T0348569

Molecular Imaging

THE AMERICAN PHYSIOLOGICAL SOCIETY
METHODS IN PHYSIOLOGY SERIES

Membrane Protein Structure: Experimental Approaches, edited by Stephen H. White (1994)

Fractal Physiology, James B. Bassingthwaighte, Larry S. Liebovitch, and Bruce J. West (1994)

Physiology of Inflammation, edited by Klaus Ley (2000)

Methods in Cellular Imaging, edited by Ammasi Periasamy (2001)

Molecular Imaging: *FRET Microscopy and Spectroscopy*, edited by Ammasi Periasamy and Richard N. Day (2005)

Molecular Imaging

FRET Microscopy and Spectroscopy

Edited by
AMMASI PERIASAMY
RICHARD N. DAY

Published for the American Physiological Society by

UNIVERSITY PRESS

2005

OXFORD
UNIVERSITY PRESS

Oxford University Press, Inc., publishes works that further
Oxford University's objective of excellence
in research, scholarship, and education.

Oxford   New York
Auckland   Cape Town   Dar es Salaam   Hong Kong   Karachi
Kuala Lumpur   Madrid   Melbourne   Mexico City   Nairobi
New Delhi   Shanghai   Taipei   Toronto

With offices in
Argentina   Austria   Brazil   Chile   Czech Republic   France   Greece
Guatemala   Hungary   Italy   Japan   Poland   Portugal   Singapore
South Korea   Switzerland   Thailand   Turkey   Ukraine   Vietnam

Published by Oxford University Press, Inc.
198 Madison Avenue, New York, New York 10016

www.oup.com

Oxford is a registered trademark of Oxford University Press

Library of Congress Cataloging-in-Publication Data
Molecular imaging : FRET microscopy and spectroscopy / edited by
Ammasi Periasamy, Richard N. Day.
p. cm. — (Methods in physiology series / American
Physiological Society)
Includes bibliographical references and index.
ISBN-13 978-0-19-517720-6
ISBN 0-19-517720-7
1. Fluorescence microscopy.   2. Fluorescence spectroscopy.   I.
Periasamy, Ammasi.   II. Day, Richard N.   III. Methods in physiology
series.
QH212.F55M64 2005
570'.28—dc22   2004061702

Printed and bound by CPI Group (UK) Ltd, Croydon, CR0 4YY

Foreword

This volume comprises an admirable collection of chapters that offer the reader a valuable and accessible overview of the broad applicability of FRET (Förster resonance energy transfer). Such applications range from the tracking of single molecules to measurements in tissue, with a main emphasis on FRET imaging. The topics have been carefully chosen to impart not just a comprehensive synopsis and up-to-date highlights, but bring to light many technical and fundamental aspects that are indispensable for practitioners.

The FRET process is familiar to scientists in many diverse disciplines, including biology, chemistry, biochemistry, biotechnology, bioengineering, chemical engineering, and physics. There are few acronyms in science that are so widely recognized; this alone is a tribute to the importance and usefulness of FRET techniques. Based in quantum mechanics, this technique continually inspires new applications in a wide range of research. One of the major goals of all physical, chemical, and biological sciences is to understand phenomena on a molecular scale. This requires information about the spatial relationships between molecules. FRET is usually the best, and often the only, way to measure quantitatively distances between molecules, on a scale of 0.2 to 20 nanometers, or to simply determine whether two molecules interact within this distance.

Over 75 years ago, Jean and Francis Perrin proposed that dipole–dipole interactions could explain the well-known experimental fact that chromophores can interact without collision at distances greater than molecular diameters. Twenty years later, Förster provided a quantitative explanation for the nonradiative energy transfer. In a few seminal papers he laid the groundwork for all later theoretical extensions, which provide the basis for modern applications of FRET to determine accurate intermolecular distances, and distance differences. Most importantly, Förster presented a theory that makes it possible to extract the relevant information from simple experimental data that allows the quantitative analysis of molecular-scale distances. Förster showed us how to relate the molecular-scale, electronic Coulombic coupling between the donor and acceptor chromophores to the transition dipoles of the interacting molecules. This allows the efficiency of transfer to be expressed simply in terms of an overlap integral, the index of refraction, the quantum yield of the donor, the dipole–dipole interaction angle (kappa), and fundamental constants. In general, the only unknown is the distance between the chromophores, which is usually the parameter sought. The overlap integral and the quantum yield are calculated from simple experiments on the separate chromophores. The kappa parameter can usually be adequately approximated, or the error estimated. Thus, the original physical insights of Förster continue to inspire deeper theoretical analyses of energy transfer that extend the applicability of FRET to new experimental situations.

FRET is the most commonly applied experimental measurement for estimating distances between molecules in solution, and more recently it has been extensively applied in optical microscopy. New experimental techniques measuring FRET efficiency, and more efficient and accurate methods for data analysis are continually being developed for use in new experimental environments. Because the molecular information lies in the spectroscopic signal, FRET can be applied to macroscopic samples as well, even to whole organisms.

The chapter titles in this edited volume give a good indication of both broad and up-to-date applications of FRET. We thank Ammasi Periasamy and Richard Day, the editors of this excellent volume, for undertaking the job of assembling chapters from internationally recognized experts, to give interested readers an extensive, insightful overview of the wide range of FRET applications, particularly in the biological and imaging areas. Although lively research is being pursued in other scientific areas, it is in biological research and biotechnological applications that FRET has been most profoundly exploited in the last few decades.

Robert M. Clegg
University of Illinois at Urbana-Champaign
Urbana, Illinois

Preface

Over the past decade, remarkable developments have occurred in the application of Förster (fluorescence) resonance energy transfer (FRET)–based microscopy and spectroscopy to the biomedical sciences. These advances are being driven by the rapid progress of imaging and data processing technologies combined with the development of novel fluorescent probes that are genetically encoded by transferable DNA sequences. The application of these technologies is shedding new light on protein–protein interactions, protein conformational changes, and the behavior of signaling molecules inside living cells. Given these rapid advances, it may be useful in this preface to offer a brief historical overview of the important early contributions that provide the framework for the modern FRET techniques covered in the following chapters.

As long ago as 1922, Cario and Franck (Z. Physik. 11, 161 (1922)) observed that illumination of a mixture of mercury and thallium vapors at a wavelength absorbed only by mercury resulted in fluorescence emission from both atoms. In 1927, Jean-Baptiste Perrin recognized that energy could be transferred (*transfert d'activation*) from an excited donor molecule to its neighbors through direct electrodynamic interactions. These near-field interactions would allow the donor to transfer excitation energy without the emission of an intermediate photon. Perrin's model, however, was based on dye molecules with precisely defined oscillator frequencies, and incorrectly predicted that energy transfer could occur over distances of up to 1000 Å (J. Perrin, *C.R. Acad. Sci. (Paris)* 184, 1097 (1927)). Several years later, Perrin's son Francis expanded on his father's work, supplying a corresponding quantum mechanical theory of excitation energy transfer (F. Perrin, *Ann. Chim, Physique* 17, 283 (1932)). In his own work, Francis recognized that "spreading of absorption and emission frequency" because of dye molecule collisions with the solvent molecules would decrease the probability of energy transfer. His calculations reduced the intermolecular distance over which efficient energy transfer could occur to 150–250 Å, which was still greater, by a factor of 3, than the experimental observations.

In 1948, Theodor Förster extended the work of the Perrins to quantitatively describe the FRET process. Förster showed that the efficiency of FRET varied as the inverse of the sixth power of the distance between the oscillating dipoles, and defined the critical molecular separation R_0, now called the Förster radius, at which the rate of energy transfer was equal to the rate of fluorescence emission. In contrast to the earlier assumptions made by the Perrins, Förster observed that "the absorption and fluorescence spectra of similar molecules are far from completely overlapping" and quantified the spectral overlap integral (T. Forster, *Ann. Phys.*, 2, 55–75 (1948)). Using this information, Förster determined an R_0 value for fluorescein of 50Å, much smaller than the values resulting from the Perrins' work, and in perfect agreement with the experimental observations. Our colleague, Dr. Robert Clegg, recently suggested that the FRET acronym should refer to Förster resonance energy transfer (FRET) to give credit to these valuable contributions, and we entirely agree (R. Clegg, *Biophotonics International*, September, 42–45 (2004)). The chapters below illustrate how the techniques pioneered by Förster have been extensively applied to measure protein proximity in biological systems.

This is the first book to address the technological developments in FRET microscopy from the viewpoint of fundamental concepts, methods, and biological applications. Each chapter includes the theory and basic principles behind the FRET microscopy or spectroscopy technique being discussed. The range of techniques considered is broad: wide-field, confocal, multiphoton, and lifetime FRET, spectral imaging FRET, photobleaching FRET, single-molecule FRET, bioluminescence FRET, time and image correlation spectroscopy. The authors describe the physics of FRET and the basics of the various light microscopic FRET techniques used to provide high spatial and temporal resolution with the goal of localizing and quantitating the protein–protein interactions in live specimens.

Each chapter provides many possible combinations of conventional and green fluorescent protein fluorophores for FRET with appropriate filter combinations and spectral configurations. This will allow readers to address specific biological questions through the choice of probes for FRET experiments, as well as the selection of the most suitable experimental approaches. More importantly, this book covers the extraction of FRET signal from the contaminated FRET images acquired with various microscopy techniques and provides a dedicated step-by-step FRET data analysis algorithm. This should make these methods of molecular imaging easier to understand for all readers, especially graduate students, postdoctoral fellows, and scientists who are new to state-of-the-art FRET microscopy imaging systems.

Charlottesville, Virginia   A.P
                                                                                     R.N.D

Acknowledgments

We would like to extend special thanks to all the contributors for the time spent in preparing the valuable manuscripts included in this book. We also want to thank Mr. Jeffrey House and Ms. Nancy J. Wolitzer, Oxford University Press, for constant feedback and timely answers to all our questions. Our special thanks go to Ms. Ye Chen for her valuable help in editing the artwork for this book. Also, we wish to thank Mr. Hal Noakes for the design of the cover for the book.

Importantly, we would also like to thank the University of Virginia and the listed organizations below for their support in bringing out this valuable book:

Carl Zeiss, Inc.

Chroma Technology Corporation

Nikon, Inc.

Olympus America, Inc.

Omega Optical, Inc.

Universal Imaging Corporation

Contents

Contributors

MARGARIDA BARROSO, PH.D.
Center for Cardiovascular Sciences
Albany Medical Center
Albany, New York

KEITH M. BERLAND, PH.D.
Physics Department
Emory University
Atlanta, Georgia

CLAIRE M. BROWN, PH.D.
Department of Cell Biology
University of Virginia
Charlottesville, Virginia

VICTORIA E. CENTONZE, PH.D.
Department of Cellular and Structural
Biology
University of Texas Health Science Center
San Antonio, Texas

YE CHEN
W.M. Keck Center for Cellular Imaging
Department of Biology
University of Virginia
Charlottesville, Virginia

ROBERT M. CLEGG, PH.D.
Physics Department
Laboratory for Fluorescence Dynamics
University of Illinois at Urbana–Champaign
Urbana, Illinois

RICHARD N. DAY, PH.D.
Departments of Medicine and Cell Biology
University of Virginia Health Systems
Charlottesville, Virginia

IGNACIO DEMARCO
Department of Medicine and Cell Biology
University of Virginia Health System
Charlottesville, Virginia

MARY E. DICKINSON, PH.D.
Division of Biology
California Institute of Technology
Pasadena, California

MASILAMANI ELANGOVAN
Carl Zeiss MicroImaging, Inc.
Thornwood, New York

SCOTT E. FRASER, PH.D.
Division of Biology
California Institute of Technology
Pasadena, California

IGNACY GRYCZYNSKI, PH.D.
Center for Fluorescence Spectroscopy
Department of Biochemistry and
Molecular Biology
University of Maryland
Baltimore, Maryland

ZYGMUNT GRYCZYNSKI, PH.D.
Center for Fluorescence Spectroscopy
Department of Biochemistry and
Molecular Biology
University of Maryland
Baltimore, Maryland

TAEKJIP HA, PH.D.
Physics Department
University of Illinois at Urbana–Champaign
Urbana, Illinois

GREGORY A. HELM, M.D., PH.D.
Department of Neurosurgery
University of Virginia Health System
Charlottesville, Virginia

BRIAN HERMAN, PH.D.
Department of Cellular and Structural
Biology
University of Texas Health Science Center
San Antonio, Texas

SUNGCHUL HOHNG, PH.D.
Physics Department
University of Illinois at Urbana–Champaign
Urbana, Illinois

CARL H. JOHNSON, PH.D.
Department of Biological Sciences
Vanderbilt University
Nashville, Tennessee

ELEONORA KEATING, PH.D.
Department of Chemistry
The University of Western Ontario
London, Ontario
Canada

ANNE K. KENWORTHY, PH.D.
Departments of Molecular Physiology &
Biophysics and Cell & Developmental
Biology
Vanderbilt University School of Medicine
Nashville Tennessee

ROBERT H. KRETSINGER, PH.D.
Commonwealth Professor of Biology
University of Virginia
Charlottesville, Virginia

JOSEPH R. LAKOWICZ, PH.D.
Center for Fluorescence Spectroscopy
Department of Biochemistry and
Molecular Biology
University of Maryland
Baltimore, Maryand

HENRY A. LESTER, PH.D.
Division of Biology
California Institute of Technology
Pasadena, California

JAMES D. MILLS, M.D.
Department of Neurosurgery
University of Virginia Health System
Charlottesville, Virginia

RAAD NASHMI, PH.D.
Division of Biology
California Institute of Technology
Pasadena, California

DAVID O. OKONKWO, M.D., PH.D.
Department of Neurosurgery
University of Virginia Health System
Charlottesville, Virginia

AMMASI PERIASAMY, PH.D.
Departments of Biology and Biomedical
Engineering
W.M. Keck Center for Cellular Imaging
University of Virginia
Charlottesville, Virginia

NILS O. PETERSEN, PH.D.
National Institute for Nanotechnology
University of Alberta
Edmonton, Alberta
Canada

V. KRISHNAN RAMANUJAN, PH.D.
Department of Cellular and Structural
Biology
University of Texas Health Science Center
San Antonio, Texas

GLEN REDFORD, PH.D.
Physics Department
Laboratory for Fluorescence Dynamics
University of Illinois at Urbana–Champaign
Urbana, Illinois

FRED SCHAUFELE, PH.D.
Metabolic Research Unit and Department
of Medicine
University of California
San Francisco, California

MOHAMMED SOUTTO, PH.D.
Department of Biological Sciences
Vanderbilt University
Nashville, Tennessee

C. MICHAEL STANLEY, PH.D.
Chroma Technology Corp.
Rockingham, Vermont

JAMES R. STONE, PH.D.
Department of Neurosurgery
University of Virginia Health System
Charlottesville Virginia

HORST WALLRABE
Department of Biology
University of Virginia
Charlottesville, Virginia

YAO XU, PH.D.
Department of Biological Sciences
Vanderbilt University
Nashville, Tennessee

JIAN-HUA ZHANG, M.D., PH.D.
Department of Cellular and Structural
Biology
University of Texas Health Science
Center
San Antonio, Texas

Molecular Imaging

1

Proteins and the Flow of Information in Cellular Function

Robert H. Kretsinger

1. Introduction

Förster resonance energy transfer (FRET) microscopy provides many beautiful and informative images, most of them focusing on protein complexes. What is the appropriate framework to understand these data? This introductory chapter argues that the flow of information provides that concept. Further, the synergism of the perspectives of molecular structure and of emergent properties of complex systems is essential to understanding this flow. This chapter should provide basic ideas and vocabulary to microscopists who are not so familiar with the concepts or jargon of protein structure, function, and evolution and of information flow.

2. Proteins and Information

Francis Crick (1958) posited in his *Central Dogma* that "DNA makes RNA makes protein." This has subsequently been elaborated to include reproduction, "DNA makes DNA," and in some viruses, "RNA makes RNA." The discovery of reverse transcriptase by Baltimore (1970) and by Temin and Mizutani (1970) necessitated adding "RNA makes DNA" (Fig. 1–1). However, the fundamental assertion that genetic information flows, "makes" in the vernacular quoted above, from nucleic acid to protein and not in the reverse sense remains one of the fundamental tenets of molecular biology and has driven much of its research over the past half century.

The strong corollary of the *Central Dogma* is that proteins represent the molecular phenotype of the cell and of the organism. Two exceptions to this approximation are especially fascinating. Ribosomal RNAs (rRNA) and transfer RNAs (tRNA) play essential roles in protein synthesis and probably reflect the dominant role of catalytic RNAs in ancient cells. In addition to other catalytic functions, for example, of spliceosomes (Rappsilber et al., 2002), there are several recently discovered classes of small, interfering RNAs (siRNA) and of microRNAs (miRNA) that can modulate the functions of messenger RNAs (mRNA).

Important as these non-messenger RNAs are, the generalization that proteins represent the cell's phenotype remains valid. Certainly, there are components of

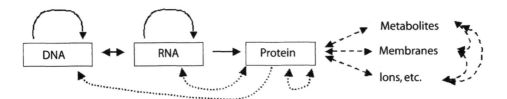

FIGURE 1–1 Flow of information. The scheme originally suggested by Crick (1958), that DNA makes RNA makes protein. The essential point is that genetic information encoded in the sequence of nucleic acids can be transduced to the sequence of amino acids in proteins but not in the reverse direction. A additional transmission of discrete sequence information: reverse transcription of RNA to DNA and replication, DNA makes DNA and RNA makes RNA, in solid lines. Proteins in turn can act on and change in a continuous manner the information content of itself or of other proteins, RNA, and DNA, either by covalent modification or simply binding reversibly, all indicated by dashed lines. Further, proteins can change the content and flow of information in a cellular mileau by catalysis, generating metabolites, altering ion fluxes, or assembling structures as membranes and cell walls, all indicated by dotted lines.

cells and extracellular materials other than proteins—lipids, carbohydrates, metabolites, and minerals—in addition to proteins that have undergone various posttranslational modifications (Nakai, 2001). Nonetheless, these components and modifications ultimately reflect the actions of proteins, hence the emphasis over the past fifty years on identifying and characterizing proteins.

3. PROPERTIES OF PROTEINS, THE REDUCTIONIST APPROACH

Scientists have long studied proteins in terms of their amino acid sequences, enzymatic activities, thermodynamic properties, and atomic structures. This has yielded a trove of empirical data from which general principles have emerged. Yet, increasingly one comes to appreciate that the character and function of a cell cannot now or in the foreseeable future be deduced or predicted from the properties of its constituent proteins. The cell's phenotype is an emergent property, certainly consistent with and explained in terms of these characteristics, but not predictable from them. It is essential that one understand these basic properties, a few of which are summarized in this section. However, one must also measure the indirect and direct interactions of these proteins over time—that is, the further flow of information in the cell beyond translation. This is the subject of the next section and of the other chapters in this book.

The (almost) complete genomic sequences of 18 archae, 141 bacteria, and 22 eukaryotes (http://www.ebi.ac.uk/genomes/) are available. From these the amino acid sequences of a million proteins, representing over a thousand homolog families, are available. Usually the codons encoding the first (N-terminal) and last (C-terminal) amino acids can be safely inferred, as can intron sites; however, some genes (in this context referring only to those genes encoding proteins) may be missed. Of greater concern, exons are sometimes spliced out of the processed mRNA prior to its translation in the ribosome (Rappsilber, 2002) to a linear sequence of amino acids.

Protein sequences are also deduced from the sequences of mRNA as copied back to complementary DNA (cDNA), using reverse transcriptase in vitro. These sequences should be free of splicing ambiguities and reflect protein that is actually being synthesized in the cell(s) of interest under specified conditions (Scamborova et al., 2004).

Having gotten the amino acid sequences of the expressed proteins, one must appreciate that the linear combinations of 20, genetically encoded amino acids generate a great variety of structures that have a remarkable range of properties. Few of the 20^N possible sequences, for a protein N residues long, have ever been synthesized (Di Lallo et al., 2003). Yet several unifying themes, with associated terminology, have emerged.

Anfinsen (1973) demonstrated that RNAase A can be denatured, then renatured in vitro, and will regain full enzymatic activity. This treatment involves no cleavage or formation of covalent bonds, no change in *configuration*. However, the *conformation* or rotation about single bonds between carbon, oxygen, nitrogen, and sulfur atoms changes a great deal. One can infer that the conformations of (most) proteins are in their global, free energy minima, and that this conformation is determined solely by the amino acid sequence of the protein in question, even though this has been explicitly demonstrated for only a few proteins.

During the folding process, some cytosolic proteins fall into a local energy minimum in which they are vulnerable to proteolysis. A misfolded protein can diffuse to the interior of a chaperone (Alberti et al., 2002; Phillips et al., 2003), where it is denatured then given another chance to find its energy minimum. If it remains misfolded, it soon has several ubiquitin proteins covalently attached and is thereby labeled for proteolysis in a 26S proteosome (Belz et al., 2002; Woodham et al., 2003).

Proteins synthesized on polyribosomes in the cytosol are released into the cytosol. Those with specific leader sequences bind to receptors on the endoplasmic reticulum and the polypeptide emerging from the ribosome is directed to the lumen of the endoplasmic reticulum. There it is modified, prior to exocytosis, to the extracellular environment or for incorporation as an integral membrane protein. Other leader sequences may target proteins for transport into the matrix or inner membrane of the mitochrondrion, into the nucleus, or to other organelles (Nakai, 2001), such as the parasitophorous vacuole of the malaria parasite, *Plasmodium falciparum* (Adisa et al., 2003).

Many proteins, regardless of destination within or outside the cell, are subject to post-translational modifications. Examples include (usually) irreversible glycosylation, characteristic of extracytosolic proteins; proteolysis involved in conversion from pre- to pro- to final isoforms often seen in the generation of peptide hormones and neurotransmitters; and reversible phosphorylation of serine, threonine, and tyrosine residues associated with cell signaling. There are scores of such specific modifications, such as ubiquination (mentioned above) and generation of green fluorescent proteins (discussed in later chapters), all of which can significantly affect the physical and functional properties of proteins (Jang and Hanash, 2003).

In addition to the changes in covalent structure, or in configuration, involved in post-translational modification, most proteins reversibly bind ligands (small molecules or ions). This often results in significant changes in conformation and in properties without further change of configuration.

Despite these and many other complications, the inference from Anfinsen's (1973) experiments remains valid; from the amino acid sequence one should in principle be able to predict the conformation(s) of the protein, excluding modifications and binding of ions or of other molecules. Indeed, if the three-dimensional structure of a homolog is known, one can usually predict the structure assumed by the sequence in question. The closer in sequence these two proteins sharing a common precursor, the more similar in structure (Wood and Pearson, 1999). Correspondingly, the more similar the sequences of the unknown and of the reference, the closer the predicted structure to the real one (Moult et al., 2003; Kretsinger et al., 2004). However, one cannot yet predict with confidence the structure of a new protein ab initio—a very humbling admission—after 30 years of intense effort by the research community. The related question of whether most chains of amino acids randomly assembled by sequence have a distinct structure associated with a distinct energy minimum remains unanswered. Just as one cannot predict ab initio the structure of a protein from its sequence, one also cannot predict the interaction site or affinity of two proteins from their respective structures (Bock and Gough, 2001).

The main chain, or backbone, of a protein can be characterized and computed to excellent approximation by two dihedral angles—ϕ CO,NH-C_α,CO) and ψ (NH$_i$,C_α-CO,N$_{i+1}$)—per residue. This is because covalent bond lengths and bond angles are almost constant among amino acids and because the dihedral angle of the peptide bond, ω ($C_{\alpha\,n}$,CO-NH, $C_{\alpha(n+1)}$), is planar and (near) constant. The conformations of amino acids as observed in proteins are restricted to small areas of the ϕ, ψ (plot (Ramachandran and Sassiekharan, 1968); this reflects the constraints imposed by van der Waals contacts between atoms of the main chain and between the main-chain and side-chain atoms, primarily C_β, when ϕ and ψ lie outside of those acceptable ranges (Fig. 1–2). When four consecutive residues have ϕ, ψ values in the H region they form an α helix, characterized by a hydrogen bond from the NH of residue $i+4$ to the CO of residue i, counting from the N-terminal residue toward the C-terminus. When three consecutive residues have ϕ, ψs in the S region they form a β strand. Two or more β strands, side by side, either parallel or antiparallel, form a β sheet, which is stabilized by strand-to-strand hydrogen bonds. About a third of residues in globular proteins are involved in α helices, a third in β sheets, and a third in *other*, imprecisely called *turn*, or *coil*, or *linker*. α Helices tend to be resistant to thermal and chemical denaturation or to proteolysis; proteins with high α-helical content are more stable. Turn regions often provide flexibility between domains, changes in conformation at ligand binding sites, and sites of proteolysis.

A domain of a globular protein, in contrast to a fibrous protein, is often the appropriate unit for analysis from structure to evolution. Most globular proteins consist of several domains, although smaller proteins, such as myoglobin or lysozyme, often consist of a single domain. Most domains under physiological conditions adopt an ensemble of closely related conformations. The axial ratio, from longest to shortest, of a domain is usually less than 3:1. Its constituent chain of residues has a single entrance at its N terminus and a single exit at its C terminus—

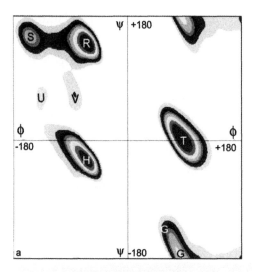

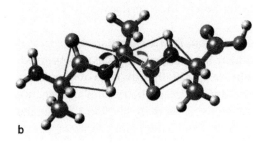

FIGURE 1–2 Ramachandran plot. *a*: Distribution of φ, ψ values observed for all amino acids in proteins determined to high resolution (Hovmöller et al., 2002). Proline is restricted to φ ~ –60° because its side chain forms a five-member pyrrolidine ring involving its α-amine nitrogen. Glycine enjoys a broader range of φ, ψ because it has only a hydrogen atom for a side chain. The other 18 amino acids have slighlty different distributions from one another, as if each had a conformational fingerprint. These seven distinct regions in the φ, ψ plot account for > 95% of observed conformations of amino acids in crystal structures determined to high resolution; they are assigned identifying symbols: H, residues in α-helices assume this conformation; T, residues in left handed α-helices assume this conformation; S, residues in β-strands assume this conformation; R, residues in polyproline helices assume this conformation; U, a conformation occasionally seen in turns; V, another conformation, distinct from U, occasionally seen in turns; G, a conformation restricted to Gly. *b*: Conformation of the backbone at φ = 180° and ψ = 180°, as found in the β strand. Arrows indicate positive or right-hand, rotation. Rotation about φ (NH-C_α) moves C_β relative to the fixed $C_{\alpha-1}$,CO_{-1}, NH,C_α plane. Rotation about ψ (C_α-CO) moves the C_α,CO,NH_{+1},$C_{\alpha+1}$ plane relative to the fixed C_β. *c*: Conformation at φ = –109° and ψ = 121°.

that is, it could be cut out with two proteolytic clips and assume essentially the same shape as it had when part of the full protein. Often a monomer protein can be cleaved to (or expressed as) its constituent domains then reassembled without formation of covalent bonds between these domains to form a functional protein. This characteristic is exploited in the yeast two-hybrid assay.

Although one speaks of conformation in the singular, nearly all proteins, except a few devoted solely to structure such as collagen or keratin, can assume several closely related conformations. Like nanomachines, these conformations reflect states in the functional cycle of a protein (Valpuesta et al., 2002; Lee et al., 2004). Within a domain the orientations of a few side chains and the conformation(s) of a turn(s) may vary among states. However, the conformations of the domains themselves are usually very similar among these various states of the entire machine. In contrast, the relationships between domains within a multidomain protein may vary significantly because of changes in conformation of the interdomain turns, often called *linkers*.

The domain is the recognizable unit of evolution. Two domains are *homologous* if they evolved from a common precursor protein. Since this process was not observed, the burden of proof, or at least of probability, rests with the advocate of homology. Often one asks what the probability is that the domain in question is similar by chance to another domain in a database. If the probability is less than, say, 1 in 10^5, then the observed similarity is inferred to reflect descent from a common precursor, as opposed to convergent evolution (http://www.ncbi.nlm.nih.gov/BLAST/). Two homologs are referred to as *orthologs* if they are related by speciation of their host organisms, and as *paralogs* if they are related by gene duplication. Dendrograms, or phylogenetic trees, of organisms based on their respective orthologs reflect the evolution of those organisms; the more orthologs involved in the analysis, ideally reflecting the entire genome, the more significant the resulting phylogeny (Stuart et al., 2002; Cheng et al., 2003). Dendrograms based on paralogs provide insight into the origins of isoforms, and possible divergence and specialization of function, within that one protein family.

The term *protein* has several meanings that should be clear from its context. Experimentalists usually encounter the protein replete with post-translational modifications and ligands bound; they deal with the properties thereby imparted. In contrast, geneticists consider the protein according to its sequence of encoded amino acids.

Some proteins function as *monomers*—a single chain of residues comprised of one or of several domains. Each of these domains in either a homochimeric (homologous domains) or a heterochimeric (different, nonhomologous domains) monomer protein has its own conformation(s) and its own evolutionary history. Most proteins are oligomers. Several identical monomers associate noncovalently to form a homo-oligomer or different monomers form a hetero-oligomer. The monomers of some oligomers associate and dissociate as part of their functional cycles; some associate with much higher affinity. Some monomers, for instance, actin, polymerize to form thin filaments of muscle. Whether *actin* refers to the monomer or to the polymer should be clear from the context.

Domain swapping describes the phenomenon in which a heterodomain protein extends (part of) one of its domains to replace its counterpart in another protein. This can lead to formation of a homodimer, a homo-oligomer, or a hetero-oligomer (Liu and Eisenberg, 2002; Juy et al., 2003). Domain swapping can be extended to form a polymer. The term may refer to part of the functional cycle of a protein(s) or be invoked in the context of a long timescale to explain the evolution of oligomeric proteins (Hakansson and Linse, 2002; Rousseau and Schymkowitz, 2003).

Many proteins are enzymes; they function as catalysts and are classified with a four number Enzyme Classification (EC; 1992) code. Often one EC code number corresponds to one homolog family; however, there are many exceptions. Proteins in one homolog family may have different substrates or mechanisms and hence several codes; conversely, one catalytic function may be performed by more than one (family of) protein(s). The binding of substrate and subsequent catalysis may occur within a single domain or at the interface of two domains or even subunits; other domains or subunits may be involved in regulation. Different isoforms of the protein may have different other domains spliced to it (Fig. 1–3).

Proteins are characterized under defined experimental conditions in vitro. This places a premium on purification procedures. One wants to determine enzyme kinetics and heat capacities, to record spectral properties (including nuclear magnetic resonance and electron spin resonance), and to grow crystals.

All of these reductionist concepts and techniques can lead one to think of proteins as isolated macromolecules interacting with selected ligands and only occa-

FIGURE 1–3 Domains of 3-deoxy-D-arabinoheptulosonic acid synthase (DAHPS). The middle panel shows the backbone of KDOPS[Ec] (3-deoxy-D-*manno*-octulosonate-8-phosphate synthase) from *Eschericia coli*, a member of the $(\beta/\alpha)_8$ homolog family. The core consists of eight parallel β strands (blue), arranged sequentially to form a barrel; they are connected by eight α helices (yellow). To the left is DAHPS[Ec] (3-deoxy-D-*arabino*-heptulosonate-7-phosphate synthase), a paralog of KDOPS[Ec], that has the same $(\beta/\alpha)_8$ core. It contains two extra elements of secondary structure, β strands β6a and β6b, that provide the binding site for the feedback inhibitors—phenylalanine, tyrosine, or tryptophan (specific to the appropriate isoform). To the right is DAHPS[Tm] from *Thermotoga maritima*. It lacks the β6a and β6b loop of DAHPS[Ec], but instead has acquired by gene splicing and fusion a flavodoxin like (FL) domain. It is also subject to feedback regulation by tyrosine and tryptophan, which bind to the FL domain. However, the pathway and mechanism of inhibition is entirely different in the two orthologs (Shumilin et al., 2004).

sionally with other proteins. This may provide an appropriate perspective for the chemist; the biologist, however, needs a broader view.

4. INFORMATION REVISITED

Most current models of information flow derive from systems engineering. They distinguish content from transmission and, in turn, transmission from processing of information, or computation. *Information* is represented by states that have sharply defined transitions and is transmitted among elements with well-defined boundaries. Those elements that send and receive information are distinct from the information itself. A *program* operates on the data, or computes. The program is a description of an algorithm; it is clearly located somewhere, and it is expressed in a formal language that can be read or interpreted by humans or by other programs. For better or worse, nature did not consult engineers when evolving life.

While genetic information stored in DNA is highly discrete and relatively static, some kinds of cellular information vary smoothly and are better described as continuous. The state of a cell, or some subsystem of it, depends on the integrated interactions over time of myriad, possibly modified and/or liganded proteins and on the metabolites, ions, and lipids that reflect the actions of proteins. Further, this cellular (protein) information, for instance, as transcription factors, can affect the replication and transcription of DNA and the translation of RNA.

Of much greater complexity, cellular computations are self-referential and self-altering. The proteins that contain information are also involved in transmission and computation. Further, these relationships change with time. In many biological information systems it is not evident where the boundaries are, if indeed there can be said to be boundaries at all in the engineering sense.

The storage, transmission, and manipulation of information at the cellular level seem to be profoundly different in biology from what they are in computer systems. Both disciplines can learn from the other; however, superficial analogies can be quite misleading.

5. PROTEOMICS AND THE FLOW OF INFORMATION

The success and power of genomics have focused new research on understanding the proteome of the cell (Chen and Xu, 2003; Zhu et al., 2003; Cho et al., 2004). A comprehensive characterization would include knowing which proteins are present and at what concentrations in various organelles and functional states of the cell, the amino acid sequences and tertiary structures of those proteins, their post-translational modifications, their ligands, enzymatic activities (if any), subcellular locations, and interactions with other proteins, and most important, their functions. Given the limitations of funding resources, time, and existing techniques, one has to set priorities in sampling this vast character space of the proteome. As already discussed, some of these properties—sequence, structure, modifications, ligand binding, and catalytic activity—can be determined in vitro using (partially) purified proteins. For many proteins these purifications and characterizations do not lend themselves to automation and can be demanding of resources and time; doing proteomics one at a time is tedious.

One should distinguish between content of proteins on the one hand and synthesis and degradation on the other. Expression of proteins can be inferred by measuring appearances and concentrations of their mRNA (Jung and Stephanopoulos, 2004). The differential degradation of selected proteins can also be measured (Woodham et al., 2003). Proteomics also asks which proteins are expressed at different stages of development and under various physiological challenges. To paraphrase the sleuth, "follow the information."

It is becoming ever more apparent that it is also essential to know the subcellular localizations and interactions of these proteins to understand the emergent property of cell function. Given all, or more realistically, a small subset of these data, how can one best analyze them with the goal of understanding the function of the proteome? Then given this framework, how does one best use FRET?

A few decades ago, most reviews of proteins and the researchers themselves would have implicitly or even explicitly given the impression that most proteins, many of them enzymes, function as isolated entities. The role in the flow of information of an enzyme involved in intermediary metabolism would have consisted of operating on substrate(s) α and producing product(s) β; the enzyme would have communicated with other computing elements by manipulating concentrations of α and β over time. At a biochemical, level this problem was addressed in the heroic studies of intermediary metabolism. Those complex charts posted in every biochemistry laboratory show the enzymes responsible for the synthesis of (almost) every small ($M_r < 500$) molecule found in most cells (www.sigmaaldrich.com/Area_of_Interest/ Biochemicals___Reagents/Enzyme_Explorer/Metabolic_Pathways_Map.html). One now appreciates that those enzymes represent <10% of the cell's armamentarium of proteins (Venter et al., 2001). Note that these metabolites are quite stable. Almost *no* chemical reactions occur within the cell, and only a few occur in the extracellular environments of metazoa, without catalysis by a specific enzyme. This dependence on catalysis permits regulation of metabolism at several levels, on different timescales, and at different points in the flow in information—feedback inhibition by products further along the pathway, post-translational modification reflecting cell signaling pathways (Miyawaki, 2003), and regulation of transcription (Chen and Xu, 2003). As will be discussed, these enzymes of intermediary metabolism often reflect the terminal nodes of very complicated networks of information.

Or an enzyme might operate directly on other proteins (or nucleic acids) by modifying their configurations, as involved in cell signaling pathways (http:// www.ariadnegenomics.com/products/pathway.html). Such charts in cell biology labs followed those adorning biochemistry labs. These pathways transduce information in the forms of molecules, metal ions, radiation, or mechanical signals from outside the cell to changes in conformation and/or configuration, and ultimately of function, of proteins inside the cell. In addition to kinase, phosphatase cycles are those involving methylation–demethylation, (de)acetylation, and (de)tyrosylation, among other processes. Each of these post-translational modifications reflects changes in both information content and the elements that compute, a distinction not necessarily captured by the arrows on the charts. In the extreme case, enzymes might be autocatalytic, using their identical brethren as substrates or even themselves—for instance, the green fluorescent protein, to be discussed below (Fig. 1–4).

One now appreciates that a large fraction of the proteins within a cell or in the intercellular space interact with other proteins (or with DNA and RNA) in a functionally significant way, other than using one another as substrates for post-translation modifications. The mean concentration of protein in the cytosol, ~200 mg/ml, is about a third that in a protein crystal, which contains approximately half solvent. Each protein can potentially interact with (nearly) all the others. To complete and extend this comprehensive understanding of the proteome, one needs to know the interactions of all of the (N) encoded and expressed proteins with one another—to first approximation, completing a (N × (N + 1) / 2) matrix of interactions with integrated and compatible data from different techniques (Giot et al., 2003; Chamrad et al., 2004; Orchard et al., 2004). Some decades hence it may appear naive when one asks for the details of all of those interactions; now, however, just completing the yes/no entries will be a milestone (Fig. 1–5). It is an essential step toward understanding the emergent property of cell function. The mapping of those interactions, even without determining affinities or functions, is recognized as an essential, albeit imperfect, first step toward understanding the role of the proteome in the overall flow of information in the cell, then the organism, and finally the community.

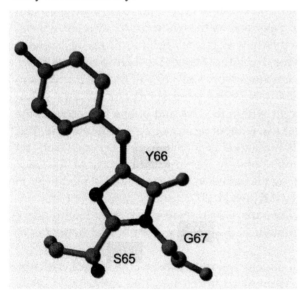

FIGURE 1–4 The active site of green fluorescent protein (GFP). The chromophore of GFP is formed by the autocatalyzed cyclization and oxidation of serine 65, tyrosine 66, and glycine 67 to form an imidazolidinone ring. The carbonyl carbon atom of S65 is bonded directly to the nitrogen of G67 with the elimination of the carbonyl oxygen of S65, as water. A double bond, indicated in black, is formed between the α and β carbon atoms of Y66 to link the conjugated bond systems of the imidazole and the tyrosinyl rings (from PDB entry 1GFL). Nitrogen atoms are blue, oxygen is red, and carbon is green. Various mutations in other residues spatially nearby tune the absorbed and emitted wavelengths to blue, cyan, yellow, or red in engineered flourescent proteins.

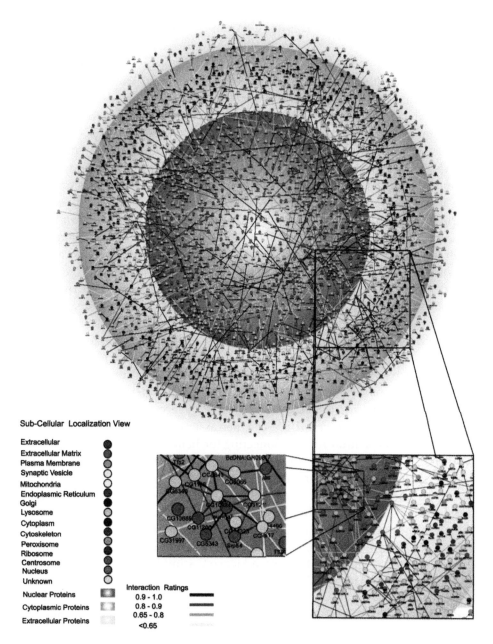

Figure 1–5 Protein interaction map of *Drosophila melanogaster*. A total of 10,623 predicted transcripts were screened against complementary DNA libraries to produce an initial map of 7048 proteins and 10,405 interactions. From this a high-confidence map of 4679 proteins and 4780 interactions was made. The short-range clustering appears to reflect multiprotein interactions; the long-range interactions may reflect interactions between complexes (Giot et al., 2003).

6. EXPERIMENTAL APPROACHES TO UNDERSTANDING THE PROTEOME

Several *in vitro* techniques have been used for estimating the protein contents and composition of the cell and its compartments. Each of these has been the subject of extensive reviews; they are mentioned here to provide perspective for the role of FRET in understanding the proteome and its role in information flow.

It is essential for the transduction of both energy and information that there be some sort of boundary between what is *inside* the living system and what is *outside*. This is another way of saying that the cell, with its encompassing membrane, is the fundamental unit of living systems, unicellular or metazoan. Even bacteria and archae have compartments or fractions, including flagella, condensed DNA, the cell membrane, and, for gram-negative bacteria, the outer (second) cell membrane. The architecture of eukaryotic cells is much more complex and includes organelles bounded by single membranes, endoplasmic reticulum, and its antecedent and derived vesicles, and the nucleus and the plastids surrounded by a double membrane reflecting their evolutionary origins by endosymbiosis (McFaddin, 1999).

The portions of the proteins that span the ~30 Å of hydrophobic lipid membrane are (almost) all α helices (DeGrado et al., 2003) or β strands that fold into a few characteristic barrel patterns (Wimley, 2003; Ryan, 2004). Sophisticated plots based on hydrophobicity and patterns of occurrence of amino acids can often identify the transmembrane helices or strands in the sequence, but errors are still frequent (Ott and Lingappa, 2002). If successful, one can then assign cytosolic and extracytosolic sides to the linkers of amino acids connecting the helices or strands. Interactions with ligands or other proteins can change the packing of the helices and strands; however, these integral membrane proteins usually spend their lives imbedded in phospholipid bilayers with no change of sidedness or topology (Jang and Hanash, 2003). Many of these integral membrane proteins function as pumps or channels; some as receptors of external messengers, others serve as anchors. For instance, the trafficking of Na, K-ATPase fused to green fluorescent protein in COS-1 renal cells in response to protein kinase C stimulation has been directly observed (Kristensen et al., 2003).

In contrast, extrinsic membrane proteins can associate with or dissociate from membranes. Sometimes these interactions are facilitated by ligands, such as the Ca^{2+} ion, bridging between the phosphate of a phospholipid and the carboxylate groups of the protein (Heim and Griesbeck, 2004). In other instances an integral membrane protein provides the anchor to the membrane. These proteins can diffuse laterally in the two-dimensional plane of the membrane but not (usually) flip sidedness. Sometimes they associate into higher-order oligomers or into clusters of unknown stoichiometry; the factors and mechanism controlling these associations are poorly understood but fundamentally important.

One should understand the distribution of proteins among these organelles as well as among the extracellular spaces of metazoa. Fractionation techniques, including filtration, differential sedimentation based on differences in density and/ or sedimentation velocity, and salt fractionation, have been refined over the past century. These fractionations precede in vitro analyses and provide reference points for in vivo investigations.

Having extracted (some of) the proteins from the entire cell or from purified or at least enriched components of the cell, one sets out to discover what's there. Two-dimensional gel electrophoresis (2DGE) has been used for over two decades and in many ways was the first technique explicitly designed to view (parts of) the proteome. Proteins are separated in the first dimension by means of isoelectric focusing and then in the second dimension by size with SDS-polyacrylamide gels. Two-dimensional gel eletrophoresis has several limitations and is cumbersome. Staining of spots with Coomassie, silver ions, fluorophores, or antibodies specific to proteins under investigation (Western blotting) has a limited dynamic range, at best 10^3. Some membrane proteins cannot be run in an isoelectric focusing gel because the detergents required for solubilization interfere with the ion gradient. Highly acidic or basic proteins cannot be focused between pH 4 and 10, as is practical in isoelectric focusing, and proteins with molecular masses below 15 kDa or over 200 kDa either run off the gel or cannot enter it. Subsequent analyses require manually cutting out plugs of gel and eluting the protein. Conceding all of this, 2DGE is still a solid reference and the workhorse of many labs. Up to 10,000 different proteins can be resolved and visualized in a single gel. Even if locally distorted, the 2D gel retains topological relations among spots; it can be scanned and digitized then computationally unbent to permit comparison with other gels. Two-dimensional gel eletrophoresis is especially valuable in detecting differences between two states of a cell or between two cell types (Lin et al., 2003).

High-pressure liquid chromatography (HPLC) provides much of the same information as that of 2DGE and potentially much more. Several chromatography resins and procedures provide resolutions in one dimension approaching that achieved in 2DGE. Liquid chromatography offers additional advantages of being readily automated and highly reproducible. The concentrations of proteins can be estimated directly by recording their absorptions at several wavelengths, e.g., 230, 260, and 280 nm. In the context of analysis of the proteome, both 2DGE and HPLC are best used in conjunction with mass spectroscopy (MS) (http://www.waters.com/WatersDivision/).

Advances in MS over the past decade have fundamentally altered the way in which one investigates the proteome (Lin et al., 2003). There are several broad categories of approaches and results. From a mixture of a few intact proteins, for instance, a spot(s) eluted from a 2D gel or a fraction from HPLC, one can determine the molecular weight of the intact protein to 1 part in 10,000. Peak heights in the spectrogram are roughly proportional to protein content; one can extend the dynamic range to 10^{10} by varying input volumes and other factors. Hence, at the end of the combined 2DGE or HPLC and MS experiment, one has the spot or elution position, molecular weight, and estimate of content of every reasonably abundant protein in the input. Note that the molecular weight of the protein, ± 0.01%, does not necessarily identify it or permit its assignment to a known gene product.

The real power of MS is realized by determining the molecular weights of fragments or peptides from the protein(s) derived from 2DGE or HPLC then using these data to identify the parent protein from its molecular weight deduced from its encoding gene. In the "top-down" strategy employing electrospray ionization Fourier transform MS, the protein(s) is (are) fragmented by means of electron capture dissociation and yields a spectrogram of mass/charge (m/z) values. These proteins or peptides are introduced into the gas phase by electrospray ionization or by matrix-assisted laser desorption/ionization.

Alternately, in the "bottom-up" strategy, the protein(s) can be digested with trypsin and (partly) separated by HPLC prior to introduction into the mass spectrometer and volatilization by electrospray ionization, so-called MudPIT (multidimensional protein identification technology). To illustrate, assume that the fraction consists primarily of three major proteins, each of which yields 20 tryptic peptides, all of which have C-terminal lysine or arginine (except the C-terminal peptides of the three proteins). This would produce 60 peptides of molecular weights ~150 to ~3000, all of which could be resolved and assigned a molecular weight on the spectrogram. The important point is that the molecular weight of each peptide reflects an almost unique amino acid composition. Many post-translationally modified residues can also be identified in this analysis. Quite remarkably, these fractured and scrambled fingerprints, from either fragmentation or proteolysis, can be deconvoluted and assigned with reasonable confidence to a protein identified thus far only by its amino acid sequence as deduced from the DNA sequence of the genome (Chamrad et al., 2004; Orchard et al., 2004). The protein has been identified by MS and its sequence deduced from the genome without having been purified, characterized, or directly sequenced.

The combination of cell fractionation and preparative procedures, 2DGE, HPLC, and MS interpreted in light of the sequence of the genome gives one a good summary of the proteins that are present in a reasonable quantity in a chosen organelle, cell, or tissue at any given time under any given growth conditions. These results are then augmented and refined by the various studies, previously mentioned, of individual purified proteins. Yet, even with all of this information, the picture of the proteosome is still incomplete.

There are several approaches to the N^2 interactions problem. One can extract and purify a subset of the N proteins and attach each individually to a substrate or chip (Seong and Choi, 2003; Jung and Stephanopoulos, 2004). Then a mixture of the N proteins is exposed to the chip and the excess removed. To score the (subset) matrix, one needs a specific identifier or label for each protein from the mixture that binds. This procedure is difficult to automate and generalize because one needs to purify each of the proteins to be attached to the chip. Even so, up to 2400 different, purified antibodies have been covalently attached to a single chip to screen cell extracts and body fluids for specific proteins or small molecules (Lueking et al., 2003).

Many if not most of the proteins in a cell have specific, functional interactions with other proteins, including, but not limited to, being enzymes that catalyze post-translational modifications in the other proteins. These interactions can be indicated by exploiting various tricks of molecular genetics, best illustrated by yeast two-hybrid displays and phage displays. One can anticipate many more clever variations and extensions of these genetic procedures.

In the idealized yeast two-hybrid display, a yeast (*Saccharomyces cerevisiae*) is genetically engineered so that a reporter gene, e.g., *lacZ* or the gene encoding the green fluorscent protein (GFP), can be expressed and monitored under the chosen growth conditions. The gene that encodes the DNA binding domain of the protein that regulates transcription of the gene encoding that reporter, *lacZ* or GFP, is fused to protein, called the *bait*, that is under investigation and incorporated into the yeast genome. The *activating domain* of the regulatory protein is fused to thousands of different genes or gene fragments encoding the population of proteins, *preys*, under investigation, these fused prey constructs are then inserted into plasmids that can infect the yeast carrying the activating domain*prey fusion gene. Ideally the yeast grows, i.e., forms a colony, only if the DNA binding domain and the activating *do*main of the regulatory protein interact and the reporter protein is then expressed. The probability that the two domains of the regulatory protein combine on the DNA in the nucleus is greatly increased if the bait under investigation interacts with one or more preys encoded in the population being screened for interaction with the bait—that is, the bait*prey interaction greatly increases the probability of a productive DNA binding domain*activating domain interaction. Hence, growth of a colony suggests, but does not prove, a functionally significant interaction between bait and prey. The plasmid can then be extracted from any yeast colony that tests positive. From its DNA sequence the identity of the prey that (might have) interacted with bait can be determined and compared against the sequence of the genome being screened. That screened genome may be from yeast or from another organism whose proteins interact like their homologs in yeast (Di Lallo et al., 2003; Kondo et al., 2003; Vazquez et al., 2003).

The yeast two-hybrid screen has many sources of error. The fusion of *bait* with the DNA binding domain might generate steric interference preventing proper interaction with the activating domain or preventing bait–prey interaction. Or the components and/or complexes regulatory domain*prey cannot pass through the nuclear envelope from their site of expression in the cytosol. Or the bait and/or its cognate preys were intended for another organelle or membrane and balk at entering the nucleus. The problems are myriad; many false positives are recorded and many real positives are missed. Nonetheless, about half of the positives, as confirmed by other experiments, reflect real, physiologically meaningful interactions.

The functions, but not necessarily physical interactions, of gene products from an alien organism can be evaluated in a yeast complementation system. One promoter, e.g., tetO (determining doxycycline repressible expression) is used to control essential yeast genes; another, e.g., pMET3 (which is switched off in the presence of methionine) is used to regulate the expression of alien cDNA clones and to determine whether their expessions are essential for growth (Zhang et al., 2003).

The random phage display technique defines the binding site, or epitope, of an antibody, usually monoclonal, that was raised against an entire protein (complex). Random fragments of the gene encoding the protein are fused to the terminus of the gene encoding the coat protein of a phage, e.g., M13. Millions of phage with different peptide fragments fused to their coat proteins are bound to a chip or column with the fixed antibody, then eluted and used to reinfect the host bacteria. Several cycles of phage binding and reinfection yield a phage population highly enriched for the high-affinity epitope(s). The encoding gene from the phage can be sequenced and the amino acid sequence of the epitope deduced (Smothers et al., 2002).

There are three advantages of these gene-based tests for functional protein–protein interaction. One need not isolate the protein; one can work from just the encoding DNA sequence. Second, one can (partially) automate and scale the screens to examine all of the proteins of the proteome under investigation. Third, and most important, one gets a strong indication of which proteins interact inside the cell. Mass spectroscopy does not yet provide these indications of interaction. Certainly the high rate of false positives and missed interactions makes one cautious. However, one can in principle and in practice repeat the screens using different regulatory proteins. Getting positives in several different two-hybrid screens gives one confidence in those positive results. The reasons for missed interactions may be common to several screens. One is then in the position of underestimating the number of protein–protein interactions, which is already much larger than most researchers had anticipated.

On the basis of these screens, tentative maps of proteome interactions have been published for several organisms (Giot et al., 2003). One can anticipate many refinements and corrections over the next decade. At this time, several generalizations seem solid. Most proteins have one to a dozen interactions with other proteins. Many of the deduced interactions make sense in that they are consistent with interactions demonstrated by *in vitro* techniques. Further, many of the interactions reflect proteins known to be associated with the same general function and with the same organelle or compartment. The availability of these data has generated the need to develop quite sophisticated programs to categorize and display the interactions in two dimension plots (Fig. 1–5).

7. Why FRET?

Given the ability to generate these proteome interaction maps and the characterizations of individual proteins and their post-translational modifications by liquid chromatography and mass spectroscopy, why should one use FRET?

Here the challenges come in at least four flavors: the physics of quite subtle and complex optics, the reduction and manipulation of vast amounts of data, the biology of inducing the right cell to do the right thing, and, finally, the tagging of the proteins. The first three of these topics are addressed in several chapters in this volume.

Turning to the fourth, proteins, as synthesized in the ribosome, do not have identifying spectral properties. A few acquire such tags by post-translational modification, for instance, the covalently attached heme group of cytochrome *c*. Others can be identified by the molecules or ions that they bind; these interactions include their (partial) incorporation into the hydrophobic interior of a phospholipid bilayer (Ritter et al., 2004).

Many proteins can be specifically labeled with antibodies or substrate analogs, both of which can carry specific, covalently attached spectral labels. These are fine for labeling extracted proteins or proteins in fixed cells with membranes made permeable to proteins. However, the permeation required for entry into a living cell can introduce artifacts (Ensenat-Waser et al., 2002). Alternatively, one can microinject the labeled, or tagged, protein (Kim et al., 2002).

As noted, there are several naturally occurring or engineered modifications of proteins that make them appropriate fluorescence energy donors or receivers. However, most researchers employ the green fluorescent protein, GFP (Day et al., 2001) or one of its genetically engineered mutants—blue, BFP; cyan, CFP; yellow, YFP; or red, RFP (Fig. 1–4). As will be discussed in detail in subsequent chapters, there are several variations on the basic experimental design. The gene encoding XFP is fused to the presumptive N terminus or C terminus of the gene encoding the first protein of interest, ONE. ZFP, whose absorption profile matches the emission of XFP, is fused to protein TWO. Both fusion proteins, XFP–ONE and ZFP–TWO, are expressed in the cell of interest behind cleverly chosen control elements. Or, one is expressed and the other is introduced via microinjection or electroporation, or both are introduced. If expressed, XFP-ONE and/or ZFP-TWO can be engineered to have peptide signals that target them to the desired cellular compartment or component. If proteins ONE and TWO interact, probably the chromophore sites of XFP and the ZFP will be brought within the ~30 Å required for fluorescence energy transfer when the fluorescent proteins are properly folded (Philipps et al., 2003). For instance, the GFP-fused 1724RolB protein from the root-inducing plasmid is localized to the nucleus in tobacco (Moriuchi et al., 2004). Another example, the GFP-encoding gene, has been introduced into *Candida albicans* cells to record the general level of gene expression of this yeast while growing in infected tissue (Barelle et al., 2004). In these various applications one must be cognizant of the same sorts of potential artifacts as those discussed for the yeast two-hybrid system.

Given a system that is well characterized by yeast two-hybrid screens (Fig. 1–5), what advantages are to be gained by also examining FRET interactions? The high levels of both false positives and false negatives in yeast two-hybrid screens make it valuable, even if only for confirmation, to have any and all measurements of interactions by methods that rely on different principles. Second, since one is using molecular genetic techniques, one could in principle scan an entire column of an $N \times N$ (protein) matrix by holding XFP–ONE constant and scanning many ZFP–TWO constructs.

Some investigations explore the physical and chemical properties of proteins with little regard for their roles in the functions of the cell or organisms. Others may describe the expressions, locations, and interactions of proteins, ignoring their physical properties. Yet time and again, one sees that the development of new techniques and the interpretations associated with proteomics involve the synergistic interaction of both the reductionist and the systems perspectives.

By far the greatest advantage of FRET is that one can observe in real time a living cell computing.

REFERENCES

Adisa, A., M. Rug, N. Klonis, M. Foley, A.F. Cowman, and L. Tilley. The signal sequence of exported protein-1 directs the green fluorescent protein to the parasitophorous vacuole of transfected malaria parasites. *J. Biol. Chem.* 278:6532–6542, 2003.

Alberti, S., J. Demand, C. Esser, N. Emmerich, H. Schild, and J. Hohfeld. Ubiquitylation of BAG-1 suggests a novel regulatory mechanism during the sorting of chaperone substrates to the proteasome. *J. Biol. Chem.* 277:45920–45927, 2002.

Anfinsen, C. Principles that govern the folding of protein chains. *Science* 181:223–227, 1973.

Baltimore, D. RNA-dependent DNA polymerase in virions of RNA tumour viruses. *Nature* 226:1209–1211, 1970.

Barelle, C.J., C.L. Manson, D.M. MacCallum, F.C. Odds, N.A.R. Gow, and A.J.P. Brown. GFP as a quantitative reporter of gene regulation in *Candida albicans*. *Yeast* 21:333–340, 2004.

Belz, T., A.D. Pham, C. Beisel, N. Anders, J. Bogin, S. Kwozynski, and F. Sauer. In vitro assays to study protein ubiquitination in transcription. *Methods* 26:233–244, 2002.

Bock, J.R. and D.A. Gough. Predicting protein–protein interactions from primary structure. *Bioinformatics* 17:455–460, 2001.

Chamrad, D.C., G. Korting, K. Stuhler, H.E. Meyer, J. Klose, and M. Bluggel. Evaluation of algorithms for protein identification from sequence databases using mass spectrometry data. *Proteomics* 4:619–628, 2004.

Chen, Y. and D. Xu. Computational analyses of high-throughput protein–protein interaction data. *Curr. Prot. Pept. Sci.* 4:159–180, 2003.

Cheng, M., G. Ruan, and Z.H. Rao. Phylogeny reconstruction based on protein phylogenetic profiles of organisms. *Prog. Nat. Sci.* 13:109–113, 2003.

Cho, S.Y., S.G. Park, D.H. Lee, and B.C. Park. Protein–protein interaction networks: from interactions to networks. *J. Biol. Chem. Mol. Biol.* 37:45–52, 2004.

Crick, F.H.C. On protein synthesis. In: *Symposium of the Society for Experimental Biology* XII, New York: Academic Press, 1958, pp. 153–163.

Day R.N., A. Periasamy, and F. Schaufele. Fluorescence resonance energy transfer microscopy of localized protein interactions in the living cell nucleus. *Methods* 25:4–18, 2001.

DeGrado, W.F., H. Gratkowski, and J.D. Lear. How do helix–helix interactions help determine the folds of membrane proteins? Perspectives from the study of homo-oligomeric helical bundles. *Protein Sci.* 12:647–665, 2003.

Di Lallo, G., M. Fagioli, D. Barionovi, P. Ghelardini, and L. Paolozzi. Use of a two-hybrid assay to study the assembly of a complex multicomponent protein machinery: bacterial septosome differentiation. *Microbiology SGM* 149:3353–3359, 2003.

Ensenat-Waser, R., F. Martin, F. Barahona, J. Vazquez, B. Soria, and J.A. Reig. Direct visualization by confocal fluorescent microscopy of the permeation of myristoylated peptides through the cell membrane. *IUBMB Life* 54:33–36, 2002.

Enzyme Nomenclature, NC-IUBMB, Academic Press, New York, 1992.

Giot L., J.S. Bader, C. Brouwer, A. Chaudhuri, B. Kuang, Y. Li, et al. A protein interaction map of *Drosophila melanogaster*. *Science* 302:1727–1736, 2003.

Hakansson, M. and S. Linse. Protein reconstitution and 3D domain swapping. *Curr. Prot. Pept. Sci.* 3:629–642, 2002.

Heim N. and O. Griesbeck. Genetically encoded indicators of cellular calcium dynamics based on troponin C and green fluorescent protein. *J. Biol. Chem.* 279:14280–14286, 2004.

Hovmöller, S., T. Zhou, and T. Ohlson. Conformations of amino acids in proteins. *Acta Crystallogy D. Biol. Crystallogy.* d58:768–776, 2002.

Jang, J.H. and S. Hanash. Profiling of the cell surface proteome. *Proteomics* 3:1947–1954, 2003.

Jung, G.Y. and G. Stephanopoulos. A functional protein chip for pathway optimization and in vitro metabolic engineering. *Science* 304:428–431, 2004.

Juy, M., F. Penin, A. Favier, A. Galinier, R. Montserret, R. Haser, J. Deutscher, and A. Bockmann. Dimerization of Crh by reversible 3D domain swapping induces structural adjustments to its monomeric homologue Hpr. *J. Mol. Biol.* 332:767–776, 2003.

Kim J.Y., Z.A. Yuan, M. Cilia, Z. Khalfan-Jagani, and D. Jackson. Intercellular trafficking of a KNOTTED1 green fluorescent protein fusion in the leaf and shoot meristem of *Arabidopsis*. *Proc. Natl. Acad. Sci. U.S.A.* 99:4103–4108, 2002.

Kondo, E., H. Suzuki, A. Horii, and S. Fukushige. A yeast two-hybrid assay provides a simple way to evaluate the vast majority of hMLH1 germ-line mutations. *Cancer Res.* 63:3302–3308, 2003.

Kretsinger, R.H., S. Hovmöller, and R.E. Ison. Prediction of protein structure. *Methods Enymole.* 383:1–27, 2004.

Kristensen, B., S. Birkelund, and P.L. Jorgensen. Trafficking of Na,K-ATPase fused to enhanced green fluorescent protein is mediated by protein kinase A or C. *J. Membr. Biol.* 191.25–36, 2003.

Lee S., M.E. Sowa, J.M. Choi, and F.T.F. Tsai. The ClpB/Hsp104 molecular chaperone—a protein disaggregating machine. *J. Struct. Biol.* 146:99–105, 2004.

Lin, D., D.L. Tabb, and J.R. Yates. Large-scale protein identification using mass spectrometry. *Biochim. Biophys. Acta* 1646:1–10, 2003.

Liu, Y. and D. Eisenberg. 3D domain swapping: as domains continue to swap. *Prot. Sci.* 11:1285–1299, 2002.

Lueking, A., A. Possling, O. Huber, A. Beveridge, M. Horn, H. Eickhoff, J. Schuchardt, H. Lehrach, and D.J. Cahill. A nonredundant human protein chip for antibody screening and serum profiling. *Mol. Cell. Proteomics* 2:1342–1349, 2003.

McFadden, G.I. Endosymbiosis and evolution of the plant cell. *Curr. Opin. Plant Biol.* 2:513–519, 1999.

Miyawaki, A. Visualization of the spatial and temporaldynamics of intracellular signaling. *Dev Cell* 4:295–305, 2003.

Moriuchi, H., C. Okamoto, R. Nishihama, I. Yamashita, Y. Machida, and N. Tanaka. Nuclear localization and interaction of RolB with plant 14-3-3 proteins correlates with induction of adventitious roots by the oncogene rolB. *Plant J.* 38:260–275, 2004.

Moult, J., K. Fidelis, A. Zemla, and T. Hubbard. Critical assessment of methods of protein structure prediction (CASP)—round V proteins. *Struct. Funct. Genet.* 53:334–339 (Suppl. 6), 2003.

Nakai, K. Review: prediction of in vivo fates of proteins in the era of genomics and proteomics. *J. Struct. Biol.* 134:103–116, 2001.

Orchard, S., H. Hermjakob, R.K. Julian, K. Runte, D. Sherman, J. Wojcik, W.M. Zhu, and R. Apweiler. Common interchange standards for proteomics data: public availability of tools and schema. *Proteomics* 4:490–491, 2004.

Ott, C.M. and V.R. Lingappa. Integral membrane protein biosynthesis: why topology is hard to predict. *J. Cell Sci.* 115:2003–2009, 2002.

Philipps, B., J. Hennecke, and R. Glockshuber. FRET-based in vivo screening for protein folding and increased protein stability. *J. Mol. Biol.* 327:239–249, 2003.

Ramachandran, G.N. and V. Sassiekharan. Conformation of polypeptides and proteins. *Adv. Prot. Chem.* 28:283–437, 1968.

Rappsilber J., U. Ryder, A.I. Lamond, and M. Mann. Large-scale proteomic analysis of the human splicesome. *Genome Res.* 12:1231–1245, 2002.

Ritter, M., A. Ravasio, J. Furst, M. Jakab, S. Chwatal, and M. Paulmichl. Self-interactions and protein-membrane interactions of ICln assessed by fluorescence resonance energy transfer (FRET). *Biophys. J.* 86:348A–348A (Part 2 Suppl. S), 2004.

Rousseau, F. and J.W.H. Schymkowitz. The unfolding story of three-dimensional domain swapping. *Structure* 11:243–251, 2003.

Ryan, M.T. Chaperones: inserting beta barrels into membranes. *Curr. Biol.* 14:R207–R209, 2004.

Scamborova P., A. Wong, and J.A. Steitz. An intronic enhancer regulates splicing of the twintron of *Drosophila melanogaster* prospero pre-mRNA by two different spliceosomes. *Mol. Cell Biol.* 24:1855–1869, 2004.

Seong, S.Y. and C.Y. Choi. Current status of protein chip development in terms of fabrication and application. *Proteomics* 3:2176–2189, 2003.

Shumilin, I.A., R. Bauerle, J. Wu, R.W. Woodard, and R.H. Kretsinger. Crystal structure of the reaction complex of 3-deoxy-D-*arabino*-heptulosonate-7-phosphate synthase from *Thermotoga maritima*. *J. Mol. Biol.* 341:455–466, 2004.

Smothers, J.F., S. Henikoff, and P. Carter. Affinity selection from biological libraries. *Science* 298:621–622, 2002.

Stuart, G.W., K. Moffett, and J.J. Leader. A comprehensive vertebrate phylogeny using vector representations of protein sequences from whole genomes. *Mol. Biol. Evol.* 19:554–562, 2002.

Temin, H.M. and S. Mizutani. RNA-dependent DNA polymerase in virions of Rous sarcoma virus. *Nature* 226:1211–1213, 1970.

Valpuesta, J.M., J. Martin-Benito, P. Gomez-Puertas, J.L. Carrascosa, and K.R. Willison. Structure and function of a protein folding machine: the eukaryotic cytosolic chaperonin CCT. *FEBS Lett.* 529:11–16, 2002.

Vazquez, A., A. Flammini, A. Maritan, and A. Vespignani. Global protein function prediction from protein–protein interaction networks. *Nat. Biotechnol.* 21:697–700, 2003.

Venter, J.C. et al. The sequence of the human genome. *Science* 291:1304–1351, 2001.

Wimley, W.C. The versatile beta-barrel membrane protein. *Curr. Opin. Struct. Biol.* 13:404–411, 2003.

Wood, T.C. and W.R. Pearson. Evolution of protein sequences and structures. *J. Mol. Biol.* 291:977–995, 1999.

Woodham, C., L. Birch, and G.S. Prins. Neonatal estrogen down-regulates prostatic androgen receptor through a proteosome-mediated protein degradation pathway. *Endocrinology* 144:4841–4850, 2003.

Zhang, N.S., M. Osborn, P. Gitsham, K.Y. Yen, J.R. Miller, and S.G. Oliver. Using yeast to place human genes in functional categories. *Gene* 303:121–129, 2003.

Zhu, H., M. Bilgin, and M. Snyder. Proteomics. *Annu. Rev. Biochem.* 72:783–812, 2003.

2

Basics of Fluorescence and FRET

ZYGMUNT GRYCZYNSKI, IGNACY GRYCZYNSKI,
AND JOSEPH R. LAKOWICZ

1. INTRODUCTION

In the last 20 years, there has been a remarkable growth of scientific interest in examining the properties of living cells and membranes using fluorescence microscopy techniques. During this time, the fluorescence microscopy field has been advanced by the application of steady-state and time-resolved digital imaging techniques, as well as by the recent development of convenient probes suitable for in vitro and in vivo labeling.

Fluorescence quickly became one of the dominant technologies in biomedicine. The rapid growth of applications to a variety of scientific fields calls for a better understanding of the basic principles of fluorescence by practitioners. Fluorescence resonance energy transfer (FRET) is probably the most widely applied technology that allows researchers to study molecular processes with subnanometer resolution. The enormous power of FRET lies in the fact that it yields not only static information but also insight into the internal mobility and flexibility of bimolecular systems.

The idea of nonradiative resonance energy transfer originated almost 80 years ago. It was stimulated by initial reports on fluorescence self-depolarization in solutions (Gaviola and Pringshein, 1924; Perrin, 1925, 1926a, 1926b; Wawilow and Lewshin, 1923; Weigert, 1920; Kawski, 1983). The first approximation of the interaction of two oscillating dipoles was presented by Perrin in 1925 (Perrin, 1925, 1927) and was based on classical physics. However, even subsequent quantum mechanical improvements of this model (Perrin, 1932) were unable to give a quantitative account of the experimental data measured for fluorophores at high concentrations. The exact theory of fluorescence (Förster) resonance energy transfer (FRET) was first correctly explained by Förster almost 20 years later (Förster, 1946, 1947, 1949).

Over the last 40 years, FRET has been widely used to study all kinds of molecular assemblies. The most common is to measure the distances between two sites on a macromolecule. Typically, a macromolecule can be covalently labeled with a donor (D) and an acceptor (A). For a fixed single donor and acceptor, the transfer efficiency can be determined from steady-state measurements of donor quenching by the acceptor. However, for many systems like protein assemblies, membranes, and DNA/RNA, a single separation does not apply and the distance distribution of D-A separations has to be used. Also, molecular dynamics of typical biological systems may lead to a significant change in D-A separation during the donor lifetime. In such cases the information from steady-state measurements is not always accurate and the more advanced technology of time-resolved measurements should be used.

In this chapter we describe the most important characteristics of fluorescence that play a fundamental role in understanding the basics and applications of Förster resonance (radiationless) energy transfer (FRET).

2. BASICS OF FLUORESCENCE SPECTROSCOPY

Luminescence is the emission of light from a substance through electronically excited states. *Photoluminescence* is the emission of light from the electronically excited state that was stimulated by the absorption of light. Depending on the nature of the excited state, photoluminescence can be divided into two categories: fluorescence and phosphorescence. *Fluorescence* is the emission of light from the excited singlet state. Because this type of transition is usually allowed within the molecular orbitals, the emission rates of fluorescence are fast, on the order of 10^8 s^{-1}, resulting in nanosecond fluorescence lifetimes. In contrast, *phosphorescence* is the emission of light from the triplet excited state and this transition is typically forbidden. The emission rates are much smaller, on the order of 10^0 to 10^3 s^{-1}, and phosphorescence lifetimes are usually much longer, typically milliseconds to seconds.

The excitation of molecules by light occurs via the interaction of molecular dipole transition moments with the electric field of the light and, to a much lesser extent, interaction with the magnetic field. A detailed description of excitation processes is provided by quantum electrodynamics, but for our purposes we will consider much simpler semiclassical descriptions in which the states of the molecules are quantified but the electromagnetic field is viewed in the classical sense of Maxwell (Michl and Thulstrup, 1986).

In this section we will provide a simplified and limited quantitative view of the basics of one-photon and two-photon absorption as well as emission. In general, the absorption of typical substances is characterized by a wavelength-dependent absorption cross section. Emission (fluorescence) is characterized by its emission spectra, quantum yield, fluorescence lifetime, and polarization. The processes following light absorption and emission are usually illustrated by a Jabłoński diagram, shown in Figure 2–1.

2.1 Absorption

To understand light absorption, one should account for two well-accepted facts. First, the transition of energy $E_0 \rightarrow E_i$ requires photons of frequency $v_i = (E_i - E_0)/h$, where h is the Planck constant. Second, the strength of the molecule–light interaction depends on the strength of the electric and magnetic field of the light wave (light intensity) and on the mutual orientation of the field vectors (electric and magnetic)

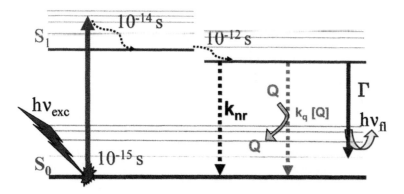

FIGURE 2–1 Jabłoński diagram with collisional quenching. The term $k_q[Q]$ represents concentration-dependent quenching. exc, excitation; fl, fluorescence; h, Planck constant; k_{nr}, rate of non-radioactive decay; Q, quencher.

and the molecular transition moments. We usually need to consider only the electric dipole transition moment, since contribution of the higher-order components of quadrupole transition moments and the magnetic dipole transition moments are negligible in normal conditions (low-light intensity). In this case, one needs to consider the electric dipole transition moment and the electric field vector of the light wave. (Mickel and Thulstrup, 1986)

To describe light absorption in such a simple case, consider a monochromatic light beam of incident intensity I_0 and a wavelength λ passing through an absorbing homogenous isotropic sample. Assume that the molar concentration of absorbing solute (molecules) is C (mol/L), so the number of molecules per cubic centimeter is $n' = CN/1000$, where N is Avogadro's number, $N = 6.023 \cdot 10^{23}$ per mol. After passing through a layer of l cm thickness, the light intensity I is given by the Lambert-Beer law:

$$I(\lambda) = I_0\,(\lambda)\cdot 10^{-\varepsilon(\lambda)Cl} = I_0\,(\lambda)\cdot e^{-\sigma(\lambda)n'l} \qquad (1)$$

where $\varepsilon(\lambda)$ is the wavelength-dependent decadic molar extinction coefficient and $\sigma(\lambda)$ is the wavelength-dependent absorption cross section for the solute in the particular solvent used. The exponent part of equation (1), $\varepsilon(\lambda)Cl$, is called the absorbance or optical density of the sample. From equation (1), one may calculate a straightforward relation between the extinction coefficient and absorption cross section, $\sigma(\lambda) = 2303\ \varepsilon(\lambda)/N$. In this relationship, the extinction coefficient is in L mol^{-1} cm^{-1} and the absorption cross section is in cm^2 per molecule.

Figure 2–2 shows the typical absorption and emission spectra for perylene and quinine sulfate. Absorption is the wavelength-dependent plot of the negative of

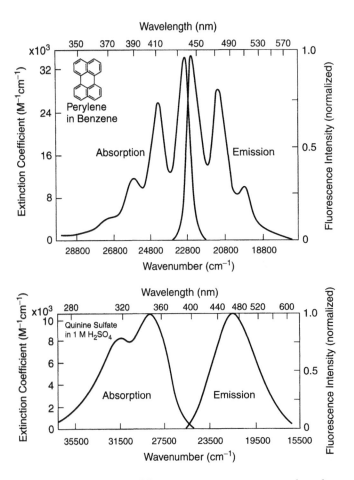

Figure 2–2 Absorption and fluorescence emission spectra of perylene and quinine. [Adapted from Lakowicz (1999).]

the exponent in equation (1), $\varepsilon(\lambda)Cl$, and the emission spectrum is the normalized wavelength-dependent intensity.

2.2 Multiphoton Absorption

Very intense radiation fields, such as those produced by ultrafast femtosecond lasers, can cause simultaneous absorption of two or more photons (two-photon, three-photon absorption, etc.). This phenomenon was originally predicted by Maria Göppert-Mayer in 1931. In general, the contributing photons do not need the same energy. We use the term *two-photon two-color absorption* (Lakowicz et al., 1996) to describe the case in which two photons have different energy (wavelength). The quantum-mechanical theory of two-photon absorption is rather complex; the interested reader is referred to original descriptions (Göppert-Mayer, 1931; Mazurenko, 1971; McClain and Harris, 1977) for a detailed account of this theory. In a simplified description, one photon is needed to mix a small amount of all excited state functions: ψ_j into the ground state ψ_0 to produce the virtual state, which is the energetically perturbed nonstationary state of the molecule. To imagine the virtual state one may also use the Heisenberg principle. For a very short well-defined period of time (in the range of femtosecods), the uncertainty of energy spreads over a large gap to satisfy the relationship $\delta t \times \delta E \geq \hbar$. The incoming low-energy photon will have a small but finite probability of being absorbed to the virtual state. The mixing of excited states ψ_j into the virtual state is favorable if the transition moments for the excitation $0 \rightarrow j$ have a large magnitude and if their orientation relative to the electric vector of the light is appropriate. The promotion from the virtual state to the final state ψ_f is stimulated with the energy of another photon that will be similarly favored if the transition moments for $j \rightarrow f$ are large and favorably oriented. Contributions to the excitation amplitude from all excited states ψ_j should be summed in a coherent fashion, appropriately weighting factors that reflect the mismatch between light-appropriate frequencies. Both photons are absorbed simultaneously, even if they have a different wave number (wavelength), because they contribute to the expression for probability of absorption in a symmetrical fashion.

An examination of absorption rates for sequential absorption steps as shown in Figure 2–3 leads to simple approximate expressions for the multiphoton cross section (Faisal, 1987):

$$\sigma_2 = \sigma_{ij}\sigma_{jf}\tau_j \qquad \text{Two Photon}$$
$$\sigma_3 = \sigma_{ij}\sigma_{jk}\sigma_{kf}\tau_j\tau_k \qquad \text{Three Photon}$$

(2)

where σ_{ij}, σ_{ik}, σ_{kf} represent appropriate one-photon absorption cross sections and τ_j, τ_k are intermediate-state lifetimes that determine the timescale for simultaneous arrival of the photons. The intermediate-state lifetime is related through the uncertainly principle to the reciprocal of the difference of transition frequency and photon frequency.

There are important physical consequences for multiphoton absorption processes that follow from quantum-mechanical predictions. First, the probability of such a process is very low and the absorption cross sections are extremely small. Typically, two-photon absorption cross sections are on the order of 10^{-50} to 10^{-48}

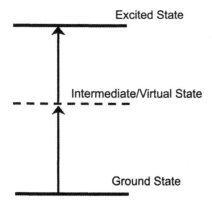

FIGURE 2–3 Schematic illustration of intermediate state.

and three-photon absorption cross sections are on the order of 10^{-83} to 10^{-81}. Such small numbers are very difficult to measure. They are usually estimated by comparison of fluorescence signal rather than by direct absorption measurement. The direct absorption measurement is very difficult because it depends on incident light intensity, spatial distribution of the incident light, and temporal coherence (pulse overlap in the time space).

It is important to realize that fluorescence intensity resulting from one-photon and two-photon excitation has a different dependence on excitation light intensity (power). For one-photon excitation with low absorption, the observed fluorescence intensity F_1 is given by (McClain and Harris, 1977)

$$F_1 = KQP\varepsilon Cl = KQP\sigma_1 n'l \tag{3}$$

where K is the experimental parameter describing the efficiency of light collection, Q is the quantum yield of the fluorophore, P is the laser power, and σ_1 is the absorption cross section.

For two- photon and higher-order excitation processes, the fluorescence intensity F_n is given by

$$F_2 = KQ\frac{P^2}{A}\sigma_2 n'l$$
$$\dots$$
$$F_n = KQ\frac{P^n}{A}\sigma_n n'l \tag{4}$$

where A is the area of the excitation beam cross section. It is important to realize that the units of absorption cross section for multiphoton excitation are different for each excitation order. For example, the unit for two-photon excitation is cm^4 s/(photon molecule).

Consequently, the fluorescence intensity depends on squared laser power for two-photon excitation, cubed (third) power for three-photon excitation, and the fourth power for four-photon excitation. This very strong dependence of fluorescence signal on the excitation power is frequently used to control the mode of excitation. The examples of power dependence for two- and three-photon excitations are shown in Figure 2–4. The upper plot shows the emission intensity changes for one-, two-, and three-photon excitations when excitation power drops by half. The lower panel shows power-dependent intensities for two- and three-photon excitations.

2.3 Emission

To understand the quantum-mechanical basics of spontaneous emission, one must go beyond the semiclassical description and introduce the perturbation of the molecular stationary state by the fluctuation of the electromagnetic field. In this chapter we shall limit our description to a presentation of the most important facts and characteristic parameters for fluorescence processes.

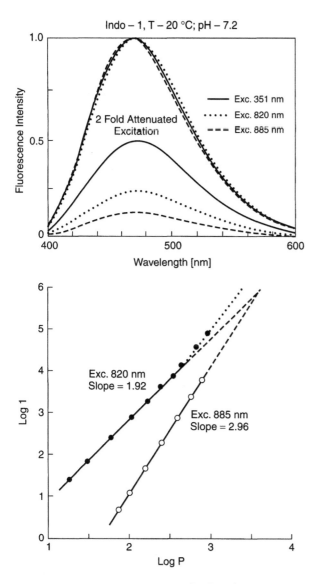

FIGURE 2–4 *Top*: Emission spectra of Indo-1 for excitation at 351, 820, and 885 nm. Also shown are the emission spectra for each excitation with twofold attenuated incident light. *Bottom*: Dependence of the emission intensity of Indo-1 on laser power (in mW) at 820 and 885 nm. [Adapted from Gryczynski and Lakowicz (1997).]

The phenomenon of fluorescence displays many general characteristics. The most common ones are emission spectra, Stokes' shift, fluorescence quantum yield and lifetime, and fluorescence polarization.

2.3.1 Stokes' shift of the emission spectrum

The Jabłoński diagram in Figure 2–1 shows that the energy of the emitted photon is typically less than that of the absorbed photon. Hence, the fluorescence occurs at lower energy (longer wavelength). This phenomenon was first described by Sir

G. G. Stokes in 1852. This characteristic energy loss between excitation and emission is universally observed for fluorescent molecules in solution (Lakowicz, 1999). The reasons for Stokes' shift are rapid transition to the lowest vibrational energy level of the excited state S_1 and decay of the fluorophore to a higher vibrational level of S_0. The excess of excitation energy is typically converted to thermal energy.

2.3.2 Emission spectra

A *fluorescence emission spectrum* is a plot of the fluorescence intensity vs. wavelength or wavenumber. Examples of some fluorescence spectra are shown in Figure 2–2. The emission spectra are dependent on the chemical structure of the fluorophore and the solvent in which the fluorophore is dissolved. Usually for rigid and planar molecules such as perylene, the emission spectrum may show significant structure corresponding to the individual vibrational energy levels of the ground state. For other fluorophores, such as quinine, a spectrum does not show vibrational structures at room temperature.

An important characteristic of the emission spectrum is that its shape is generally independent of the excitation wavelength. This is known as Kasha's rule (Kasha, 1950). Upon excitation into higher electronic and vibrational levels, the excess energy is quickly dissipated (in 10^{-12} s) via internal processes and interaction with the solvent, leaving the fluorophore in the lowest vibrational level of S_1 (also called the fluorescent state). Because of this rapid relaxation, the fluorophore always emits from the lowest excited state and observed emission spectra are independent of the excitation wavelength. Very rare exceptions are some molecules such as azulenes that may emit from the S_2 level (Birks, 1973).

2.3.3 Fluorescence quantum yield

The *fluorescence quantum yield* is the number of emitting photons relative to the number of absorbed photons. One should realize that even a 100% quantum yield is less than 100% of energetic yield (energy emitted relative to energy absorbed) because of energetic differences between absorbed and emitted photons (Stokes' losses). The meaning of quantum yield is best represented by a simplified Jabłoński diagram (Fig. 2–1). The emission process is governed by the rate constants (emissive rate of fluorophore) and k_{nr} (rate of nonradiative decay), which both depopulate the excited state. The fraction of fluorophores that decay through emission rate Γ directly contributes to the quantum yield:

$$Q = \frac{\Gamma}{\Gamma + k_{nr}} \tag{5}$$

The quantum yield approaches unity when $\Gamma \gg k_{nr}$. For convenience we have grouped all possible nonradiative decay processes with the single rate constant k_{nr}. Usually the nonradiative rates are difficult to separate and predict. The radiative rate constant can be calculated from the fluorophore spectral properties (Birks, 1973; Strickler and Berg, 1962):

$$\Gamma = 2.88 \cdot 10^{-9} n^2 \frac{\int_0^\infty F(v)\,dv}{\int_0^\infty \frac{F(v)}{v^3}\,dv} \int_0^\infty \frac{\varepsilon(v)}{v}\,dv = 2.88 \cdot 10^{-9} n^2 \langle v^3 \rangle^{-1} \int_0^\infty \frac{\varepsilon(v)}{v}\,dv \qquad (6)$$

where $F(n)$ is the wavenumber dependent emission spectrum, $\varepsilon(n)$ is the extinction coefficient for the absorption spectrum, and n is the refractive index of the medium. The integrals are calculated over the $S_0 \leftrightarrow S_1$ absorption and emission spectra. Although equation (6) does not account for solvent interaction and refractive index change, it works rather well in many occasions. In particular, it is well suited for rigid fluorophores of high quantum yield such as perylene (Birks, 1973).

2.3.4 Fluorescence quenching

Many processes can decrease the intensity (or quench) fluorescence. There are a number of mechanisms by which fluorescence quenching may occur. The most common are collisional quenching, static quenching, and electron transfer.

Collisional quenching is the deactivation of excited fluorophores by contact with some other molecule in solution that increases the nonradiative rate, shown in the Jabłoński diagram in Figure 2–1. This other molecule is called a *quencher*. In this case the fluorophore is returned to the ground state during a collision encounter with the quencher. The fluorophore molecules are not chemically altered in the process. Typical colissional quenchers include oxygen, iodide, halogens, amines, and acrylamide. For collisional quenching, fluorescence intensity can be described by the Stern-Volmer equation:

$$F = \frac{F_0}{1 + K[Q]} = \frac{F_0}{1 + k_q \tau_0 [Q]} \qquad (7)$$

where K is the Stern-Volmer quenching constant, k_q is the bimolecular quenching constant, τ_0 is the lifetime for the unquenched fluorophore, and $[Q]$ is the quencher concentration. A more detailed description of quenching processes and the effect on fluorescence can be found in Lakowicz (1999).

2.4 Fluorescence Lifetime

The fluorescence lifetime is perhaps the most important characteristic of fluorescence. The measurements of fluorescence lifetimes provide much more information about molecular systems than typically is available from steady-state intensity measurements.

The lifetime of the excited state (fluorescence lifetime) is the average time the molecule spends in the excited state prior to returning to the ground state. The meaning of the fluorescence lifetime can be represented by the Jabłoński diagram shown in Figure 2–1. After excitation, the deactivation process is governed by the emissive rate constant Γ and nonradiative rate constant k_{nr} that depopulates the excited state. For the system in Figure 2–1 the lifetime is

$$\tau = \frac{1}{\Gamma + k_{nr}} \tag{8}$$

It is important to have a firm understanding of the physical meaning of the fluorescence lifetime, τ. Emission is a statistical process and τ represents the average time the molecules spend in the excited state. Suppose a sample containing a large number of fluorophore molecules is excited with an infinitely sharp pulse (δ function) of light. The pulse excites an initial population (N_0) of fluorophores into the excited state. The excited state population decays with the cumulative rate $\Gamma + k_{nr}$ according to the relation

$$\frac{dN(t)}{dt} = -(\Gamma + k_{nr}) \cdot N(t) \tag{9}$$

where $N(t)$ is the number of excited molecules ($N(t_0) = N_0$). Emission is a random process, and each excited fluorophore has an equal probability of emitting a photon in a given period of time. Since fluorescence intensity is proportional to the number of molecules in the excited state, $N(t)$, equation (9) can be written in terms of time-dependent intensity $I(t)$. Solution of equation (9) yields the following expression for a single exponential decay:

$$I(t) = I_0 \cdot e^{-t(\Gamma + k_{nr})} = I_0 \cdot e^{\frac{t}{\tau}} \tag{10}$$

where I_0 is the initial intensity. For a single exponential decay, 63% of the molecules have decayed prior to $t = \tau$ and 37% decay after $t > \tau$.

It is important to understand statistical meaning of excited state decay. For example, a fluorescence lifetime of 10 ns means that the population of the excited state decreases exponentially e-fold during the first 10 ns, but it is a distinct probability that excited molecules will be found after a much longer time.

Fluorescence lifetime in the absence of nonradiative processes is called the *intrinsic* (natural) *fluorescence lifetime* and can be calculated from the absorption spectra, excitation coefficient, and emission spectra:

$$\tau_n = \frac{1}{\Gamma} \tag{11}$$

where Γ is given by the equation (6). Natural lifetime can always be calculated from the measured lifetime, τ, and quantum yield, Q:

$$\tau_n = \frac{\tau}{Q} \tag{12}$$

For most fluorophores, lifetimes are on the order of 1 to 10 ns. There are also a significant number of molecules with shorter lifetimes in the picosecond range, and a limited number of molecules with fluorescence lifetimes in the microsecond or submillisecond range.

The fluorescence lifetime and the quantum yield can be modified by factors that affect either rate constant (Γ or k_{nr}). There are a number of processes that may change rate constants. The most common ones are a variation of the nonradiative rate due to interaction with the solvent, collisional quenching by other molecules or ions, electron transfer, or radiationless energy transfer to a suitable acceptor. The exclusive sensitivity of the lifetimes to the fluorophore external conditions leads to the valuable concept of detecting different processes in solution, or in macromolecules, by means of measuring the fluorescence lifetime. Fortunately, numerous biological processes, such as diffusion in a solution, have a timescale comparable to fluorescence lifetimes. For example, the diffusion coefficient of oxygen in water at 25°C is about 2.5 10^{-5} cm^2 s, which gives the average distance an oxygen molecule can diffuse in 1 ns as about 7 Å. Typical fluorescence lifetimes easily allow diffusion throughout most proteins or membranes. Measurement of fluorescence lifetime provides a powerful means to study FRET in solution and in macromolecules. FRET from donor (D) to acceptor (A) is a dynamic quenching process that mostly affects donor lifetime. Measurements of donor lifetime may not only provide information about average D–A separation but also give an accurate depiction of dynamic motion of macromolecular components (Haas et al., 1975, 1978a; Amir and Haas, 1986; Lakowicz, 1999).

2.4.1 Time-resolved measurements

Time-resolved measurements provide much more information about molecular systems than what is available from steady-state measurements, and sophisticated time-resolved measurements are widely used in fluorescence spectroscopy (and more recently in microscopy) to study physicochemical processes and biological macromolecules. For instance, by measuring fluorescence lifetimes of the system one may detect multiple molecular species with characteristic lifetimes, distinguish static from dynamic quenching, and evaluate the extent of resonance energy transfer.

The most commonly used methods for time-resolved fluorescence measurements are time-domain, streak camera method, and frequency-domain. In the time-domain and streak camera methods, the sample is excited with a pulse of light. The time-dependent intensity is measured following the excitation pulse, and the decay time is calculated from the slope of a plot of log $I(t)$ vs. t (Fig. 2–5, right). In the frequency-domain method, the sample is excited with intensity-modulated light (Fig. 2–6). The frequency of light intensity modulation is comparable to the reciprocal of the decay time τ, which is much smaller than the frequency of the electric component of an electromagnetic wave of the light. Because of the lifetime of the fluorophore, the emission is delayed in time and modulation decreases relative to the excitation. This delay, measured as a phase shift (φ) and demodulation (m), can be used to calculate the decay time.

Typically, intensity measurements (and especially intensity decay measurements) should be done under magic angle conditions to avoid the effects of rotational diffusion and/or anisotropy change on the intensity decay. For vertical polarization of the excitation light, measurements should be made through a polarizer oriented at 54.7 from the vertical axis (for details see Emission Anisotropy, below). Typical examples of time-resolved fluorescence measurements (time-domain

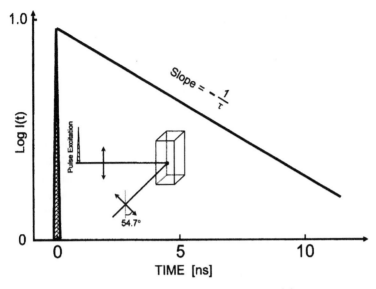

FIGURE 2–5 Concept of time-domain (pulse excitation) lifetime measurements.

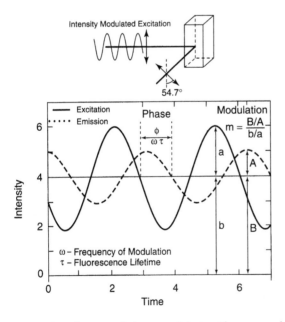

FIGURE 2–6 Concept of phase-modulation (frequency-domain) lifetime measurements.

and frequency-domain modes) for N-Acetyl Tryptophan Anide (NATA) are shown in Figure 2–7 (Lakowicz, 1999).

2.5 Emission Anisotropy

Linearly polarized light selectively excites fluorophore molecules that have a dipole of absorption transition moments parallel to the electric vector of the excitation beam. The photoselective excitation process (photoselection) is the origin for observed emission anisotropy in the isotropic media. Anisotropy measurements are commonly used in the biological applications of fluorescence. Such measurements provide information on the size and shape of biomolecules and/or the rigidity of various molecular environments.

Theoretical consideration of emission anisotropy is based on three assumptions:

1. There is no correlation between the excitation light and light emitted by the fluorophore. In fact, the absorption process is very quick (10^{-15} s) compared to the average fluorescence lifetime (which is in the nanosecond range). This assumption is generally true for low-light intensities when the stimulated emission is negligible.

2. The emitting molecules are incoherent, and there is no dependence between photons emitted by different fluorophores.

3. The direction of emission transition moments is fixed within the molecular coordinates and is independent of the excitation mode.

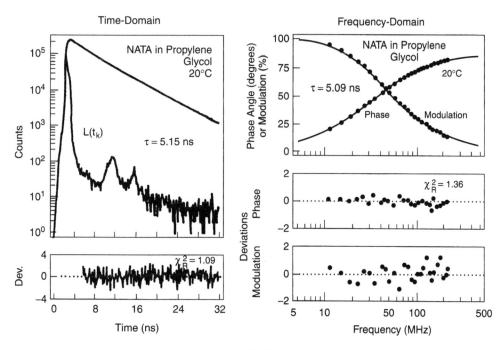

FIGURE 2–7 Time-resolved measurements of NATA in propylene glycol. *Left*: time-domain method. *Right*: Frequency-domain method. [Adapted from Lakowicz (1999).]

The anisotropy of the emission field is described by the following relationship first presented by Jabłoński (Jabłoński, 1960, 1961, 1967, Kawski, 1993):

$$r^2 = \frac{(I_x - I_y)^2 + (I_x - I_z)^2 + (I_y - I_z)^2}{2I^2} \tag{13}$$

where $I = I_x + I_y + I_z$ is the total intensity and $I_z \geq I_x \geq I_y$ are relative light intensities with the electric vector aligned along the appropriate axis of the coordinate system in Figure 2–8.

Assume that a rigid isotropic solution of fluorophores is excited by linearly polarized light coming in on the sample along the y-axis as in Figure 2–8. The axial symmetry of the system around the excitation electric vector requires $I_x = I_y$ and one may obtain

$$r^2 = \left(\frac{I_z - I_y}{I}\right)^2 \tag{14}$$

For linearly polarized excitation light, I_z and I_y are called $I_\parallel$ and $I_\perp$ (parallel and orthogonal component):

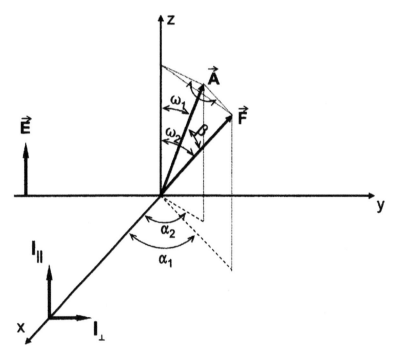

FIGURE 2–8 Coordinate system for emission anisotropy calculation. $\vec{A}$, absorption transition moment; $\vec{E}$, electric vector; $\vec{F}$, emission transition moment.

$$r = \frac{I_{\parallel} - I_{\perp}}{I} = \frac{I_{\parallel} - I_{\perp}}{I_{\parallel} + 2I_{\perp}} \tag{15}$$

Since

$$I_{\perp} = \frac{I - I_{\parallel}}{2} \tag{16}$$

equation (15) can be written as follows:

$$r = \frac{3}{2}\frac{I_{\parallel}}{I} - \frac{1}{2} \tag{17}$$

For the case in Figure 2–8, the electric vector of the excitation light is oriented along the z-axis. The projection of the electric vector onto the arbitrary absorption transition moment that forms an angle ω_1 with the z-axis is proportional to $\cos \omega_1$. Since the light intensity is $I \sim E^2$, the probability of light absorption by such an absorption vector is proportional to $\cos^2 \omega_1$. Similarly, emission intensity parallel to the z-axis for a single emission vector that forms angle ω_2 with the z-axis is proportional to $\cos^2 \omega_2$, and for assemblies of fluorophores to the weighted average (Kawski and Gryczynski, 1986),

$$I_{\parallel} = \langle \cos^2 \omega_2 \rangle \quad \text{and} \quad \frac{I_{\parallel}}{I} = \langle \cos^2 \omega_2 \rangle \tag{18}$$

Inserting equation (18) into equation (17) results in:

$$r = \frac{3}{2}\frac{I_{\parallel}}{I} - \frac{1}{2} = \frac{3}{2}\langle \cos^2 \omega_2 \rangle - \frac{1}{2} \tag{19}$$

Taking into account the simple geometric relation between the angles ω_1, ω_2, and δ: $\cos \omega_2 = \cos \omega_1 \cos \beta + \sin \omega_1 \sin \beta \cos \delta$ one may calculate the following:

$$\langle \cos^2 \omega_2 \rangle = \cos^2 \omega_1 \cos^2 \beta + \frac{1}{2}\sin^2 \omega_1 \sin^2 \beta = \left(\frac{3}{2}\cos^2 \omega_1 - \frac{1}{2}\right)\left(\cos^2 \beta - \frac{1}{3}\right) + \frac{2}{6} \tag{20}$$

where β is the fixed angle between absorption and emission transition moments. To calculate anisotropy, one will need to average the $\cos^2 \omega_1$ functions:

$$\langle \cos^2 \omega_2 \rangle = \frac{\int_0^{\pi/2} \cos^2 \omega_1 f(\omega_1) d\omega_1}{\int_0^{\pi/2} f(\omega_1) d\omega_1} \tag{21}$$

where $f(\omega_1)$ is the distribution of the absorption transition moments owing to the polarized absorption (photoselection). For one-photon excitation,

$$f_1(\omega_1) = \cos^2 \omega_2 \sin \omega_1 \tag{22}$$

One will then calculate

$$r = \frac{2}{5}\left(\frac{3}{2}\cos^2 \beta - \frac{1}{2}\right) \tag{23}$$

The relationship expressed in equation (23) is known as a *Perrin product*. It is didactic to consider possible values for emission anisotropy for different angles between the absorption and emission transition moments. For $\beta = 0°$ (collinear absorption and emission transition moments), $r = 2/5 = 0.4$; for $\beta = 30°$, $r = 0.25$; and for $\beta = 90°$, $r = -0.2$.

It is interesting to consider emission anisotropy for multiphoton absorption phenomena. In two-photon absorption, two photons simultaneously interact with and are absorbed by the fluorophore. For our purposes, we can simply assume that the second photon interacts only with the ensemble of molecules, which were preselected by the first photon. Therefore, the only nonzero elements in the two-photon transition tensor (Johnson and Wan, 1997) are the diagonal elements, and one element is significantly larger than the other. The orientation distribution in the excited state followed by two-photon absorption is given by (Lakowicz et al., 1992; Kawski et al., 1993)

$$f_2(\omega_1) = \cos^2 \omega_1 f_1(\omega_1) = \cos^4 \omega_1 \sin \omega_1 \tag{24}$$

and the emission anisotropy for the two-photon process is given by

$$r = \frac{4}{7}\left(\frac{3}{2}\cos^2 \beta - \frac{1}{2}\right) \tag{25}$$

The limiting value for anisotropy ($\beta = 0$) for two-photon excitation is $r = 4/7 = 0.571$, and for $\beta = 30°$, $r = 5/14 = 0.357$, and for $\beta = 90°$, $r = -2/7 = -0.286$. For three-photon excitation one will obtain limiting anisotropy ($\beta = 0$) $r = 7/9$. The experimental values of anisotropy close to these limits have already been reported (Lakowicz and Gryczynski, 1997; Malak et al., 1997).

In earlier publications the term *polarization* was frequently used. The emission polarization is defined

$$P = \frac{I_{\parallel} - I_{\perp}}{I_{\parallel} + I_{\perp}} \tag{26}$$

For an experimental system such as that in Figure 2–8 (with cylindrical symmetry along the z-axis), the anisotropy and polarization values can be interchanged using

$$P = \frac{3r}{2+r} \quad \text{and} \quad r = \frac{2P}{3-P} \tag{27}$$

Use of anisotropy is preferred because most theoretical expressions are much simpler when expressed in term of emission anisotropy. An example is the expression for the average anisotropy of a mixture of fluorophores, each with anisotropy r_i (polarization P_i) and its fractional fluorescence intensity f_i:

$$\bar{r} = \sum_i f_i r_i \tag{28}$$

when in contrast, average polarization is given by

$$\left(\frac{1}{P} - \frac{1}{3}\right) = \sum \frac{f_i}{\left(\frac{1}{P} - \frac{1}{3}\right)} \tag{29}$$

The previous expression using anisotropy is clearly preferable.

Fluorescent molecules are often observed under conditions in which rotational motions can occur during the excited state lifetime. Examination of anisotropy decay may reveal the rates and amplitudes of such fluorophore motions. The decay of fluorescence anisotropy can be analyzed following the pulse excitation:

$$r(t) = \sum r_{0j} e^{\frac{-t}{\theta_j}} \tag{30}$$

where r_{0j} is the amplitudes (anisotropy at $t = 0$) associated with the molecule correlation time θ_j. It is interesting to note how the decay time determines what motion can be observed using fluorescence. The steady-state anisotropy (r) is given by the average of $r(t)$ weighted by $I(t)$:

$$r = \frac{\int_0^{\pi/2} r(t) I(t) dt}{\int_0^{\pi/2} f(\omega_1) d\omega_1} \tag{31}$$

In this equation, the denominator serves to normalize the anisotropy to be independent of total intensity. In the numerator the anisotropy at any time t contributes to steady-state anisotropy according to the intensity at time t. For a single correlation time θ and an exponential decay of fluorescence intensity with the rate constant $1/\tau$, one will obtain from equation (31)

$$r = \frac{r_0}{1 + \left(\tau/\theta\right)} \tag{32}$$

where r_0 is the limiting anisotropy, which would be measured in the absence of rotational diffusion. The relationship expressed in equation (32) is frequently called the *Perrin equation* and was originally derived by F. Perrin in 1932 through use of complex polarization representation. The Perrin equation can be used to follow binding and association reactions of biomolecules. For example, binding of the fluorophore to the protein would slow the probe's rate of rotation, increasing the observed steady-state anisotropy. Similarly, the association of two small proteins would also lead to an increase of observed anisotropy. The essential point is that the rotational correlation times for most proteins are comparable to typical fluorescence lifetimes, and measurements of fluorescence anisotropy are sensitive to any factors that affect the rate of rotational diffusion.

3. ENERGY TRANSFER

Förster resonance energy transfer (FRET) is the transfer of electronic excitation energy between isolated donor D and acceptor A of suitable spectroscopic properties (Fig. 2–9). The donor molecules typically emit at shorter wavelengths, which overlap with the absorption spectrum of the acceptor. This energy transfer occurs without the appearance of the photon and is the result of long-range interactions between D and A dipoles. The efficiency of this process is dependent on the extent of spectral overlap of the emission spectrum of the donor with the absorption spectrum of the acceptor, the quantum yield of the donor, the relative orientation of the donor and acceptor transition moments, and the distance between donor and acceptor molecules. The sixth-power distance dependence of FRET has resulted in its widespread applications to measure separations between donor and acceptor molecules (Steinberg, 1971; Clegg, 1992; Clegg et al., 1994; Van Der Meer et al., 1994; Wu and Brand, 1999; Lakowicz, 1999; Shih et al., 2000).

It is now well established that for D–A separations greater than a few Å, the Förster dipole–dipole mechanism of FRET is the only one of importance. The theoretical concept for transfer of excitation energy by the resonance mechanism has been developed along both the classical model (Perrin, 1925) and the quantum-mechanical model (Perrin, 1932; Förster, 1946, 1948, 1949, 1965). The theory for resonance energy transfer (RET) is rather complex and in this chapter only the final results will be discussed. Readers interested in the physical basis and mathemati-

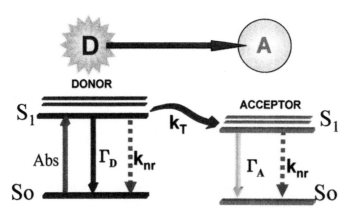

FIGURE 2–9 Jabłoński diagram of representation of FRET. Abs, absorption; k_{nr}, rate of nonradioactive decay; k_t, rate of transfer.

cal derivations of FRET are referred to excellent reviews by Clegg (1992, 1996) and to some original papers by Förster (1948, 1999). For the very weak coupling, where the energy of interaction between donor and acceptor is small compared to the vibrational splitting of the donor energy levels, the classical and quantum-mechanical approaches yield an essentially identical dependence. Considering a single donor and acceptor separated by the distance r, one may calculate the rate of transfer (probability of transfer of energy quantum from donor to acceptor per unit time):

$$k_T = \frac{Q_D \kappa^2}{\tau_D r^6} \left(\frac{9000 \ln 10}{128 \pi N n^4} \right) \int_0^\infty F_D(\lambda) \varepsilon(\lambda) \lambda^4 d\lambda \tag{33}$$

where Q_D is the quantum yield of the donor in the absence of acceptor; τ_D is the lifetime of the donor in the absence of acceptor; n is the refractive index of the medium; N is Avogadro's number; F_D is the normalized fluorescence intensity of the donor (under the curve normalized to unity); $\varepsilon(\lambda)$ is the extinction coefficient of the acceptor at λ; and k^2 is the orientational factor describing the relative orientation in space of the transition moment of the donor and acceptor. The integral in equation (33), referred to as the *overlap integral* $J(\lambda)$, expresses the extent of spectral overlap between the donor emission and acceptor absorption and is given by

$$J(\lambda) = \frac{\int_0^\infty F_D(\lambda) \varepsilon(\lambda) \lambda^4 d\lambda}{\int_0^\infty F_D(\lambda) d\lambda} \tag{34}$$

where $F_D(\lambda)$ is dimensionless. If the extinction coefficient $\varepsilon(\lambda)$ is expressed in units of $M^{-1} cm^{-1}$ and λ in nanometers, then $J(\lambda)$ is in units of $M^{-1} cm^{-1} nm^4$. The overlap integral has been defined in several ways with different units. This sometimes causes confusion if one tries to calculate the so-called R_0 value for a specific donor–acceptor system. We usually recommend use of nanometers or centimeters for the wavelength and $M^{-1} cm^{-1}$ for the extinction coefficient.

R_0 is a characteristic Förster distance and its meaning can be understood from the frequently used form of equation (33), where the right-hand side can be written in the form

$$k_T = \frac{1}{\tau} \left(\frac{R_0}{r} \right)^6$$

and

$$R_0 = 8.79 \times 10^3 \left[Q_D \kappa^2 n^{-4} J(\lambda) \right]^{1/6} \tag{35}$$

In this expression R_0 is given in Å and the overlap integral is given in $M^{-1} cm^{-1}$. It is important to realize that the energy transfer process competes with the spontaneous decay of the donor, which proceeds with the rate constant $1/\tau_D$. The probability p that the donor will not lose its energy at a time t after excitation is given by

$$-\frac{1}{p} \frac{dp}{dt} = \frac{1}{\tau_D} + \frac{1}{\tau_D} \left(\frac{R_0}{r} \right)^6 \tag{36}$$

From this relationship one may understand the phenomenological meaning of R_0, that it is the distance between donor and acceptor at which half of the excitation energy of the donor is transferred to the acceptor, while the other half is dissipated by all other processes, including emission. In other words, for $r = R_0$, 50% of donor excitation energy is transferred to the acceptor and 50% deactivates in all other processes (radiative and nonradiative).

If the transfer rate is much faster than the decay rate ($1/\tau$), then energy transfer will be efficient. If the transfer rate is slower than the decay rate, then little transfer will occur during the excited-state lifetime.

The efficiency of energy transfer (E) is the fraction of photons absorbed by the donor that are transferred to the acceptor. According to Figure 2–9, this fraction is given by

$$E = \frac{k_T}{k_T + k_{nr} + \Gamma} = \frac{k_T}{k_T + 1/\tau_D} \tag{37}$$

which is the ratio of the transfer rate to the total decay rate of the donor. Taking into account equation (35), one would obtain

$$E = \frac{R_0^6}{R_0^6 + r^6} = \frac{1}{1 + (r/R_0)^6} \tag{38}$$

This equation shows that energy transfer efficiency is strongly dependent on distance when the D-A separation is near R_0, as shown in Figure 2–10. The efficiency quickly increases to 1.0 as the D-A separation decreases below R_0. For instance, if

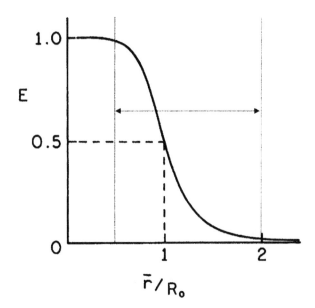

FIGURE 2–10 Dependence of energy transfer (E) efficiency on D–A separation. R_0 is the Förster distance. [Adapted from Lakowicz 1999).]

$r = 0.5\,R_0$ the energy transfer efficiency is 98.5%, and if $r = 2R_0$ the energy transfer efficiency is only 1.5%. This outstanding dependence on $1/r^6$ of FRET has been extensively tested experimentally (Stryer and Haugland, 1967; Steinberg, 1971; Stryer, 1978).

Transfer efficiency is typically measured using the relative fluorescence intensity of the donor, in the absence (F_D) and presence (F_{DA}) of acceptor:

$$E = \frac{F_D - F_{DA}}{F_D} = 1 - \frac{F_{DA}}{F_D} \tag{39}$$

Usually it is more convenient to calculate transfer efficiency from the measured fluorescence lifetimes of the donor:

$$E = \frac{\tau_D - \tau_{DA}}{\tau_D} = 1 - \frac{\tau_{DA}}{\tau_D} \tag{40}$$

where τ_D and τ_{DA} are the fluorescence lifetimes measured in the absence and presence of the acceptor, respectively. Equation (40) is true for a single exponential decay. However, single exponential decays are rare in biomolecules. In general, the intensity average decay time should be used for multiexponential intensity decay.

3.1 Factors Affecting FRET

A crucial step in the practical use of energy transfer is to calculate the R_0 value for a given donor–acceptor pair. According to equation (33), the most important factors to consider are the overlap integral, $J(\lambda)$, the quantum yield of the donor in the absence of acceptor, Q_D, and the orientation factor, κ^2.

3.1.1 Overlap Integral

To calculate the overlap integral one should use equation (34). Neverless, it is important to remember that one use proper units and spectra normalization to obtain appropriate values, as discussed earlier. Typically, absorption spectra are not very sensitive to the probe environment (solvent) or temperature. However, fluorescence spectra are usually sensitive to the fluorophore environment and proper donor emission should be used. The value of the overlap integral thus depends on the medium in which the chromophore is embedded. For example, if one uses tryptophan residues in proteins as a donor, the emission spectrum will strongly depend on the tryptophan localization within the protein structure. A highly buried tryptophan will typically have short wavelength emission (330 nm) in comparison with a surface tryptophan that may have an emission spectrum red-shifted up to 360 nm. A desirable characteristic to consider for donor fluorophores is minimal sensitivity to the environment (solvent). Examples of overlap integrals for some typical D-A systems are presented in Table 2–1.

According to equation (33), the transfer rate is directly proportional to the overlap integral. Haugland and collaborators (1969) tested this dependence extensively

TABLE 2–1. Examples of Overlap Integrals and R_0 Values Calculated for Common Donor–Acceptor Pairs

Donor (Quantum Yield)		Acceptor	$J(M^{-1} cm^{-3})$	$R_0(\text{Å})$	Reference
Indol	(0.34) PG	Dansylamide	4.29×10^{-15}	24.3	Tyson et al., 2000
1,5-DPE	(0.37)	EPE	2.36×10^{-13}	48.7	Lakowicz, 1999
2,5-DPE	(0.76)	EPE	1.54×10^{-13}	51.2	Lakowicz, 1999
2,6-DPE	(0.71)	2,5-DPE	1.3×10^{-15}	22.8	Lakowicz, 1999
AMCA	(0.49)	Cy3	1.66×10^{-13}	49.6	Malicka et al., 2003a
Cy3	(0.24)	Cy5	5.77×10^{-13}	54.2	Malicka et al., 2003b
Fluorescein	(0.40)	Cy5	1.49×10^{-15}	47	Brewer et al., 2004
C-460	(0.58) glycerol	$[Ru(pby)_3]^{2+}$	4.04×10^{-14}	38	Tyson et al., 2000
C-460	(1.00) CH_3CN	$[Ru(pby)_3]^{2+}$	4.06×10^{-14}	43.8	Tyson et al., 2000

by changing the overlap integral over a 40-fold range and proved the validity of Förster's theory.

3.1.2 Donor quantum yield

Similar precautions should be considered for the donor quantum yield. The quantum yield can also be strongly dependent on the surrounding environment. It is generally difficult to find the quantum yield of a donor fluorophore in a protein or peptide. For example, the quantum yields of tyrosine or tryptophan in different proteins vary significantly (Lakowicz, 1999), sometimes by almost an entire order of magnitude. Some authors have suggested using an average value of quantum yield for R_0 calculations. It is rather fortunate that R_0 depends on the sixth root of Q and is therefore not very sensitive to small changes in donor quantum yield.

3.1.3 Index of refraction

In general, the refractive index does not vary much between typical solvents or proteins. However, in the R_0 calculation, this parameter appears as $1/n^4$ and even a relatively small change will have a much bigger effect than quantum yields or overlap integral. In general, the value of n may vary from 1.33 to 1.6 for biological media, which in practice means over 100% changes in $1/n^4$ factor. The refractive index for a typical buffer is only slightly higher than that for water (1.33), but in a protein environment may vary from 1.4 to 1.6. Also, dependence of refractive index on wavelength and temperature should not be ignored.

3.1.4 Orientation factor κ^2

The orientation factor κ^2 gives the dependence of the interaction between two dipoles on their orientation and can be defined by

$$\kappa^2 = \left(\cos \theta_T - 3\cos \theta_D \cos \theta_A \right)^2$$

and

$$\cos \theta_T = \sin \theta_D \sin \theta_A \cos \phi + \cos \theta_D \cos \theta_A$$

(41)

where θ_D and θ_A are the angles between the separation vector R and donor and acceptor vectors, respectively; θ_T is the angle between donor and acceptor transition moments, as shown in Figure 2–11. Depending on the relative orientation of the donor and acceptor, κ^2 can range from 0 to 4. It is important to realize that $\kappa^2 = 4$ only for collinear transition dipole orientation and is significantly smaller for any other orientation. For example, two parallel dipoles will have a much smaller orientation factor, $\kappa^2 = 1$. In fact, a dipole orientation resulting in $\kappa^2 = 0$ is much more probable than one of $\kappa^2 = 4$. For an ensemble of D-A pairs of high mobility (the orientation of transition moments can average before the emission), the average orientation factor $\kappa^2 = 2/3$ and is even smaller for rigid systems with no mobility, $\kappa^2 = 0.476$ (Steinberg, 1968; Dale and Eisinger, 1979). It is important to realize that energy transfer efficiency is directly proportional to κ^2:

$$E = \frac{k_T}{1/\tau_D + k_T} = \frac{\kappa^2}{R^6/C_T + \kappa^2} \tag{42}$$

where C_T is the constant for the given D-A pairs made up of all parameters in equation (33) describing R_0 (overlap integral, the quantum yield, and refractive index). It has been emphasized by Dale and Eisinger (Dale and Eisinger, 1974, 1979; Eisinger et al., 1981) that only when each D-A pair in the experiment has the same orientation or, equivalently, the same extent of dynamic averaging (fast compare with the fluorescence and transfer rates), is it valid to substitute a single or average value for κ^2 in equation (42). This will also be a good approximation for low transfer efficiency in the static averaging regime.

The orientation factor (more precisely, angular dependence of energy transfer efficiency) is rather difficult to establish (control) in a real experimental system. This may result in controversy over experimental data interpretation. Real biological systems have intrinsically significant conformational freedom that results in a wide range of possible D-A orientations leading to a big spread of possible κ^2 values. In this case, one would calculate an average value for the orientation factor, $\langle \kappa^2 \rangle$. In practice, the averaging of expression (41) is usually a complicated procedure, which depends on a number of factors. The situation is only simple for a system with fast

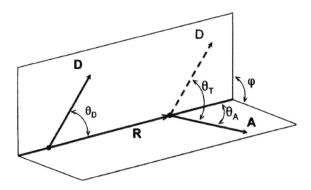

FIGURE 2–11 Representation of angles used to define relative orientation of the donor and acceptor transition moments. R is the separation for donor (D) and acceptor (A). transition moments. R is the separation for donor (D) and acceptor (A). [Revised from Dale and Eisinger (1979).]

rotational dynamics where one may expect full dipole randomization before the transfer occurs. In such cases the average value of the orientation factor $\langle \kappa^2 \rangle = 2/3$. This is the most commonly used value for orientation factor that is true for only this specific case. For example, the same system with isotropic orientational distribution of donor and acceptor transition moments, but with no rotational dynamics (frozen), yields an average value of orientation factor $\langle \kappa^2 \rangle = 0.476$.

Unfortunately, both of these well-defined values for the average orientation factor $\langle \kappa^2 \rangle$ only rarely apply to real experimental systems. For many labeled systems, the mobility of donor and acceptor is significantly limited, precluding full randomization. Also, efficient energy transfer results in a short donor fluorescence lifetime that prevents efficient randomization. In this case, the orientation factor may have a significant effect on observed transfer, especially for small values of κ^2 when the relative dipole orientation forbids the transfer. Astoundingly, it is more probable to have relative dipole orientations that significantly slow ($\kappa^2 < 0.1$) the transfer than those that significantly enhance ($\kappa^2 > 2$) it. For a system in which donor and acceptor may independently adopt any random orientation, the probability of k^2 being < 0.1 is almost 25% (Eisinger et al., 1981).

It may be informative to discuss the effect of relative chromophore orientation on the orientation factor and energy transfer. Some authors have considered k^2 distributions to assess the error in a determination of D-A separation caused by a particular choice of κ^2. This consideration has statistical validity for cases in which D and A have a large number of entirely independent possible orientations. The most rational strategy for delimiting the possible values of k^2 relies on results of depolarization measurements to quantify the possible range for the orientation factor (Cheueng 1991; Dale and Eisinger 1974, 1979). Measured depolarization (anisotropy drop) provides direct information on the reorientational mobility of the fluorophores.

Dale and co-workers have shown that the extent of orientational effects on energy transfer can be calculated from the series of depolarizing events that intervene between the light absorption by donor and the emission by acceptor. Figure 2–12 illustrates the model for calculating depolarization resulting from energy transfer between two fluorophores that may freely rotate around a symmetry axis lying along the macromolecule's radius. The cones indicate the axially symmetric orientational distributions of the donor and acceptor transition moments. In principle, there are three types of depolarizing steps: those due to reorientation of donor, $\langle d_D^x \rangle$, those due to reorientation of acceptor, $\langle d_A^x \rangle$, and those due to energy transfer between donor and acceptor, d_T^x. The average depolarization factor resulting from the energy transfer, $\langle d_T \rangle$, is given by the product of these three axial depolarization factors:

$$\langle d_T \rangle = \langle d_D^x \rangle d_T^x \langle d_A^x \rangle \tag{43}$$

$\langle d_D^x \rangle$ and $\langle d_A^x \rangle$ refer to the axial depolarization factors for donor and acceptor, respectively; d_T^x refers to the axial transfer depolarization factor associated with their axial orientation factor, given by

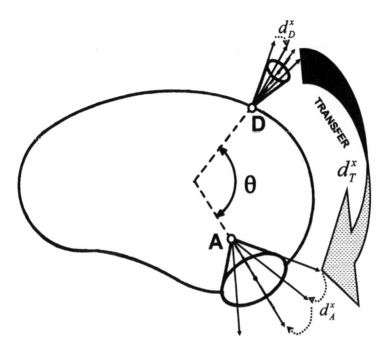

FIGURE 2–12 Schematic representation of the three depolarizing steps after photon absorption by the donor (D): donor depolarization (d_D^x), transfer depolarization (d_T^x) and acceptor (A) depolarization (d_A^x).

$$d_T^x = \frac{3}{2}\cos^2\theta - \frac{1}{2} \tag{44}$$

where θ is the angle between the mean orientation of donor and acceptor (Fig. 2–12). Derivation of equation (44) is analogous to that of equation (25). The axial depolarization factors are related to observable depolarization of donor and acceptor by

$$\langle d_D^x \rangle = \left(\frac{r_0^D}{r_f^D}\right)^{1/2}$$

$$\langle d_A^x \rangle = \left(\frac{r_0^A}{r_f^A}\right)^{1/2} \tag{45}$$

where r_0 is the experimentally determined limiting anisotropy of donor and acceptor and r_f is the observed anisotropy for donor and acceptor. In practice one could determine limiting anisotropy for the system from anisotropy measurements in a highly viscous solvent at low temperature, when all depolarizing rotations are frozen. On many occasions these values are close to 0.4 for one-photon excitation. The observed anisotropy r_f can be measured for the system at normal experimental conditions. In both cases special care must be taken to observe exclusively the donor emission for donor depolarization factor measurements (r_f^D and r_0^D) and to excite only the acceptor when measuring acceptor depolarizing factor (r_f^A and r_0^A).

To illustrate the extent of possible changes for average orientation factors, we shall consider a simple example of an ensemble of identical D and A pairs attached to identical spherical macromolecules, as schematically shown in Figure 2–13. For simplicity, only axially symmetric reorientations (rotation) are considered and a symmetry axis is assumed to lie along the sphere's radius. Figure 2–13 shows the schematics of a spherical macromolecule, (globular protein, for example) with D and A attached to the macromolecule surface. If the fluorophore (donor and acceptor) rotates much faster than the measured fluorescence lifetime, we may assume that orientation of transition moments will average before the energy transfer occurs. To assess the effect of the orientation factor, we shall consider two simple cases. In the first case, the transition moment is oriented along the rotation axis; chromophore rotation does not change (displace) the transition moment orientation. In the second case, the transition moment is orthogonal to the rotation axis; chromophore rotation leads to planar distribution of transition moments.

Using equation (41), we can compute the average orientation factor for the systems shown in Figure 2–13 as a function of relative probe (D and A) localization on the macromolecular surface (angle θ in Fig. 2–13), assuming full independent reorientation of each chromophore around the rotation axis. To illustrate this, we shall consider three cases.

In case 1, both D and A transition moments are oriented along the rotation axis. The average value of orientational factor is

$$\left\langle \kappa^2 \right\rangle_{\|,\|} = \frac{(3 - \cos\theta)^2}{4} \tag{46}$$

In case 2, one chromophore has a transition moment oriented along the rotation axis and another orthogonal to the rotation axis. The average value of κ^2 is

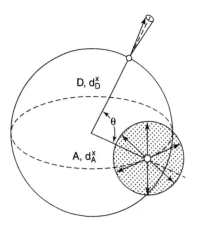

FIGURE 2–13 Schematic representation of the specific case of transition moment distribution. The linear–transition moment is oriented along the sphere's radius, and the planar–transition moment is orthogonal to the sphere's radius.

independent of which fluorophore has the orthogonal transition moment orientation. The average orientational factor is given by

$$\langle \kappa^2 \rangle_{\perp,\parallel} = \langle \kappa^2 \rangle_{\parallel,\perp} \frac{\sin^2 \theta}{8} \tag{47}$$

In case 3, both D and A have a transition moment oriented orthogonal to the rotation axis, yielding the average orientation factor:

$$\langle \kappa^2 \rangle_{\perp,\perp} = \frac{13 + 6\cos\theta + \cos^2\theta}{16} \tag{48}$$

Figure 2–14 shows the dependence of the average orientation factor on angle θ for all three cases. Even in this simple case, the expected average value of $\langle \kappa^2 \rangle$ is drastically different from the usually assumed 2/3 value. For the first case, the possible values of $\langle \kappa^2 \rangle$ are quite large, significantly higher than the usual assumed average of 2/3. Case 2 is even more dramatic. The average orientation factor is much smaller than the usual 2/3 value—very close to zero. Both cases will lead to significant overestimation and underestimation of real D–A separations. Some detailed examples of hemoprotein systems with significantly smaller values of $\langle \kappa^2 \rangle$ have been discussed by Gryczynski et al. (1995, 1997). More detailed analysis of the orientation factor is elegantly presented by Dale and Eisinger (1974, 1979) and by Cheueng (1991).

It is important to stress that with the availability of crystal structures for many proteins, investigators can make good estimates of probe localization. With the

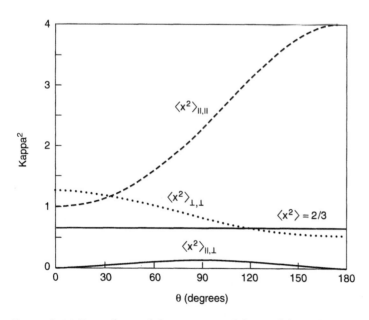

FIGURE 2–14 Dependence of the orientational factor κ^2 from the doner–acceptor (D-A) pair on the surface of the spherical molecule, as a function of angle θ in Figure 2–13.

knowledge of the transition moment orientation within the molecular framework for donor and acceptor, and the availability of polarization measurements, it should be possible to precisely estimate the expected value of $<\kappa^2>$. In many cases, this value could be significantly different from the typical value of 2/3.

3.2 Methods for Determination of Energy Transfer Efficiency

The efficiency of energy transfer can be determined from analysis of donor and/or acceptor steady-state spectra (Schiller, 1975; Mugnier et al., 1985, Lakowicz, 1999) and of fluorescence intensity decay (fluorescence lifetime) (Conrad and Brand, 1968; Schiller, 1975; Van Der Meer et al., 1994; Lakowicz, 1999).

3.2.1 Steady-State Spectra Analysis

There are a few available steady-state approaches to determine FRET efficiency. The most important ones are enhanced acceptor fluorescence and/or quenching of donor fluorescence. An important assumption underlying most of the steady-state methods is the fact that the donor and acceptor alone should have absorption and emissive properties identical to those of the donor and acceptor in the energy transfer system. For example, consider a protein labeled with a single donor and a single acceptor. One may measure absorption (excitation) and emission spectra as well as quantum yields for donor only–labeled protein and acceptor only–labeled protein. The assumption is that these properties are identical when both donor and acceptor are present on the protein. Common examples of fluorescent molecules used for protein labeling are fluorescein (donor) and rhodamine (acceptor). Figure 2–15 shows the absorption (excitation) and emission spectra for donor and acceptor alone in water. Presence of FRET will have a profound effect on donor and acceptor fluorescence. The fluorescence of donor only– and acceptor only–labeled systems will be

$$
\begin{aligned}
F_D^0(\lambda_{Obs}) &= I_{Exc}\varepsilon_D(\lambda_{Exc})C_D Q_D \\
F_A^0(\lambda_{Obs}) &= I_{Exc}\varepsilon_A(\lambda_{Exc})C_A Q_A
\end{aligned}
\tag{49}
$$

where I_{Exc} is the intensity of excitation light; $\varepsilon_D(\lambda_{Exc})$ and $\varepsilon_A(\lambda_{Exc})$ are the extinction coefficients for D and A at the excitation wavelength, λ_{Exc}; QD and QA are the quantum yields for D and A, respectively; and C_D and C_A are respective concentrations of D and A. We want to stress that for a perfectly labeled system, $C_D = C_A$. However, donor or acceptor fractional labeling (underlabeling) is often a concern when dealing with biological systems. In the presence of FRET, part of the excitation energy of the donor will be transferred to the acceptor. If the efficiency of energy transfer is E, one will get

$$
\begin{aligned}
F_D(\lambda_{Obs}) &= I_{Exc}\varepsilon_D(\lambda_{Exc})C_D Q_D(1-E) = F_D^0(\lambda_{Obs})(1-E) \\
F_A(\lambda_{Obs}) &= I_{Exc}\varepsilon_A(\lambda_{Exc})C_A Q_A + I_{Exc}\varepsilon_D(\lambda_{Exc})C_D Q_A E = F_A^0(\lambda_{Obs})\left(1+\frac{\varepsilon_D(\lambda_{Exc})}{\varepsilon_A(\lambda_{Exc})}\frac{C_D}{C_A}E\right)
\end{aligned}
\tag{50}
$$

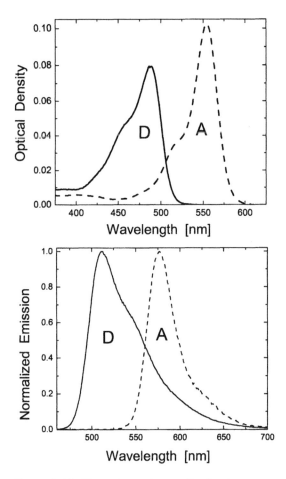

FIGURE 2–15 Absorption and normalized emission spectra for fluorescein and rhodamine in water.

The fluorescence spectrum of the donor is directly proportional to transfer efficiency, decreasing as transfer efficiency increases. The emission of the acceptor also depends on transfer efficiency. However, it is weighted by the ratio of extinction coefficients at a given excitation wavelength and relative concentrations of donor and acceptor. This feature can be an important complication when one is dealing with an imperfectly labeled system, which is frequently the case for biological systems.

In general, FRET can be calculated from the donor emission or from the acceptor emission when both donor and acceptor are excited. It is important to note that one should measure emission spectra of donor and acceptor when no energy transfer is present. In practice, this can be done when the system is labeled with only donor or only acceptor. It is easy for the D-A pair to have a separate wavelength range when exclusively donor emission can be seen; however, it is practically impossible to find a spectral range when only acceptor emission is present. For common fluorophores, the red edge of the donor emission extends over the acceptor emission (see Fig. 2–15, where emission of fluorescein contributes at all wavelengths to the emission of rhodamine). This is not a significant limitation in today's fluorescence spectroscopy and fluorescence microscopy. Collecting emission spectra in a wide observation range for different excitation wavelengths, one may deconvolyte emission (excitation) spectra down to two contributing components (donor and acceptor) related by energy transfer efficiency (equations (50)). Figure 2–16 shows simulated emission spectra for a fluorescein–rhodamine system in which energy transfer efficiency is assumed to change from 0 to 1. The analysis is simple when the labeling efficiency is 100% for both donor and acceptor. The case is much more complicated when one deals with imperfect labeling. The excess of donor or acceptor (presence of free donor or acceptor in the solution) will contribute to the overall emission spectrum of the system. The contribution of extra dye is highly correlated with transfer efficiency and a unique solution cannot be found. In some cases, measurements of emission anisotropy of acceptor (Kawski, 1983) with few different excitation wavelengths may help to solve ambiguity of imperfect labeling. As discussed earlier, one step of energy transfer usually highly depolarizes the acceptor emission.

We want to stress that the set of equations (49–50) also applies to the system of two macromolecules (proteins, for example), one labeled with the donor fluorophore and the other with the acceptor chromophore that can be brought together during a specific biological process. An example can be two complementary DNA strands labeled with donor and acceptor, respectively, or two proteins labeled with donor and acceptor that assemble on a ribosome.

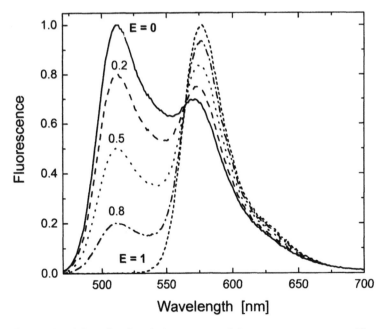

FIGURE 2–16 Simulated emission spectra of donor–acceptor system (fluorescein–rhodamine) excited at 442 nm for transfer efficiency (E) changing from 0 to 1.0.

3.2.2 Time-resolved FRET analysis

According to Figure 2–9, the energy transfer process contributes to the depopulation process of excited donor chromophores. In effect, the observed fluorescence lifetime of the donor is particularly sensitive to the presence of energy transfer. For D-A systems, time-resolved data can be used to recover the distance distribution between donor and acceptor. Since the time scales of molecular motions are comparable to fluorescence lifetimes, donor-to-acceptor motions also influence the extent of energy transfer and can be used to recover the mutual diffusion coefficient.

In flexible biological systems, a range of D-A distances is usually possible. This significantly affects the FRET process. The concept of a distribution of D-A distances was originally introduced by Haas and Steinberg (Grinvald et al., 1972, Haas et al., 1975, 1978a) for flexible polypeptides. Figure 2–17 shows an example of a macromolecule labeled at unique sites by a single donor and a single acceptor. In the native state one expects a well-defined conformation with a single D-A distance. The mean distance is sharply localized at a particular separation r. This unique distance is represented by the probability function $P(r)$, which is narrowly distributed around the mean separation r. If the protein is unfolded to the random-coil state by the addition of denaturant, a broader range of D-A distances can be expected (Lakwricz et al., 1990a, 1990b).

This distribution of distances has a profound effect on the time-resolved decay of the donor. For a well-folded protein, a well-defined D-A distance results in a single transfer rate for all the donors. Since there is one transfer rate, the single exponential donor decay becomes shorter but remains a single exponential decay (Fig. 2–17, middle).

For the unfolded system, the range of D-A distances are much larger. Some of the D-A pairs are closely spaced and display shorter decay times, and other pairs are further apart and display much longer decay times. The range of D-A separation results in a range of decay times, so the single decay of the donor becomes more complex than the initial single exponential decay (Fig. 2–17, bottom). Time-resolved analysis of donor decay is the most common method of determining distance distributions in the biological system. Detailed examples of distance distribution analysis for flexible molecules can be found in Lakowicz (1999). However, under special circumstances, it is possible to recover distance distribution without the use of time-resolved measurements (Cantor and Pechukas, 1971; Gryczynski et al., 1988a, 1988b).

3.2.3 Distance distribution from steady-state measurements

Analysis of equations (49) and (50) shows that there is a strong correlation between all parameters and it is impossible to recover a unique solution from simple emission measurements. However, by using different R_0 values, it is possible to use steady-state measurements (spectral analysis) to recover distribution of distances in flexible molecules. This is possible because steady-state energy transfer efficiency strongly depends on the ratio (R_0/r), and radiationless interaction samples different regions of the distance distribution when R_0 changes. The steady-state transfer efficiency described by equation (38) can be related to distance distribution by

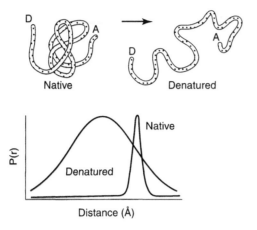

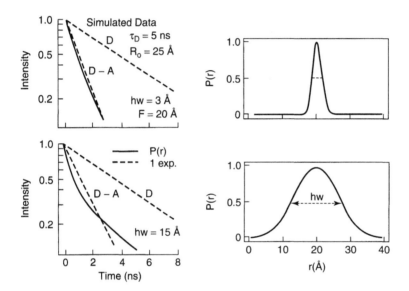

FIGURE 2–17 *Top:* Distribution of donor–acceptor (D-A) distances for a native and denatured protein. *Lower panels:* Simulated data illustrating the the effect of D-A distance distribution of different half-widths (hw) on time-domain intensity decay of the donor. $P(r)$, probability function; solid lines, in the presence of acceptor; dashed lines, single exponential fit for donor intensity decay in the absence and presence of acceptor.

$$E(R_0) = \int_0^\infty \frac{P(r)R_0^6}{R_0^6 + r^6}\, dr \tag{51}$$

where $P(r)$ is the distance distribution function and describes the normalized probability of finding the specific D-A separation. Small R_0 values probe only the closely spaced D-A pairs, and as R_0 increases, the contribution of all D-A pairs increases. Originally, the proposal was made to synthesize the number of analogous D-A pairs with different R_0 values (Cantor and Pechuka, 1971). This is a very laborious procedure that is not always possible to do, as shown by Wiczk et al. (1991).

There is an alternative way to obtain different values of R_0, by using the dynamic quenching of the donor (Gryczynski et al., 1988a, 1988b). As shown in equation (35), the R_0 value directly depends on quantum yield of the donor. The process of collisional quenching (dynamic quenching) of the fluorophore increases the nonradiative rates in equation (5) and results in the apparent change of fluorophore quantum yield. A decrease in quantum yield results in a smaller value of R_0, as seen in equation (35). The new value of the Förster distance upon quenching will be

$$R_0^Q = R_0 \left(\frac{Q_D^Q}{Q_D^0} \right)^{1/6} \tag{52}$$

where Q_D^0 and Q_D^Q are the initial quantum yield of the donor and the quantum yield of the donor in the presence of the quencher, respectively. Unfortunately, via this method the value of R_0 can be changed to a limited extent for a reasonable amount of quenching. Quenching is a process that competes with energy transfer. Very high quenching decreases the signal and makes fluorescence hard to detect.

4. Conclusion

In this chapter, we have presented basic physical parameters that regulate FRET. With a well-established foundation of FRET, one has a powerful tool for quantitative analysis of macromolecules and biological systems. Quantitative analysis of steady-state and time-resolved FRET measurements provides information on global structures and conformational dynamics and may reveal thermodynamic parameters for conformational transition. This information can be essential for the understanding of biological functions of proteins, DNA/RNA, and other biological assemblies that are frequently mediated by transitions between alternative conformations.

Recent developments in microscopy, detection systems, and excitation sources have made FRET measurements possible on a single-molecule level. This will provide a wealth of new information on structural dynamics of individual macromolecules. Time-resolved measurement of a single donor–acceptor pair will provide new opportunities for studies on conformational dynamics of molecular assemblies, without assumptions about the shape of conformational distribution.

This chapter is dedicated to Professor Alfons Kawski on the occasion of his 77th birthday.

References

Amir, D. and E. Haas. Determination of intramolecular distance distributions in a globular protein by nonradiative excitation energy transfer measurements. *Biopolymers* 25:235–240, 1986.

Berlman, I.B. Azulene and its derivatives. In: *Handbook of Fluorescence Spectra of Aromatic Molecules*, 2nd edition, New York: Academic Press, 1971, pp. 94–95.

Birks, J.B. *Organic Molecular Photophysics*. New York: John Wiley and Sons, 1973.

Brewer, B., J. Malicka, P. Blackshear, and G. Wilson. RNA sequence elements required for high affinity binding by the zinc finger domain of tristetraprolin. *J. Biol. Chem.* 279:27870–27877, 2004.

Cantor, C.R. and P. Pechukas. Determination of distance distribution functions by singlet-singlet energy transfer. *Proc. Natl. Acad. Sci. U.S.A.* 68:2099–2101, 1971.

Cheueng, H.C. Resonance energy transfer. In: *Topics in Fluorescence Spectroscopy, Vol. 2: Principles*, edited by J.R. Lakowicz, New York: Plenum Press, 1991, pp. 127–176.

Clegg, R.M. Fluorescence resonance energy transfer and nucleic acid. *Methods Enzymol.* 211:353–388, 1992.

Clegg, R.M. Fluorescence resonance energy transfer. In: *Fluorescence Imaging Spectroscopy in Microscopy*, edited by X.F. Wang, and B. Herman, New York: John Wiley and Sons, 1996, pp. 179–252.

Clegg, R.M., A.I.H. Murchie, and D.M. Lilley. The solution structure of the four way DNA junction at low salt concentration: a fluorescence resonance energy transfer analysis. *Biophys. J.* 66:99–109, 1994.

Conrad, R.H., and Brand, L. Intermolecular transfer of excitation from tryptophan to 1-dimethylaminonaphthalene-sulfonamide in a series of model compounds. *Biochemistry* 7; 777–783, 1968.

Dale, R.E. and J. Eisinger. Intermolecular distances determined by energy transfer. Dependence on orientational freedom of donor and acceptor. *Biopolymers* 13:1573–1605, 1974.

Dale, R.E. and J. Eisinger. The orientation freedom of molecular probes. The orientation factor in intermolecular energy transfer. *Biophys. J.* 26:161–194, 1979.

Eisinger, J., Blumberg, W.E., Dale, R.E. Orientational effects in intra- and intermolecular long range excitation energy transfer. *Ann. N.Y. Acad. Scio* 366, 155–175, 1981.

Faisal, F.H.M. *Theory of Multiphoton Processes*. New York: Plenum Press, 1987.

Förster, T. Energiewanderung und Fluoreszenz. *Naturwissenschaften* 6:166–175, 1946.

Förster, T. Fluoreszenzversuche an Farbstoffmischungen. *Angew. Chem. A* 59:181–187, 1947.

Förster, T. Zwischenmolekulare Energiewanderung und Fluoreszenz. *Ann. Phys.* 2:55–75, 1948.

Förster, T. Experimentalle und theoretische Undersuchung des zwischengmolekularen Ubergangs von Elektronenanregungsenergie. *Naturforschung* 4A:321–327, 1949.

Förster, T. Delocalized excitation and excitation transfer. In: *Modern Quantum Chemistry*, Part III, edited by O. Sinanoglu, New York: Academic Press, 1965, pp. 93–137.

Gaviola, E. and P. Prinsgsheim. Über den einfluss der konzentration auf die polarisation der fluoreszenz von farbstoffösungen. *Z. Phys.* 24:24–36, 1924.

Göppert-Mayer M. Ueber Elementarakte mit Quantensprengungen. *Ann. Phys.* 9:273–295, 1931.

Grinvald, A., E. Haas, and I.Z. Steinberg. Evolution of the distribution of distances between energy donors and acceptors by fluorescence decay. *Proc. Natl. Acad. Sci. U.S.A.* 69:2273–2277, 1972.

Gryczynski, Z., J. Lubkowski, and E. Bucci. Heme–protein interactions in horse heart myoglobin at neutral pH and exposed to acid invastigated by time-resolved fluorescence in the pico- to nanosecond time range. *J. Biol. Chem.* 270:19232–19237, 1995.

Gryczynski, Z., J. Lubkowski, and E. Bucci. Fluorescence of myoglobin and hemoglobin. *Methods Enzymol.* 78:538–569, 1997.

Gryczynski, I., W. Wiczk, M. Johnson, and J.R. Lakowicz. End-to-end distance distribution of flexible molecules from steady-state fluorescence energy transfer and quenching induced change in the Förster distance. *Chem. Phys. Lett.* 145:439–446, 1988a.

Gryczynski, I., W. Wiczk, M. Johnson, H. Cheueng, C. Wang, and J.R. Lakowicz. Resolution of the end-to-end distance distribution of flexible molecules using quenching-induced variation of the Förster distance for fluorescence energy transfer. *Biophys. J.* 54:577–586, 1988b.

Haas, E., E. Katchalski-Katzir, and I. Stainberg. Brownian motions of the ends oligopeptides chains in solution as estimated by energy transfer between chain ends. *Biopolymers* 17:11–31, 1978a.

Haas, E., E. Katchalski-Katzir, and I. Stainberg. Effect of the orientation of donor and acceptor on the probability of energy transfer involving electronic transitions of mixed polarizations. *Biochemistry* 17:5064–5070, 1978b.

Haas, E., M. Wilchek, E. Katchalski-Katzir, and I. Steinberg. Distribution of end-to-end distance of oligopeptides in solution as estimated by energy transfer. *Proc. Natl. Acad. Sci. U.S.A.* 72:1807–1811, 1975.

Haugland, R.P., J. Yguerabide, and L. Stryer. Dependence of kinetics of singlet-singlet energy transfer on spectral overlap. *Proc. Natl. Acad. Sci. U.S.A.* 63:23–30, 1969.

Jabłoński, A. On the notion of emission anisotropy. *Bull. Acad. Pol. Sci. Ser. A* 8:259–264, 1960.

Jabłoński, A. Über die abklingungsvorgänge polarisierter photolumineszenz. *Z. Naturforsch.* 16a:1–4, 1961.

Jabłoński, A. Emission anisotropy. *Bull. Acad. Pol. Sci. Ser. A* 15:885–892, 1967.

Johnson, C.K. and C. Wan. Anisotropy decays induced by two-photon excitation. In: *Topics in Fluorescence Spectroscopy.* Vol. 5, *Nonlinear and Two-Photon Induced Fluorescence,* edited by J.R. Lakowicz, 1997, pp. 43–56.

Kasha, M. Characterization of electronic transitions in complex molecules. *Disc. Faraday Soc.* 9:14–19, 1950.

Kawski, A. Excitation energy transfer and its manifestation in isotropic media. *Photochem. Photobiol.* 38:487–508, 1983.

Kawski, A. Fluorescence anisotropy: theory and application of rotational depolarization. *Crit. Rev. Anal. Chem.* 23:459–529, 1993.

Kawski, A., I. Gryczynski, and Z. Gryczynski. Anisotropy of 4-dimethylamino-ω-diphenyl-phosphinyl-trans-styrene in isotropic media in the case of one- and two-photon excitation. *Naturforschung* 48a:551–556, 1993.

Kawski, A. and Z. Gryczynski. On the determination of transition-moment directions from emission anisotropy measurements. *Z. Naturforsch.,* 41a:1195–1199, 1986.

Lakowicz, J.R. *Principles of Fluorescence Spectroscopy,* 2nd edition. New York: Kluwer Academic/Plenum Press, 1999.

Lakowicz, J.R. and I. Gryczynski. Multiphoton excitation of biochemical fluorophores. In: *Topics in Fluorescence Spectroscopy.* Vol. 5, *Nonlinear and Two-Photon Induced Fluorescence* edited by J.R. Lakowicz, 1997, pp. 87–99.

Lakowicz, J.R., I. Gryczynski, Z. Gryczynski, E. Danielsen, and M.J. Wirth. Time-resolved fluorescence intensity and anisotropy decays of 2,3-diphenyloxazole by two-photon and frequency-domain fluorometry. *J. Phys. Chem.* 96:3000–3006, 1992.

Lakowicz, J.R., I. Gryczynski, H. Malak, and Z. Gryczynski. Two-color two-photon excitation of fluorescence. *Photochem. Photobiol.* 64:632–635, 1996.

Lakowicz, J.R., I. Gryczynski, W. Wiczk, G. Laczko, F. Prendergrastand, and M. Johnson. Conformational distribution of melitin in water/methanol mixtures from frequency-domain measurements of nonradiative energy transfer. *Biophys. Chem.* 36:300–334, 1990a.

Lakowicz, J.R., H. Szmacinski, I. Gryczynski, W. Wiczk, and M. Johnson. Influence of diffusion on excitation energy transfer in solutions by gigahertz harmonic content frequency-domain fluorometry. *J. Phys. Chem.* 94:8413–8416, 1990b.

Lewshin, W. L. Über polarisiertes fluoreszenzlicht von farbstofflösungen. III. *Z. Phys.* 26:274–284, 1924.

Malak, H., I. Gryczynski, J. Dattelbaum, and J.R. Lakowicz. Three-photon induced fluorescence of diphenylhexatriene in solvent and lipid bilayers. *J. Fluoresc.* 7:99–106, 1997.

Malicka, J., I. Gryczynski, J. Fang, J. Kusba, and J.R. Lakowicz. Increased resonance energy transfer between fluorophores bound to DNA in proximity to metallic silver particles. *Anal. Biochem.* 315:160–169, 2003a.

Malicka, J., I. Gryczynski, J. Kusba, and J.R. Lakowicz. Effects of metallic silver island films on resonance energy transfer between N,N'-(dipropyl)-tetramethyl-indocarbocyanine (Cy3)- and N,N'-(dipropyl)-tetramethyl-indodicarbocyanine (Cy5) labeled DNA. *Biopolymers* 70:595–603, 2003b.

Mazurenko, Y.T. Polarization of luminescence from complex molecules with two-photon excitation. Dichroism of two-photon light absorption. *Opt. Spectrosc. (USSR)* 31:413–414, 1971.

McClain, W.M. and R.A. Harris. Two-photon spectroscopy in liquids and gases. In: *Excited States 3*, edited by E. C. Lim, New York: Academic Press, 1977, pp. 2–77.

Michl J. and E.W. Thulstrup. *Spectroscopy with Polarized Light. Solute Alignment by Photoselection, in Liquid Crystals, Polymers, and Membranes.* New York: VCH Publisher, 1986.

Mugnier, J., J., Pouget, J. Bourson, and B. Valeur, Efficiency of intermolecular electronic energy transfer in coumarin biochromophoric molecules. *J. Luminesc.* 33:273–300, 1985.

Perrin, J. *2me Consell de Chim. Solvey.* Paris: Gauthier-Villars, 1925, pp. 322–324.

Perrin, F. Polarisation de lumiére de fluorescence. Vie moyenne des molecules dans l'etat excité. J. Phys. Radium 7:390–401, 1926a.

Perrin, F. La fluorescence des solutions. Induction moléculaire-polarisation et durée d'émission-photochimie. *Ann. Phys. (Paris)* 12:169–275, 1926b.

Perrin, J. Fluorescence et induction moleculaire par resonance. *CR Hebd. Seances Acad. Sci.* 184:1097–1100, 1927.

Perrin, F. Theorie quantique des transferts d'activation entre molecules de meme espece. Cas des solutions fluorescents. *Ann. Phys. (Paris)* 17:283–314, 1932.

Perrin, F. Diminution de la polarization de l. fluorescence des solutions résultant du movement brownien de rotation. *Acta Phys. Plon.* 5:335–345, 1936.

Schiller, P.W. In: *Biochemical Fluorescence Concepts*, edited by R.F. Chen and H. Edelhoch, New York: Marcel Dekker, 1975, pp. 115–284.

Shih, W., Z. Gryczynski, L. Lakowicz, and J. Spudich. A FRET-based sensor reveals ATP hydrolysis dependent large conformational changes and three distinct states of the molecular motor myosin. *Cell* 102:683–694, 2000.

Steinberg, I.Z. Long-range nonradiative transfer of electronic excitation energy in proteins and polypeptides. *Annu. Rev. Biochem.* 40:83–114, 1971.

Steinberg, I.Z. Nonradiative energy transfer in systems in which rotatory Brownian motion is frozen. *J. Chem. Phys.* 48:2411–2413, 1968.

Stokes, G.G. On the change of refrangibility of light. *Phil. Trans. R. Soc. Lond.* 142:463–562, 1852.

Strickler, S.J. and R.A. Berg. Relationship between absorption intensity and fluorescence lifetime of molecules. *J. Chem. Phys.* 37:814–822, 1962.

Stryer, L. Fluorescence energy transfer as a spectroscopic ruler. *Annu. Rev. Biochem.* 47:819–846, 1978.

Stryer, L. and R.P. Haugland. Energy transfer: a spectroscopic ruler. *Proc. Natl. Acad. Sci. U.S.A.* 58:719–726, 1967.

Tyson, D., I. Gryczynski, and F. Castellano. Long-range resonance energy transfer to $[Ru(pby)_3]^{2+}$. *J. Phys. Chem.* 104:2919–2924, 2000.

Van Der Meer, W., B.G. Cooker III, and S.-Y. Chen. *Resonance Energy Transfer: Theory and Data.* New York: John Wiley and Sons, 1994.

Wawilow, S.L. and W.L. Lewshin. Beiträge zur frage über polarisiertes. Fluoreszenzlicht von farbstofflösungen. II. *Z. Phys.* 16:135–154, 1923.

Weigert, F. Über polarisierte fluoreszenz, verhandb. *D. Deutsch. Phys. Ges.* 23:100–106, 1920 and Phys. Zeitschr. 23:232–233, 1922.

Weigert, F. and Käpler. Polarisierte fluoreszenz in farbstofflösungen. *Z. Phys.* 25:99–107, 1924.

Wiczk, W., P.S. Eis, M.N. Fishman, M.L. Johnson, and J.R. Lakowicz. Distance distribution recovered from steady-state fluorescence measurements on thirteen donor–acceptor pairs with different Förster distance. *J. Fluoresc.* 1:273–286, 1991.

Wu, P. and L. Brand. Review-resonance energy transfer: methods and applications. *Ann. Biochem.* 218:1–13, 1994.

3

An Introduction to Filters
and Mirrors for FRET

C. Michael Stanley

1. Introduction

There have been striking advances in the development of microscopy techniques over the years, each addressing specific issues in biomedical research. A short, noninclusive list of landmark papers would include those by Zernike (1958), Minsky (1957), Ploem (1967), Allen et al. (1969), Göppert-Meyer (1931), and Webb (Denk et al., 1990). Perhaps more importantly, the introduction of green fluorescent proteins (GFP) (Heim and Tsien, 1996) has revolutionized imaging in biological specimens with most of these microscopic techniques. The fluorescence microscope has been very effectively used to identify cellular organelles, noninvasively, by conjugation or expression of various fluorpohore/protein molecules in the form of GFPs. Currently, biologists and biophysicists are very much interested in localizing protein molecules in living specimens with wide-field, confocal, multiphoton, and lifetime imaging microscopy techniques as described in various chapters of this book (see also Pawley, 1998; Periasamy, 2001). Each of these advanced fluorescence-based techniques requires precision-made interference filters to record the emission signals from the biological molecules and to discriminate the excitation and scattered light from various components of the microscope optics and specimen (Stanley, 2003). In this chapter we describe some of the filters used for protein imaging of various Förster resonance energy transfer (FRET) pairs (fluorphores or chromophores) with high-resolution FRET techniques.

2. FRET

Förster resonant energy transfer is a process that involves the direct transfer of excitation energy between two fluorochromes (molecules that have the potential to fluoresce) with overlapping emission and excitation spectra (see Uster and Pagno, 1986; Clegg, 1992). Because energy transfer can occur over only a limited distance, FRET is useful in determining intramolecular and intermolecular reactions both spatially and temporally. It can be used to follow and record the interactions of proteins, DNA, and RNA well below the 200 nm (10^{-9} meter) resolution limit of the optical microscope at its highest magnification and with the highest numerical-aperture (NA) optics. In fact, FRET only allows resolution of interactions between 1 and 10 nm, which is its beauty as well as its curse. This application note is intended as a basic introduction only; we will focus instead on the optics required for the protocol.

If one fluorochrome has an emission spectrum that overlaps significantly with the absorption spectrum of a second fluorochrome, and if the two fluorochromes are physically very close to each other (generally between 2 and 6 nm [the Förster distance]), then energy from the first (donor) molecule may be transferred to the second (acceptor) molecule. This is a nonradiative phenomenon (no photons are emitted from the donor) in which energy from the donor (shorter wavelength emission) fluorochrome is transferred to the acceptor (longer wavelength emission) fluorochrome. When this happens, the donor fluorochrome's fluorescence is quenched (reduced in total light output). The acceptor's emission is triggered by energy absorbance or transfer from the donor and results in a wavelength-appropriate emission from the acceptor. As noted above, this transfer is exquisitely distance sensitive, and because accidental FRET is an extremely rare event (effectively never happening), it is a reliable indicator of molecular interaction.

In theory, this is a fairly simple optical arrangement: an excitation filter and dichroic mirror match the wavelength of the donor molecule, and an emission filter matchs the fluorescence emitted by the acceptor. The microscopy would involve exciting only the donor, and imaging or recording only the emission of the acceptor. The presence of an emission from the acceptor, without its being directly illuminated or excited, should indicate FRET.

However, the technique is not simple at all. Its complexity is due to several confounding issues, including overlapping absorption and emission wavelengths (spectral bleedthrough or cross-talk), the potential for motion, noise inherent in wide-field microscopy, and autofluorescence within the cells being imaged.

3. OVERLAPPING SPECTRAL WAVELENGTHS

The emission and absorption characteristics of two different fluorochromes are not necessarily distinct. Take, for example, one of the most commonly used FRET pairs used in biological investigations, the cyan and yellow fluorescent proteins (CFP, YFP), which are two mutants of green fluorescent protein (GFP); see Chapters 4, 7, and 12–13. The CFP has an absorption maximum of 435 nm with emission at 480 nm (Fig. 3–1). The YFP has a maximum absorption of 505 nm with emission at 530 nm (Fig. 3–2).

The emission curve (at least at the maxima) for the cyan protein almost completely overlaps with the absorption curve for the yellow protein, which is one of the prerequisites for FRET (Fig. 3–3), and would seem to make the two molecules an excellent FRET pair. Intuitively, FRET should be possible by exciting the CFP with light of approximately 436 nm (a large spike in the mercury emission spectra) and recording or imaging any emission of approximately 530 nm. Unfortunately, all the fluorescent proteins, including cyan and yellow, have coincident emission

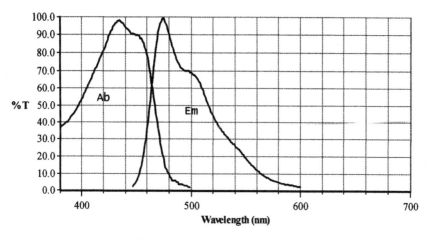

FIGURE 3–1 Absorption (Ab) and emission (Em) spectra of cyan fluorescent protein.

tails, such that there is considerable overlap (minimum of about 25%) between the emissions of the CFP and the YFP, therefore the emission from cyan may be read or recorded in the yellow channel. This emission may then be misinterpreted as FRET when, in fact, it is simply spectral bleedthrough of the extended emission from the donor protein (see Chapters 4 and 7). Furthermore, while the excitation maxima of cyan and yellow fluorescent proteins are spectrally quite distinct (435 nm and 505 nm), the absorption spectrum of YFP is broad and has some over-lap with the absorption spectrum of CFP. Thus it becomes possible that YFP can be directly excited by 436 nm light. This raises the question of whether the YFP emission is caused by FRET or by direct excitation of 436 nm light. Any and all possible FRET pairs must be tested to determine if the excitation of the donor will directly excite the acceptor.

As described below, in ratio methods (dividing CFP emission by YFP emission) will resolve many of these issues. However, it is still critical to correct for the

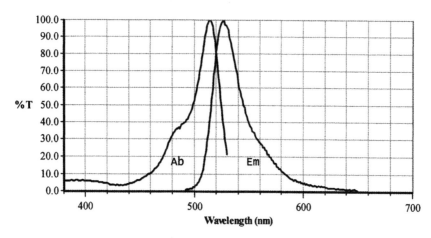

FIGURE 3–2 Absorption (Ab) and emission (Em) spectra of yellow fluorescent protein.

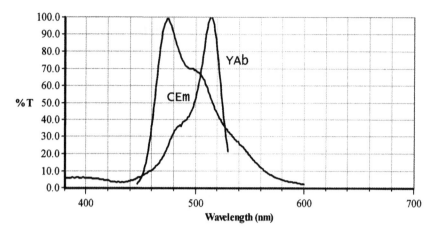

FIGURE 3–3 Cyan fluorescent protein emission (CEm) and Yellow fluorescent protein absorption (YAb) spectra.

emission overlap between cyan and yellow fluorescent proteins and/or any pair being used for FRET (see Chapter 7; Tron et al., 1984; Gordon et al., 1998; Chamberlain et al., 2000; Xia and Liu, 2001; Elangovan et al., 2003; Wallrabe et al., 2003a, 2003b).

Obviously, this problem would not occur if there were no free cyan molecules. Unfortunately, this will never occur when intermolecular FRET is being measured for stoichiometric reasons, and is extremely unlikely to occur when measuring intramolecular FRET events. Thus the cyan will be excited at 436 nm and emit with a 480 nm maxima, but there will be a concomitant tail out to 520–550 nm that will show up in the yellow emission spectra. Thus it is necessary to subtract the donor emission and divide the two emissions to determine if a FRET reaction has occurred. There are also now available (Chapter 7) sophisticated algorithms to incorporate subtraction, division, and enhanced correction techniques.

Subtraction of this cyan emission signal from the total FRET-plus-cyan signal seems simple and straightforward, except for the fact that this contamination by the cyan emission tail can, in some situations, be a very large proportion of the FRET signal.

4. FILTER SET BASICS

A typical epifluorescence microscope uses three different optics to constitute a full "set" of filters for imaging: excitation filter, dichromatic beamsplitter, and emission filter (see Chapter 4). Each of these components is very specialized in its own regard, but there are some common ingredients.

The *excitation filter* is designed to block all light from the source (usually mercury, xenon, or merury–xenon combination), except those wavelengths that will be used to "excite" the fluorochrome of interest. In our example below, the excitation filter for CFP shows a band of transmission from 426 to 446 nm (436/20x), for YFP the band is from 490 to 510 nm (500/20x). These wavelengths correspond to the maximal absorption of the two proteins. This excitation optic blocks all light outside this band from deep ultraviolet (UV) to 1200 nm. This optic is traditionally made of float glass, is used at 0° angle of incidence (AOI), and is not typically ground and polished.

The second optic, the *dichromatic beamsplitter*, also called the *dichroic*, or simply the *mirror*, is the least understood optic in the system. This dichroic has two functions: it must effectively reflect the excitation wavelengths of the excitation filter listed above, and must at the same time transmit the emission wavelengths listed below. This optic is not glass but is made from quartz or fused silica. It must be made to very exacting tolerances of both physical flatness and optical characteristics. The 455dclp shown in Figure 3–4 is designed to reflect over 90% of all the light from the 436/20x excitation filter. It also transmits over 90% from 460 to 550 nm,

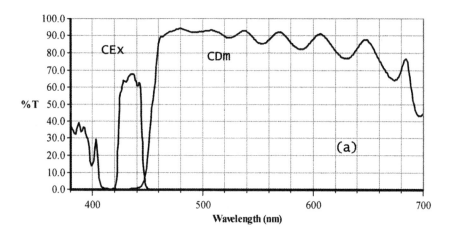

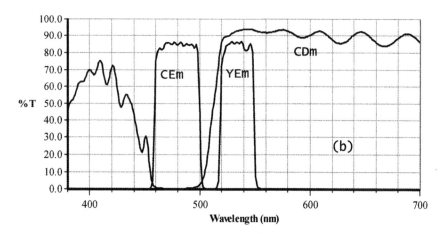

FIGURE 3–4 *a*: These filter sets are typically mounted into a microscope. Cex, cyan fluorescent protein (CFP) exciter filter cube d436/20x, CDM, CFP dichroic mirror 455dclp; *b*: The splitter unit filter sets are CFP emission filter d480/40m (CEm), dichroic 510dclp (CDm), and yellow fluorescent protein (YFP) emission filter d535/30m (YEm).

which are the required transmission bands listed below. The dichroic beamsplitter is used at 45° AOI.

The emission/barrier filter in the set has the primary function of blocking, to an optical density of 6, the excitation wavelengths transmitted by the excitation filter of the set. The reason that this must occur is that in a standard epifluorescent system, the excitation light can easily be 40,000 times brighter than the emission signal. If even a small percentage of this excitation light reaches the detector it can saturate the system, or in the very least, drastically reduce the signal-to-noise ratio. The second function of this optic is to transmit the wavelengths of light corresponding to the maximum emission of the fluorochrome. This optic is typically used at 3°–5° AOI to reduce internal reflections, and is ground and polished to optical flatness specifications. One example below shows an emission filter of 460–500 nm (480/40m).

For FRET experiments there is typically one cube designed for the donor and one cube designed for the acceptor. There is then a third cube (sometimes referred to as the *FRET cube*) made using the excitation filter and dichroic mirror appropriate for the donor but with the emission filter designed for the acceptor emission. This emission filter has to be specifically designed to block the shorter wavelengths of the donor, since it typically would be blocking the longer light of the acceptor excitation. In many modern microscopes the three-cube design has been replaced with emission filter wheels and/or motorized turrets.

The choice of excitation and emission bands for any or all fluorochromes is debatable. The fluorochrome characteristics must be considered, especially pH sensitivities, as well as the complete beam path of the microscope. The choice of light sources will also affect the decision. The emission filter in the FRET set must be designed to minimize the cross-talk from the donor emission while maximizing the transmission of the acceptor emission light. All of these decisions may translate into several different filter sets for the same FRET pair, under slightly different conditions. It is not unusual at all for the researcher to try several permutations before finding the optimal set.

5. MOTION

To collect FRET data, two images are needed—one that defines the CFP emission intensity, and one that defines the YFP emission intensity. A ratio of the two images is then calculated. Consequently, there must be no movement during the acquisition of the two images. Obviously, the best way to avoid motion is to acquire both images simultaneously, using a dual detection system. Several devices, including those from Optical Insights, Hamamatsu, and PTI, are currently available. The optical configuration for these devices would appear similar to that shown in Figure 3–4.

However, simultaneous acquisition is not always practical. Use of a fairly rapid emission filter wheel is the next best alternative. The investigator acquires the image or data from the excited donor (CFP), and then moves the wheel as quickly as possible to acquire the image from the acceptor emission (YFP). See Figure 3–5 for the optics involved with this configuration.

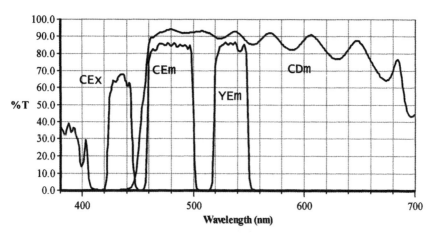

FIGURE 3–5 Cyan fluorescent protein (CFP) excitation/beamsplitter with CFP and yellow fluorescent protein (YFP) single emitters. CEx, CFP exciter d436/20x; CDM, CFP dichroic mirror 455dclp; Cem, CFP emitter d480/40m; YEM-YFP emitter d535/30m.

In addition, the acquisition time of the camera or detector must be rapid enough to minimize any movement. This means that there must be enough photons present at the detector to form an image, and high-sensitivity detectors should be used to collect the photons.

It is important to keep in mind that much of this work is done on live cells in aqueous solution, a fact that presents other issues involving motion, such as the motion arising from the use of perfusion pumps. It is imperative that the microscope and all attachments be kept very stable. Vibration isolation tables are recommended and quite common for these applications.

6. NOISE AND REGISTRATION

Signals involved in FRET are so small that they can easily become lost in the noise (background brightness) of the microscope. One logical solution is to determine the ratio between the emissions of the donor and acceptor molecules, instead of trying to measure only the FRET emission. The donor emission should decrease at the same time that the acceptor emission increases (at least for intramolecular FRET). This division process makes it possible to isolate much of the noise, thus enhancing the true FRET emission. While division of two images is considerably more difficult than addition or subtraction, most morphometry software packages now include a module that allows for ratiometric analysis. It must be kept in mind that the division of two images requires that the two be in pixel registration, which requires very accurate alignment of the emission beam path, involving the dichroic mirror and the emission filter. In many cases the emission filters for the ratio must be ground and polished to matching thickness and wedge specifications.

Authors of some very early papers (Hogg, 1871; Jones, 1997) claimed to detect FRET signals by excitation of the donor and image/calculation of emission at the acceptor only, without ratiometry. This approach is now considered naive.

Mercury light sources are notoriously noisy. Many experts consider them unacceptable for ratio imaging, yet they remain the most common light source in current use. Xenon light sources are considered more stable and less noisy. However, xenon doesn't offer the advantage of energy spikes, which are present in the mercury source. The combination mercury–xenon lamp provides more stable illumination and a potentially longer lifetime of the bulb (www.hamamatsuphotonics. com). Furthermore, even a properly aligned and cleaned microscope is likely to provide several percentage points of noise before the image reaches the camera or detector. There is also the added noise of the detector itself to consider. If the measured ratio for a common FRET relationship were around 12%, it would be easy to lose much of that in the noise of the microscope, making it unlikely that those statistics would be reliable.

Some researchers do not have emission filter wheels or dual detectors and cannot employ either of the ratio techniques listed above. Some try to use two separate cubes to measure FRET. One cube would be a typical donor (e.g., CFP) cube: 436/20x, 455dclp, and 480/40 nm (Fig. 3–6). The other cube would have the same exciter and dichroic, with the acceptor (e.g., YFP) emission, for example, 436/20x, 455dclp, and 535/30 nm (Fig. 3–7).

This approach presents several problems. One is the constant threat of motion, as the two cubes must be moved into and out of position. This threat can be lessened, but not eliminated, by using a turret system. Another potential problem is the time required to move the cubes, during which the cells may also move. Furthermore, the microscope and cubes must be nearly perfectly machined and aligned, or else the two images will not be in alignment at the camera or detector. In this case, the image from one cube may go to a slightly different location on the detector, because of beam deflection and wedge (deviation from perfect parallelism) associated with either the mirror or the emission filter or with machining of the parts within the microscope. Many software packages now provide an algorithm that will realign two images, but it is still an extra step and not something that happens automatically.

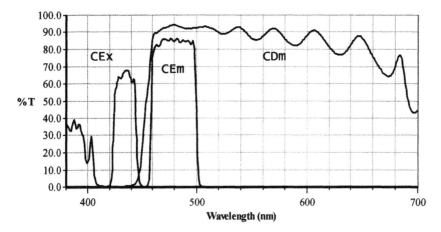

FIGURE 3–6 Cyan fluorescent protein (CFP) filter sets for imaging the cell expressed with CFP. CEx, CFP exciter d436/20x; CDm, CFP dichroic mirror 455dclp; CEm, CFP emitter d480/40m.

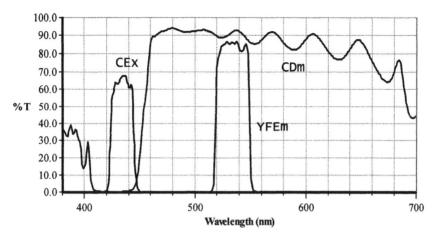

FIGURE 3–7 This filter set can be used for cyan fluorescent protein (CFP) excitation and yellow fluorescent protein (YFP) emission. CEx, CFP exciter d436/20x; CDm, CFP dichroic 455dclp; YFEm, YFP/FRET emission d535/30m.

7. AUTOFLUORESCENCE

As with all fluorescence techniques, endogenous fluorescence in a cell or tissue (the autofluorescence) can rear its ugly head. Some samples will have little autofluorescence, whereas others such as liver or brain slices may have a large autofluorescence signal. In fact, virtually every sample will have some level of autofluorescence. This should always be evaluated before the experiment by using the optics necessary for all the FRET applications to determine what the cells look like, without any external fluorescence added. There are a surprising number of researchers who go through all the steps necessary to do fluorescence work without ever evaluating their preparations to determine the background levels of light emission. This is a particular problem with any cell that is highly pigmented, such as plant tissue.

Usually the autofluorescence can be subtracted from the signal of interest, because this unwanted emission tends to be broader and less specific in both intensity and wavelength. Occasionally special steps have to be taken, such as moving the bandpass of the emission filter, to maximize the wanted signal and minimize the autofluorescence. Unfortunately, there are no optics that can determine what is "wanted" and what is considered "noise." This has to be determined by the researcher with a variety of control experiments.

8. MITIGATING TECHNIQUES

Use of a laser scanning confocal or multiphoton microscope for FRET can minimize most of the problems described above. The diffraction limited spot, coupled with the photomultiplier tubes (PMT) detection of only that one spot in time, greatly reduces the overall noise in the system and makes it much easier to isolate the FRET relationship. Confocal microscopes also commonly have multiple detectors so that the two images may be acquired simultaneously (Fig. 3–8).

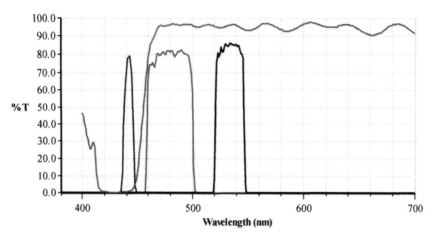

FIGURE 3–8 This filter set can be used for a laser–based microscopy system for cyan fluorescent protein (CFP)/yellow fluorescent protein (YFP) FRET emission ratio. Blue, laser clean-up filter, z442/10x; green, dichroic, z442rdc; red, CFP emission filter hq480/40m; black, YFP/FRET emission filters hq535/30m.

In either type of microscopy, there is great advantage in breaking or establishing the FRET relationship on demand by means of a chemical, electrical, or mechanical challenge to the cell or tissue. This allows for built-in control of FRET vs. non-FRET using the same field of view in the microscope. This technique is much better than relying on the average intensity of single probe samples (donor-only cells and acceptor-only cells) for the multiple images to be used in the ratio pairs. This technique is very paradigm- and experiment-specific, however, and is not always possible.

9. ACCEPTOR PHOTOBLEACHING

Acceptor bleaching is a method that offers great promise for wide-field and confocal FRET microscopy (see Chapters 4, 5, and 8; Kenworthy et al., 2000; Karpova et al., 2003). In this protocol, a ratio is determined for the emission of the donor by use of images from before and after bleaching the acceptor. Since FRET is a nonradiative energy transfer process, the absence (or destruction) of the acceptor should result in a slightly increased emission from the donor.

Again, using CFP/YFP as our example, one would acquire an image of the cyan emission (d480/40 nm) using cyan excitation (d436/20x) just as if one were imaging cyan protein alone (set 31044, Fig. 3–6). Then a special cube or filter set would be used to completely (95%+) photobleach (destroy) the YFP (Figs. 3–9a, 3–9b, and 3–9c show three different designs for wide-field; Fig. 3–9d is a laser set). The CFP filters would be used again to acquire the cyan emission. If there had been a FRET relationship between the two fluorochromes, the cyan emission would have increased slightly after bleaching the yellow protein.

This technique provides a built-in control, since the same field of view can be ratioed before and after bleaching. There is no longer a need to "break" the FRET bond, nor is there an issue with the cyan emission bleeding into the yellow emission channel. Also eliminated is the necessity of using average cyan and yellow intensities from cells with single transfections; in those cases in which the "average" cyan and "average" yellow emission intensities must be used from cells with only cyan or yellow protein, the standard deviation of intensities may be 15%–20%.

One difficulty with this procedure is the need to ensure that the yellow (acceptor) fluorescence is completely bleached (destroyed) without affecting the fluorescence of the (cyan) donor protein. This must be carefully checked with adequate controls. Issues may also arise with regard to the complete bleaching of the yellow (acceptor) molecule in a time frame that is suitable for the particular application and to minimization of motion artifacts.

Another potential problem is that the resultant ratio will be very small, and therefore its accuracy is difficult to ascertain. Even so, this technique offers great hope for wide-field and confocal FRET applications, with its built-in control mechanism.

Elsewhere in this volume (Chapter 7) are specific algorithms used in the calculations necessary for dissecting the FRET signal from the various backgrounds. Also listed are linear unmixing software packages that are coming onto the market (Chapter 10).

9. CONCLUSION

Although FRET is a difficult technique, when done properly it can be extremely valuable, as it allows measurement of protein–protein interaction in living cells. Because the technique works at the nanometer level, and if fluorescent proteins are used as the reporters, care should be taken to define where the fluorescent protein gene sequence is inserted within the protein (generally carboxyl or amino terminus) to maximize the chance of interaction. Furthermore, to avoid most of the issues described above and to ensure that there actually is a relationship, initial FRET measurements should be performed in a fluorimeter in free solution. If the fluorimeter results are positive, the next set of experiments should occur with a confocal microscope if at all possible.

Despite its associated problems, FRET has been shown to work quite successfully in many cases. The molecular constructs and manipulations are actually the hardest part. The optical arrangement of filters and mirror is fairly straightforward. The exciter filter should match the absorption maximum of the donor molecule, with the mirror's reflection band coinciding with the exciter filter. The transmission band of the mirror and the emission filters should match the emission maximum of the acceptor molecule and donor emission for ratio analysis. A common mistake is use of the dichroic mirror from the acceptor set instead of the dichroic for the donor. For our CFP/YFP example, the optics would include 440/20x, 455dclp, 480/40 nm, and 535/30 nm (Fig. 3–5).

As with all applications, however, there is more than one approach and multiple techniques are available. The choice of technique can sometimes depend a great deal on the hardware available to the researcher. In the cases shown above, dual

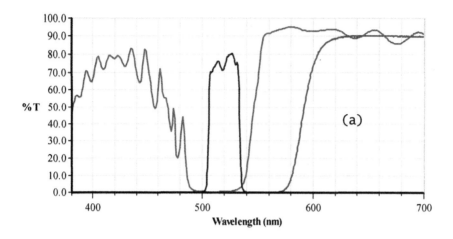

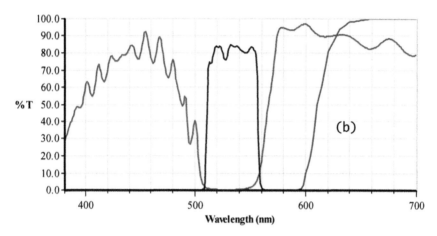

FIGURE 3–9 Yellow fluorescent protein (YFP) photobleaching filter sets. *a*: Blue, YFP bleaching exciter d520/30x; green, dichroic 545dclp; red, emission-blocking glass for safety, og590lp. *b*: Blue, YFP bleaching excitation d535/50x; green, dichroic 565dclp; red, emission-blocking glass for safety, rg610lp. *c*: This set is for use with an excitation filter wheel, without changing the dichroic. Blue, cyan fluorescent protein (CFP) exciter d436/20x; green, custom bandpass dichroic; red, CFP emitter d480/40m; blue, YFP bleaching filter d535/50x (Note: care must be taken in using this set without proper blocking optics in the eyepieces.) *d*: YFP bleaching set for 442/514 nm laser systems. This set is designed to leave the mirror in position and move the clean-up filters and the emitters. Blue, helium-cadmium clean-up filter, CFP exciter z442/10x; green, polychroic to reflect 442 nm and 514 nm laser line z442/514rpc; red, CFP emitter hq480/40m; blue, argon clean-up filter to bleach YFP z514/10x; red, absorption glass emitter for safety only, og550lp.

detectors were used simultaneously, sequential detection was done with emission filter wheels, and single acquisitions were made by use of moving single-cube designs.

There is also the possibility of using both excitation and emission filter wheels for special applications. This equipment makes it possible to leave a polychroic (dual dichroic) in place in the microscope while sequentially selecting single excitation

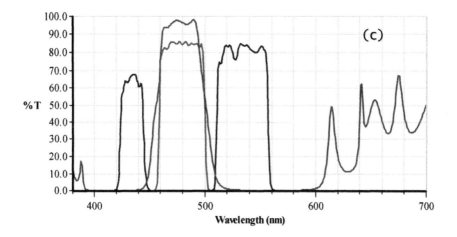

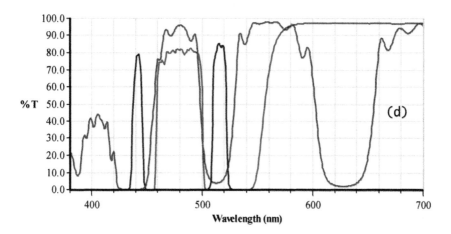

FIGURE 3–9 *(continued)*

filters and single emission filters. Even though FRET, in absolute terms, only requires the excitation of the donor, many researchers want to determine the presence of the acceptor. A dual set is shown in Figure 3–10 for a single-band exciter,
single-band emitter set. Table 3–1 lists FRET filter combinations from selected FRET
pairs for wide-field microscopy systems.

When all is said and done, probably the best way to localize or follow protein
molecules in biological systems is with bioluminescence resonance energy transfer (BRET; Chapter 14), fluorescence lifetime imaging (FLI; see Chapters 2 and 11–
13; Bastiaens and Squire, 1999), or correlation spectroscopy (see Chapters 15–16),
but that is another story.

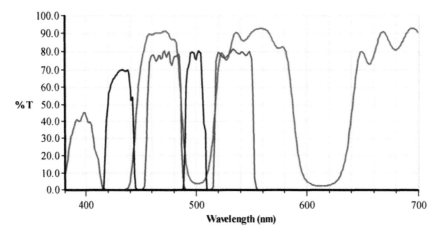

FIGURE 3–10 This filter set is good for FRET imaging in the wide-field FRET microscopy system. This set has a cyan fluorescent protein/yellow fluorescent protein (CFP/YFP) dual mirror with single exciters and emitters (see also Table 3–1). Blue, CFP exciter s430/25x; blue, YFP exciter s500/20x; green, dual dichroic, polychoic; red, CFP emitter s470/30m; red, YFP emitter s535/30m.

TABLE 3–1 Filter Configuration for Wide-Field FRET Imaging for Selected Fluorophore Combinations

FRET Pair		Excitation			Emission	
Donor	Acceptor	Donor	Acceptor	Beamsplitter Lot No.	Donor	Acceptor
CFP	YFP	436/20x	500/20x	82006	470/20m	535/30m
Alexa488	Cy3	480/20x	545/30x	86007	525/40m	620/60m
Alexa488	Alexa555	484/15x	545/30x	86109	517/30m	605/40m
Cy3	Cy5	540/25x	620/60x	86022	590/30m	695/55m
BFP	GFP	380/30x	484/15x	86003	445/40m	535/50m
BFP	YFP	380/30x	500/20x	86005	465/40m	535/50m
BFP	dsRED	375/20x	545/30x	86010	465/30m	630/60m
CFP	dsRED	430/25x	545/30x	86011	480/40m	620/60m
GFP	Rhod-2	484/15x	540/25x	86019	517/30m	605/40m
FITC	Rhod-2	484/15x	540/25x	86019	517/30m	605/40m
FITC	Cy3	480/20x	545/30x	86007	525/40m	620/60m

BFP, blue fluorescent protein; CFP, cyan fluorescent protein; Cy, cyanate; dsRED, *Discosoma sp.* red fluorescent protein; FITC, fluorescein isothiocyanate; GFP, green fluorescent protein; Rhod-2, rhodamine-2; YFP, yellow fluorescent protein.

A special thanks goes to Dr. Simon Watkins, at the University of Pittsburgh, for invaluable input, corrections, and deletions. Thanks also go to Drs. George Patterson and Dave Piston, at Vanderbilt University, and Rich Day, at the University of Virginia, for the use of the CFP and YFP spectral data. This work was made possible by the employee–owners of Chroma Technology Corporation.

REFERENCES

Allen, R.D., G.B. David, and G. Nomarski. The Zeiss-Nomarski differential interference equipment for transmitted light microscopy. Z. *Wiss. Mikrosk.* 69:193–221, 1969.

Bastiaens, P.I.H. and A. Squire. Fluorescence lifetime imaging microscopy: spatial resolution of biochemical processes in the cell. *Trends Cell Biol.* 9:48–52, 1999.

Chamberlain, C.E., V.S. Kraynov, and K.M. Hahn. Imaging spatiotemporal dynamics of
Rac activation in vivo with FLAIR. *Methods Enzymol.* 325:389–400, 2000.

Clegg, R.M. Fluorescence resonance energy transfer and nucleic acids. *Methods Enzymol.*
211:353–389, 1992.

Denk, W., J.H. Strickler, and W.W. Webb. Two-photon laser scanning fluorescence microscopy. *Science* 248:73–76, 1990.

Elangovan, M., W. Horst, Y. Chen, R.N. Day, M. Barroso, and A. Periasamy. Characterization of one- and two-photon excitation energy transfer microscopy. *Methods* 29:58–73,
2003.

Göppert-Mayer, M. Ueber Elementarakte mit Quantenspreungen. *Ann. Phys.* 9:273–295,
1931.

Gordon, G.W., G. Berry, X.H. Liang, B. Levine, and B. Herman. Quantitative fluorescence
resonance energy transfer measurements using fluorescence microscopy. *Biophys. J.*
74:2702–2713, 1998.

Heim, R. and R.Y. Tsien. Engineering green fluorescent protein for improved brightness,
longer wavelengths and fluorescence resonance energy transfer. *Curr. Biol.* 6:178–82,
1996.

Hogg, J. *The Microscope: History, Construction, and Application.* London: George Routledge
and Sons, 1871.

Jones, T. *History of the Light Microscope.* Memphis, TN, http://www.utmem.edu/personal/
thjones/hist/cl.htm., 1997.

Karpova, T.S., C.T. Baumann, L. He, X. Wu, A. Grammer, P. Lipsky, G.L. Hager, and J.G.
McNally. Fluorescence resonance energy transfer from cyan to yellow fluorescent pro-
tein detected by acceptor photobleaching using confocal microscopy and a single laser.
J. Microsc. 209:56–63, 2003.

Kenworthy, A.K., N. Petranova, M. Edidin. High-resolution FRET microscopy of cholera
toxin B-subunit and GPI-anchored proteins in cell plasma membranes. *Mol. Biol. Cell*
11:1645–1655, 2000.

Minsky, M. Microscopy apparatus. U.S. Patent 3,013,467, 1957.

Pawley, J.B. *Handbook of Biological Confocal Microscopy.* New York: Plenum Press, 1998.

Periasamy, A. *Methods in Cellular Imaging.* New York: Oxford University Press, 2001.

Ploem, J.S. The use of a vertical illuminator with interchangeable dielectric mirrors for fluorescence microscopy with incident light. *Z. Wiss. Mikrosk.* 68:129–142, 1967.

Stanley, C.M. Filters and mirrors for applications in fluorescence microscopy. *Methods
Enzymol.* 60:394–415, 2003.

Tron, L., J. Szollosi, S. Damjanovich, S.H. Helliwell, D.J. Arndt-Jovin, and T.M. Jovin. Flow
cytometric measurements of fluorescence resonance energy transfer on cell surfaces.
Quantitative evaluation of the transfer efficiency on a cell-by-cell basis. *Biophys. J.* 45:939–
946, 1984.

Uster, P.S. and R.E. Pagano. Resonance energy transfer microscopy: observations of membrane-bound fluorescent probes in model membranes and in living cells. *J. Cell Biol.*
103:1221–1234, 1986.

Wallrabe, H., M. Elangovan, A. Burchard, A. Periasamy, and M. Barroso. Confocal FRET
microscopy to measure clustering of ligand-receptor complexes in endocytic membranes.
Biophys. J. 85:559–571, 2003a.

Wallrabe, H., M. Stanley, A. Periasamy, and M. Barroso. One- and two-photon fluorescence
resonance energy transfer microscopy to establish a clustered distribution of receptor-
ligand complexes in endocytic membranes. *J. Biomed. Opt.* 8:339–346, 2003b.

Xia, Z. and Y. Liu. Reliable and global measurement of fluorescence resonance energy transfer using fluorescence microscope. *Biophys. J.* 81:2385–2402, 2001.

Zernike, F. The wave theory of microscopic image formation. In: *Concepts of Classical Optics*, edited by J. Strong, San Francisco: W.H. Freeman and Co., 1958, pp. 525–536.

4

FRET Imaging in the Wide-Field Microscope

Fred Schaufele, Ignacio Demarco, and Richard N. Day

1. Introduction

We have witnessed remarkable advances over the past decade in the application of light microscopy to visualize dynamic processes inside the living cell. The development of new fluorescent probes, coupled with advances in digital image acquisition and analysis, has dramatically improved our ability to obtain quantitative measurements from living cells. For example, the cloning of the jellyfish green fluorescent protein (GFP) afforded the enormous benefit of fluorescence labeling that is genetically encoded by transferable DNA sequences (reviewed by van Roessel and Brand, 2002; Zhang et al., 2002). This has transformed studies in cell biology by allowing the behavior of proteins to be tracked by fluorescence microscopy in their natural environment within the living cell. Strategies of mutagenesis of GFP, and the discovery of new GFP-like proteins are yielding an assortment of fluorescent proteins (FPs) that emit from the blue to red range of the visible spectrum (discussed below; Patterson et al., 2001; Matz et al., 2002; Zhang et al., 2002; Verkhusha and Lukyanov, 2004).

The FPs are widely used as noninvasive markers in living cells because their fluorescence does not require the addition of cofactors, and they are very stable and well tolerated by most cell types. The minimal toxicity of these probes is best illustrated by the many examples of healthy transgenic mice that express the FP markers (reviewed by Feng et al., 2000; Hadjantonakis and Nagy, 2001; Walsh and Lichtman, 2003). For instance, Brewer et al. (2002) generated "knockin" mice that express the GFP-glucocorticoid receptor under endogenous regulatory control. The expression and function of the chimeric receptor were indistinguishable from the endogenous counterpart it replaced, demonstrating that these imaging probes can be truly noninvasive. Further, since the FPs have no intrinsic intracellular targeting, they're most useful for monitoring the subcellular localization and trafficking properties of proteins.

Importantly, the spectral properties of the FPs allow them to be used as probes in Förster resonance energy transfer (FRET) microscopy. FRET microscopy can detect the result of the direct transfer of energy from a donor to an acceptor FP that labels proteins of interest in the living cell. Since energy transfer can only occur over distances of less than about 8 nm, the detection of FRET provides information about the spatial relationships of those proteins on the scale of angstroms (Stryer, 1978; Selvin, 1995; Periasamy and Day, 1999). In this chapter, we describe the characteristics and discuss the limitations of the FPs as labels for proteins expressed in living cells. We review the principals of acquiring FRET measurements using wide-field microscopy (WFM) of cells expressing the FPs. We then demonstrate three different methods for detecting FRET signals from living cells and consider the limitations of each of these approaches. Finally, we will discuss potential problems associated with the expression of proteins fused to the FPs for FRET-based measurements from living cells.

2. OVERVIEW OF THE FLUORESCENT PROTEINS

In 1962, Shimomura et al. characterized a GFP in the jellyfish *Aequorea victoria* that was a companion to the bioluminescent protein aequorin. The *Aequorea* GFP was cloned in 1992 (Prasher et al., 1992), and its expression in bacteria, mammalian and plant cells, and transgenic organisms proved its utility for in vivo fluorescence labeling (Chalfie et al., 1994; Inouye and Tsuji, 1994; Plautz et al., 1996; van Roessel and Brand, 2002). To be fluorescent, GFP must adopt an 11-strand β-barrel structure with the tripeptide serine^{65}–tyrosine^{66}–glycine^{67} positioned properly at its core to create the chromophore (reviewed by Tsien, 1998). Nearly the entire peptide sequence is required to achieve the correct folding that is necessary for fluorescence. This means that when a protein is expressed as a fusion to a FP, the fused protein must permit the efficient folding of the β-barrel. Similarly, in order for experimental results to be interpreted, the 27 kDa FP linked to the protein of interest must not interfere with the normal cellular functions of that protein (discussed in Section 6, below).

2.1 Spectral Variants Based on Aequorea Green Fluorescent Protein

The extensive mutagenesis of the *Aequorea* GFP has yielded proteins with improved characteristics that emit light from the blue to yellowish-green range of the visible spectrum (reviewed in Patterson et al., 2001; Zhang et al., 2002; see Table 4–1). For example, changing the fluorochrome serine^{65} to threonine (GFP^{S65T}) stabilized the fluorochrome in a permanently ionized form with a single peak absorbance at 489 nm, and resulted in a more rapidly maturing protein (Heim et al., 1994; Brejc et al., 1997; Cubitt et al., 1999). A commercially available modification of GFP^{S65T}, known as enhanced GFP (EGFP; BD Biosciences Clontech, Palo Alto, CA) incorporated preferred human codon usage and silent mutations to improve protein expression in mammalian cells. A blue color variant (BFP) resulted from substitution of tyrosine^{66} with histidine (Heim and Tsien, 1996; Cubitt et al., 1999). While this protein has a low quantum yield and is susceptible to photobleaching, it is the most blue-shifted FP, and has proven useful in multispectral and FRET imaging (Day et al., 1999, 2001; Table 4–1).

The identification of cyan color variants (CFP) provided alternative blue-shifted proteins that overcame some of the deficiencies of BFP. Enhanced CFP (ECFP)

TABLE 4–1 Spectral Characteristics of Fluorescent Proteins and Filter Combinations Commonly Used to Detect Them

	FP Characteristics			Filter Set		
FP	Excitation Maximum (nm)	Emission Maximum (nm)	Photobleach Sensitivity	Exciter	Beam Splitter	Emitter
BFP	382	446	High	380/30	420	445/40
ECFP	434	476	Low	436/20	455	470/20
EGFP	489	509	Low	480/20	495	510/20
EYFP	514	527	Moderate	500/20	515LP	535/30
mRFP	558	583	Moderate	545/30	570LP	620/60

FP FRET Pair		Exciter Filters		Multiband Beam Splitter (Regions of High Transmission)	Emitter Filters	
Donor	Acceptor	Donor	Acceptor		Donor	Acceptor
CFP	YFP	430/25	500/20	445–485/520–580	470/30	535/30
BFP	GFP	380/30	485/25	425–460/510–580	445/40	535/50

BFP, blue fluorescent protein; CFP, cyan fluorescent protein; ECFP, enhanced CFP; EGFP, enhanced GFP; EYFP, enhanced YFP; GFP, green fluorescent protein; YFP, yellow fluorescent protein.

resulted from substitution of tyrosine[66] with tryptophan in combination with mutations in several other residues within the surrounding β-barrel structure (Heim et al., 1994; Cubitt et al., 1999). The excitation spectra for ECFP is complex, and fluorescence lifetime measurements for ECFP indicated that it adopted more than one excited state in living cells (Tramier et al., 2002). Recently, Rizzo et al. (2004) reported that substitution of two hydrophobic residues on the solvent-exposed surface of ECFP (tyrosine[145] and histidine[148]) stabilized ECFP in a single conformation. This mutant variant, called "cerulean" (sky blue), also had increased quantum yield, making it a better FRET donor for the yellowish color variant, YFP. The YFP resulted from the targeted mutation of the β-barrel threonine residue at position 203, which was identified in the crystal structure of GFP[S65T] as being directly adjacent to the chromophore (Ormö et al., 1996). An enhanced version of YFP (EYFP; BD Biosciences Clontech, Palo Alto, CA) with improved protein expression is available. Starting with EYFP, Nagai et al. (2002) discovered an improved YFP variant during the development of new calcium-sensing proteins, which they called "Venus." This protein has phenylalanine[46] substituted with leucine, which improves the maturation efficiency and reduces the halide sensitivity of YFP. Venus is the brightest and most red-shifted of the mutant variants based on the *Aequorea* GFP that is currently available. The limitations of these FP variants as probes for FRET imaging studies will be discussed below.

2.2 Seeing Red

It was also recognized that most of the colors in reef corals result from GFP-like proteins, and the cloning of several different red fluorescent proteins (RFP) from Indo-Pacific corals has added to the spectrum of available fluorescent proteins (Matz et al., 1999, 2002). The first commercially available RFP (DsRed), cloned from the mushroom anemone *Discosoma striata*, was a very slow-maturing protein that was an obligate tetramer with a strong tendency to form oligomers *in vivo* (Baird et al., 2000). Bevis and Glick (2002) overcame the problem of slow maturation by using both random and directed mutagenesis to generate a rapid-maturing DsRed variant (<1 h) with increased stability and solubility. Starting with this rapidly maturing DsRed variant, Campbell et al. (2002) applied directed mutagenesis to break the tetramer. They then followed this with rounds of random mutagenesis to evolve proteins with improved red fluorescence. The result was the monomeric RFP (mRFP), which matures rapidly and has a peak fluorescence emission at 607 nm, making it very suitable for multispectral imaging applications (Voss et al., 2004; Table 4–1). The current version of mRFP is not yet optimal as a FRET acceptor because of its tendency to form an absorbing but non–fluorescent green species (Campbell et al., 2002). It is likely that new variants of the RFPs will soon yield FRET acceptors with improved characteristics.

3. Overview of Basic Requirements for FRET

Multicolor fluorescence microscopy experiments using the FPs to track different proteins in the same living cells allow the direct visualization of protein co-localized within subcellular compartments (for example, Schaufele et al., 2001; Platani et al., 2002; Enwright et al., 2003; Gerlich et al., 2003). The ability to define the spatial relationships between the FP-labeled proteins, however, is limited by the diffraction of light to a resolution of approximately 200 nm. Objects that are closer together will appear as a single object, so considerable distances may actually separate proteins that appear colocalized by fluorescence microscopy. FRET microscopy can dramatically improve this spatial resolution by detecting the direct transfer of excitation energy from donor to acceptor fluorochromes. There are several requirements that must be met, however, before FRET can occur. The first is the strong *distance* dependence of the direct energy transfer from a donor to acceptor fluorochromes, which is limited to the scale of angstroms. Second, there must be a significant *spectral overlap* of the donor fluorochrome emission with the absorption spectrum of the acceptor. Finally, both donor and acceptor fluorochromes must be in favorable mutual *orientation*. Each of these requirements is discussed below.

3.1 Distance

FRET was first described theoretically by Perrin (1932), who recognized that a fluorochrome in the excited state behaves as an oscillating dipole that creates its own characteristic electric field. If an acceptor fluorochrome enters the electric field, energy can be transferred from the donor by direct electrodynamic interactions to induce transitions in the acceptor—there is no intermediate photon involved. The FRET theory was developed further and given a quantum-mechanical basis by Theodor Förster (1948). The efficiency of energy transfer depends on the square of the electric field produced by the donor, and this field decays as the inverse of the third power of distance (reviewed by Stryer, 1978; Selvin, 1995; see Chapter 2). Therefore, the efficiency of FRET will vary as the inverse of the sixth power of the distance that separates the donor and acceptor fluorochromes, as described by

$$\text{FRET efficiency, } E = 1/[1 + (r/R_0)^6]$$

where r is the distance separating the two fluorochromes, and R_0 is the Förster distance, the critical molecular separation at which energy transfer is 50% efficient. Figure 4–1 shows the relationship of FRET efficiency to the distance separating the most commonly used FPs in FRET studies, CFP and YFP. This illustrates the dramatic decrease in FRET efficiency over the range of 0.5 R_0 to 1.5 R_0 (shaded area in Fig. 4–1), and shows why it is very difficult to accurately measure FRET when the distance separating the fluorochromes is > 8 nm.

3.2 Spectral Overlap

An efficient interaction between the electronic systems of the donor and acceptor requires a donor electric field at a resonance frequency that can drive the electrons in the acceptor to oscillate and induce transitions. The rate of energy transfer improves with increasing overlap of the donor emission with the absorption spectrum of the acceptor, and this is described by the spectral overlap integral, J_{DA} (see Chapter 2). The Förster distance, R_0, for a particular FRET pair is derived on the basis of their spectral overlap integral (Förster, 1948; see Patterson et al., 2000). This means that, provided the quantum yield of the donor is high, fluorochrome pairs sharing more spectral overlap will be more efficient FRET partners, and this will be represented by a larger R_0 value for the pair. For instance, there is a greater spectral

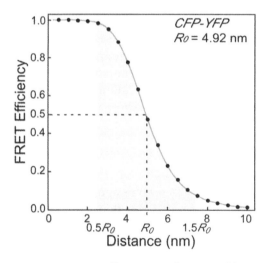

FIGURE 4–1 FRET efficiency as a function of distance separating cyan fluorescent protein (CFP) and yellow fluorescent protein (YFP), which are commonly used in FRET studies. The Förster distance, R_0, for this fluorochrome pair is 4.92 nm (Patterson et al., 2000), and the FRET efficiency drops to less than 10% at 1.5R_0 (shaded area).

overlap between GFP and YFP (R_0 = 5.6 nm) than there is for CFP and YFP (R_0 = 4.9 nm) (Patterson et al., 2000, 2001).

In theory, one should strive for the maximum overlap integral and use GFP and YFP in FRET experiments. However, there is a significant consequence of increasing the spectral overlap of the FRET partners: a dramatic increase in background fluorescence in the FRET image occurs because of spectral bleedthrough (SBT) signals contributed by both the donor and acceptor. The absorption and emission spectra for both CFP and YFP are shown in Figure 4–2, illustrating the significant spectral overlap (shaded area in Fig. 4–2) that makes CFP and YFP good FRET partners. However, these fluorescent proteins are not ideal FRET partners because they have broad spectra and small Stokes shifts. Consequently, even with optimized filter sets (see Chapter 3, for example), there will be a significant SBT signal contaminating the FRET image. This SBT signal comes primarily from two sources (illustrated in Fig. 4–2); acceptor bleedthrough because of excitation of YFP at the wavelengths used to excite CFP, and donor cross talk resulting from the strong CFP emission into the FRET channel. In practice, few fluorochromes can be considered ideal FRET pairs. It is likely that new FPs, either cloned from corals or generated through mutagenesis, will soon yield FRET pairs with improved characteristics. Until then, we must exploit the attributes of the currently available FPs and use methods, which are discussed below (Section 5), to acquire accurate measurement of FRET signals.

3.3 Orientation

The efficiency of the dipole–dipole interaction between the donor and acceptor depends on the alignment of the two dipoles. Thus, a third condition for FRET is

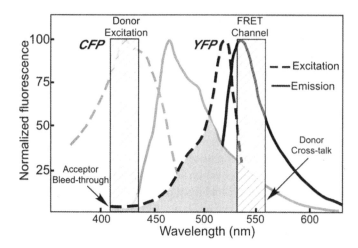

FIGURE 4–2 Spectral overlap of cyan fluorescent protein (CFP) and yellow fluorescent protein (YFP). The excitation (---) and emission (—) for both CFP and YFP are shown, illustrating the overlap (shaded area) of CFP emission with YFP absorption, a critical requirement of FRET. The wavelengths of light that are passed by typical filter sets used to excite the donor and detect signals in the FRET channel are shown by hatched boxes. The sources of spectral bleedthrough, acceptor bleedthrough, and donor cross talk are indicated.

that the donor and acceptor fluorochromes must be in a favorable mutual orientation. The orientation factor, κ^2 describes the alignment of the donor emission dipole moment with the transition dipole of the acceptor. The values for κ^2 can vary from 0, when all dipoles are perpendicular to one another, to 4, if all angles are parallel. If the dipoles are rapidly moving on timescales similar to the donor excited-state lifetime (see Chapter 9), then the dipole orientations will be random and the value for κ^2 will be 0.67.

In biological systems, it is typically assumed that the fluorochromes labeling individual proteins are free to adopt random orientations. Additionally, the protein environment within the FPs also allows some rotational freedom of the chromophore (Zimmer, 2002), which may ease constraints on the dipole orientations necessary for efficient FRET. In this regard, Riven and colleagues (2003) used fluorescence anisotropy to determine the rotational freedom of CFP- and YFP-fused potassium channels, which were inserted in the membranes of living cells. Their results showed that the FPs tethered to the membrane proteins could still randomize on the timescale of FRET, indicating that restrictions on dipole orientation were not a significant determinant of FRET efficiency in that system. Still, it is important to realize that rotational freedom can vary depending on the intracellular environment surrounding individual FP-labeled proteins, and when κ^2 is not known, there will be uncertainty in estimates of the distance separating the fluorochromes based on FRET results (van Thor and Hellingwerf, 2002).

4. FRET IMAGING IN THE WIDEFIELD MICROSCOPE

There are several different light microscopy methods that can be used to detect FRET from living cells, and each has its own inherent strengths and weaknesses. The chapters in this book cover each of these different imaging methods, and should provide insight into how to choose the best system for one's imaging needs. Frequently, no single approach will be optimal, and it is important to choose the method that fulfills the experimental needs by providing the best contrast, sensitivity, and reproducibility.

4.1 Imaging System Requirements

Most fluorescence microscope systems can be adapted to make FRET measurements, but there are several factors to consider when setting up a system for FRET. The schematic in Figure 4–3 illustrates the typical light path in a WFM, and shows several features that deserve mention. First, the arc lamp light source in a fluorescence microscope usually uses either a mercury or xenon bulb. It is important to recognize that the mercury source does not provide even intensity across the spectrum, but rather has peaks of intensity at specific wavelengths. Thus, if this source is used, the spectral intensity peaks should match the excitation wavelengths required for optimal excitation of the donor and acceptor fluorochromes. The xenon light source, by contrast, has more even intensity across the visible spectrum, but is deficient in the ultraviolet range of the spectrum. There are combination lamps

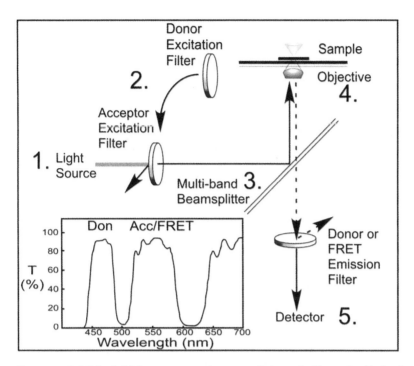

FIGURE 4–3 Wide-field fluorescence microscopy light path. The path of light through the wide-field fluorescence microscope is illustrated to show the position of (1) the arc lamp, (2) the exciter filter wheel, (3) the beam splitter mirror, (4) the obective, and (5) the detector. The multiband beam splitter has two regions of high transmission to accommodate the emission of both the donor (Don) and the acceptor (Acc) fluorochromes (inset).

available that exhibit the best characteristics of both the xenon and mercury light source. These mercury–xenon lamps provide a continuous spectrum from the ultraviolet to infrared and include the strong mercury line spectrum.

Second, it is very important to select filter combinations to detect the fluorescence signals that reduce SBT and optimize signal-to-noise ratio for each fluorochrome (Fig. 4–3). There are many different commercially available filter combinations that have been designed specifically for FRET microscopy using the FPs (see Chapter 3). Third, an often-overlooked feature in the fluorescence microscope light path is the beam splitter, or dichroic mirror. These mirrors function to reflect the excitation wavelengths of light, while transmitting the emitted fluorescence from the sample. For FRET microscopy, we recommend a single multiband beam splitter mirror used in combination with a filter wheel to select the specific excitation wavelengths (Fig. 4–3). The multiband beam splitter is designed to transmit light at both the donor and acceptor (FRET) emission wavelengths, while reflecting light at other wavelengths (inset, Fig. 4–3). The use of the single mirror eliminates potential shifts in image registration that can result from slight variations in positioning of the mirror in different filter cubes, and ensures that the efficiency of the mirror is the same for all acquired images. The filter combination and beam splitter we used for FRET measurements with CFP and YFP fluorochrome pair are shown in Table 4–1.

A fourth important consideration is the type of objective used to gather the image. The detection of weak FRET signals requires the efficient collection of fluorescence from the sample, and high numerical-aperture objectives have increased light collection efficiency. The quality of the image can be improved further by using a water-immersion objective, which will help reduce spherical aberration. Alternatively, if an oil-immersion objective is used, the oil should have a refractive index that matches the aqueous media. The final step in the light path, the imaging device, determines whether the fluorescence signals originating from the sample will be detected. The most popular imaging devices are charge-coupled device (CCD) detectors, and recent improvements in their design have resulted in greatly increased sensitivity (Spring, 2001). Significant improvements in the signal-to-noise ratio for CCD detectors have been achieved through cooling of the CCD chip, which reduces the dark current noise and increases the charge transfer efficiency. When the digitized image is read out at 12 bits, each pixel in the resulting digital image can represent up to 4096 (2^{12}) gray levels of signal intensity. When the signals from different fluorochromes are quantified, such as in FRET imaging, it is important that the quantum efficiency of the CCD detector is uniform over the range of light wavelengths measured.

4.2 Why Use Wide-Field Microscopy to Measure FRET?

Although WFM is the most basic of the imaging methods covered in this book, it can be an excellent choice for detecting FRET signals from living cells. Experiments involving light microscopy of living cells are a balance between acquiring images with optimal signal-to-noise ratio while limiting potential photodamage to the sample under observation. Wide-field microscopy is among the most sensitive methods available, allowing the illumination energy necessary to acquire significant data to be minimized (Wallace et al., 2001; Swedlow et al., 2002). The major limitation of WFM, especially under low light conditions, is the out-of-focus signal from above and below the focal plane, which reduces image contrast. Since a monolayer of cells is typically < 15 μm thick, the effect on image contrast is less than it would be for thicker and more complex specimens.

For these thin specimens, the images acquired by WFM can have less noise and greater sensitivity than those obtained by laser scanning confocal microscopy (LSCM), for several reasons. First, although all illumination sources suffer from power fluctuations, the flat-field illumination provided by a stabilized arc lamp to all regions of the sample does not have the pixel-to-pixel fluctuations in illumination found with laser scanning. Second, the use of back-illuminated or amplified CCD detector can provide greater sensitivity than photomultiplier tubes typically used in LSCM (Swedlow et al., 2002). In addition, care must be used in acquiring LSCM images because the spectral response and gain can be tuned for each of the two or more individual detectors, which will change their relative sensitivities and SBT coefficients. Finally, LSCM has a limited number of available excitation laser lines. The standard 458 nm argon ion laser line commonly used to excite CFP (peak excitation of 434 nm; Table 4–1) is not optimal for this task and overlaps substantially with the excitation spectrum of YFP, resulting in significant SBT signals in

FRET studies. The optimal excitation of CFP for FRET occurs at 430 nm, which can be achieved using a frequency-doubled diode laser (van Rheenen et al., 2004). In contrast, the arc lamp and filter wheel systems used in WFM provide uniform illumination in an unlimited choice of excitation wavelengths. As discussed in Section 5, the accurate quantification of FRET signals is critical to interpreting the spatial relationships between proteins labeled with the FPs. Use of the combination of an optimized WFM and sensitive CCD camera to acquire images of thin specimens, such as cells grown in a monolayer, can provide the most accurate measurements of dim fluorescence signals (Wallace et al., 2001). For thicker samples, LSCM or multiphoton excitation microscopy are the preferred imaging methods (see Chapters 5 and 6).

4.3 The Biological Model

The biological model used in our laboratories is the transcription factor CCAAT/ enhancer binding protein alpha (C/EBPα). The C/EBP family proteins bind to specific DNA elements as obligate dimers, where they function to direct programs of cell differentiation and play key roles in the regulation of genes involved in energy metabolism (Wedel and Ziegler-Heitbrock, 1995). Immunocytochemical staining has showed that the endogenous C/EBPα protein bound to satellite DNA repeat sequences located in regions of centromeric heterochromatin in differentiated mouse adipocyte cells (Tang and Lane, 1999). When the FP-labeled C/EBPα was expressed in mouse cells, it also preferentially localized to the regions of centromeric heterochromatin (Day et al., 2001; Schaufele et al., 2001; Enwright et al., 2003). This subnuclear positioning of C/EBPα is demonstrated in the images shown in Figure 4–4. These three-dimensional (3-D) volume renderings were generated from optical sections acquired through a cell expressing the GFP-labeled C/EBPα

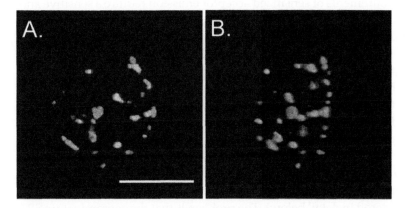

FIGURE 4–4 Three-dimensional images show the green fluorescent protein–CCAAT/enhancer binding protein alpha (GFP-C/EBPα) DNA-binding domain (DBD) in the cell nucleus. Optical sections at 0.2 µm steps were acquired from a living cell expressing the GFP-C/EBPα DBD. This protein was exclusively located in the cell nucleus, where it was bound to regions of heterochromatin. *A:* Three–dimensional volume rendering of the cell nucleus showing the protein bound to regions of heterochromatin. *B:* The same image rotated 90°; the bar indicates 10 µm.

DNA-binding domain (DBD), and show the association of C/EBPα with regions of centromeric heterochromatin in the nucleus (Fig. 4–4). We are using FRET-based imaging approaches to study the biochemical and structural origins of this subcellular distribution (Schaufele et al., 2003) and to understand how other nuclear proteins interact with C/EBPα (Day et al., 2003).

5. TECHNIQUES FOR MAKING FRET MEASUREMENTS WITH WIDEFIELD MICROSCOPY

When energy transfer occurs, the donor signal is quenched, and there is increased (sensitized) emission from the acceptor. As shown in Figure 4–2, the detection of FRET by measuring sensitized acceptor emission is complicated by SBT background signals in the FRET image. One approach for measuring FRET is to acquire an image in the FRET channel and then use an *SBT correction* method to remove the contaminating background signal from the sensitized acceptor emission. Alternatively, the effect of FRET on the donor fluorochrome can be measured by detecting the increased (or dequenched) donor signal after *acceptor photobleaching*. Because the acceptor photobleaching approach measures only the donor signal, it does not require correction for the SBT background signals. A third approach is to use ratiometric analysis to obtain relative FRET efficiencies for different protein interactions in populations of cells. We will consider each of these approaches in turn and provide examples of FRET data for each obtained using our biological model system.

5.1 Detecting Protein Interactions by FRET Using Spectral Bleedthrough Correction

Quantitative measurements of sensitized acceptor emission require a sensitive and accurate method to identify and remove SBT signals in the FRET image. Several different computer algorithms have been developed that remove SBT to yield a corrected FRET signal (reviewed by Berney and Danuser, 2003). The common approach is to acquire reference images of control cells expressing either the donor- or the acceptor-labeled proteins alone, and then to use these data to define the SBT components in the FRET channel. This information is then used to correct FRET signals acquired of experimental cells coexpressing both the donor- and acceptor-labeled proteins on a pixel-by-pixel basis.

   The approach works on the assumption that the SBT signals in the experimental cells will be the same as the combination of signals determined for the individual donor- and acceptor-labeled reference cells imaged under identical conditions (this will also be considered in detail below). The problem is that the contaminating SBT signal is determined through use of images acquired from three different cells, the two different reference cells expressing either the donor or acceptor alone, and experimental cells expressing both proteins. Thus, the individual pixel locations in these different images cannot be directly compared. What can be compared, however, are the pixel-by-pixel intensities of the SBT signals in the single-labeled reference cells, which can then be applied as a correction factor to the pixels with similar gray-level intensities in the double-labeled experimental cells. Here, we demonstrate a computer algorithm (pFRET, described in Chapter 7) that uses pixel intensity matrices to partition the pixel-by-pixel intensity values for the signals from each of the reference (donor or acceptor alone) images into particular ranges. These same pixel ranges are then identified in the experimental images, and the SBT contributions determined in each range from the reference cells are then removed by pixel-to-pixel subtraction.

For the FRET experiment illustrated in Figure 4–5, images were first acquired of cells that expressed either CFP-C/EBPα DBD or YFP-C/EBPα DBD alone to serve as intensity-matched controls for the experimental cells that co-expressed both CFP- and YFP-labeled C/EBPα DBD. All images were acquired under identical conditions with optimal signal-to-noise ratio and no saturation of pixel intensity. The same filter set (Table 4–1) was used to collect both the control cells for SBT correction and the experimental images for FRET measurements. The CFP- and YFP-specific excitation and emission filters were mounted in the filter wheels, and a single beam splitter was used to pass both the CFP and YFP emissions (Fig. 4–3). For the example shown here, the acceptor (Fig. 4–5A) and quenched donor images (Fig. 4–5B) were acquired, and the signal in the FRET channel was obtained using donor excitation (Fig. 4–5C). The uncorrected FRET (uFRET) image contained both the sensitized acceptor emission and the contaminating SBT signals. This digital image was then processed using the pFRET correction algorithm (Chapter 7) to remove the SBT signals from each pixel, resulting in the corrected FRET, or pFRET 8-bit image, that is shown in false color in Figure 4–5D.

5.2 Verifying Protein Interactions with Acceptor Photobleaching FRET

When energy transfer occurs, the donor emission will be quenched because of the direct transfer of excitation energy to the acceptor. For instance, at the R_0 for a FRET pair, 50% of the donor emission will be quenched because of direct energy transfer (see Fig. 4–1). If the acceptor fluorochrome is destroyed, FRET will be eliminated and the donor signal will be dequenched. The increase in the donor signal resulting from dequenching is a direct measure of the efficiency of FRET. The technique of acceptor photobleaching FRET exploits this characteristic and measures the dequenching of the donor signal in the regions of the cell where FRET had occurred. Importantly, this approach provides a method to verify FRET measurements made using other techniques. Here, we used the acceptor photobleaching approach to verify the previous FRET results for the C/EBPα DBD obtained using the SBT correction method.

The photobleaching approach requires the selective bleaching of the acceptor, because any bleaching of the donor fluorochrome will lead to an underestimation of the dequenching. Further, bleaching of the acceptor should be as close to completion as possible, since any remaining acceptor will still be available for FRET, again resulting in an underestimation of the donor dequenching. The YFP variant is sensitive to photobleaching, making it useful for FRET measurements by acceptor photobleaching (Miyawaki and Tsien, 2000). The results shown in Figure 4–6A

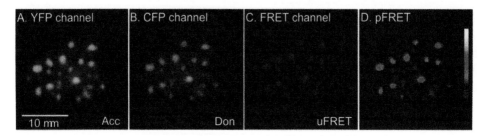

FIGURE 4–5 Use of spectral bleedthrough (SBT) correction computer algorithm to measure FRET. *A, B:* Fluorescence signals from the nucleus of a cell co-expressing yellow fluorescent protein–CCAAT/enhancer binding protein alpha (YFP-C/EBPα) DNA-binding domain (DBD) (acceptor [Acc]) and CFP-C/EBPα DBD (donor [Don]) are shown. The uncorrected signal in the FRET channel (uFRET) was detected using donor excitation and acceptor emission (*C*). This image contained both the FRET signals and the SBT signals contributed by both the donor and acceptor. The SBT signal in the FRET channel was determined using intensity-matched control cells expressing either the YFP- or CFP-C/EBPα DBD, and this contaminating background signal was removed from each pixel of the FRET image to generate the corrected (pFRET) signal (*D*). The pixel-by-pixel distribution of the corrected FRET signal in the 8-bit image is shown in the false color.

demonstrate the selective photobleaching of YFP in cells that also co-expressed CFP, using excitation at 510 nm. Greater than 90% bleaching of YFP was achieved by 5 min of exposure, whereas there was less than 5% bleaching of the co-expressed CFP under these conditions (Fig. 4–6A).

Images of a cell co-expressing the nuclear-localized CFP- and YFP-C/EBPα DBD were obtained using the donor (Fig. 4–6B) and acceptor (Fig. 4–6C) filter sets. The acceptor (YFP) was then selectively photobleached by exposure to 510 nm light to reduce the acceptor signal by more than 90%. A second donor image (Don2) was then acquired under identical conditions to those for the first donor image. Comparison of the gray-level intensity of the donor signal in the pre- and post-bleach donor images revealed an increase in the donor signal following the acceptor bleaching (indicated by the numbers in Figs. 4–6B and 4–6D). This increase in the donor signal was quantified by digital subtraction of the Don1 image from the Don2 image, and the change in gray-level intensity was mapped using a look-up table, with black indicating no change and yellow showing the maximum change in the donor signal (Fig. 4–6E). These results verified the earlier FRET results obtained with the SBT correction method, revealing the steady-state interactions of C/EBPα DBD on the angstrom scale in the intact pituitary cell nucleus.

The limitation of this approach is that it is an end-point assay and cannot be repeated on the same cells. Further, as was mentioned above, the acceptor bleaching method is dependent on complete destruction of the acceptor; any remaining acceptor will still be available to quench donor, resulting in an underestimation of FRET. We have compared the SBT correction method with the acceptor bleaching method and found that the average efficiencies of FRET estimated by means of the acceptor bleaching method were somewhat lower than average efficiencies estimated by software correction, which may reflect incomplete acceptor photobleaching. The application of acceptor photobleaching to living cells also requires that protein complexes be relatively stable over the bleach period. In this regard,

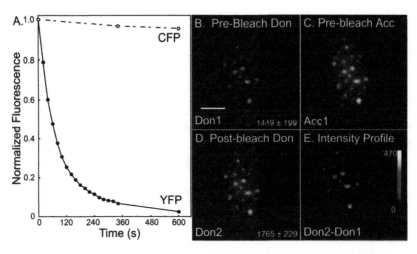

FIGURE 4–6 Use of acceptor photobleaching to measure donor dequenching. The acceptor photobleaching approach requires selective bleaching of the acceptor. *A:* Selective bleaching of yellow fluorescence protein (YFP) in cells co-expressing cyan fluorescent protein (CFP). Excitation at 510/20 nm efficiently bleached the YFP with a minimal effect on the CFP signal. *B–D:* The acceptor photobleaching FRET approach was then applied to a cell co-expressing YFP- and CFP-CCAAT/enhancer binding protein alpha (C/EBPα) DNA-binding domain (DBD). Prebleach images were acquired using the donor (Don1) and acceptor (Acc1) channels (*B, C*). The calibration bar indicates 10 μm. A second post-bleach donor image (Don2) was obtained after selective acceptor photobleaching under identical conditions to those used for the Don1 image (*D*). Gray-level intensities of the Don1 and Don2 images are shown (*B, D*). *E:* The de-quenched donor signal was then determined by digital subtraction (Don 2 – Don 1). The change in gray-level intensity of the donor signal is shown in the intensity profile, with black indicating 0 and yellow indicating the maximum increase in gray level in the 12-bit image.

the 3-D positioning of the protein complexes formed by the C/EBPα DBD is very stable over the time periods required for acceptor photobleaching; other live-cell experimental models may not be as amenable to this approach.

5.3 Extending Analysis of Protein Interactions by Means of Ratiometric Measurements

Intermolecular FRET also can be measured using ratiometric methods, provided the SBT signals are accounted for. Although SBT signals will vary from pixel to pixel (see Chapter 7), the proportion of SBT signal in the FRET image is a function of the imaging system optics and filter combinations, and will remain constant with changing donor and acceptor levels. Here, the SBT signals in the FRET channel were determined for many reference cells expressing a wide range of either the acceptor (GFP)- or the donor (BFP)-labeled C/EBPα proteins. These results are shown in Figure 4–7A and clearly demonstrate that the contributions of donor cross talk (open triangles, 0.472) and the acceptor bleedthrough (filled triangles, 0.088) to the signal in the FRET channel were constant over a wide range of expression levels. There were minimal contributions of donor excitation at the acceptor excitation wavelengths (filled circles, 0.001) and acceptor emission bleedthrough into the donor

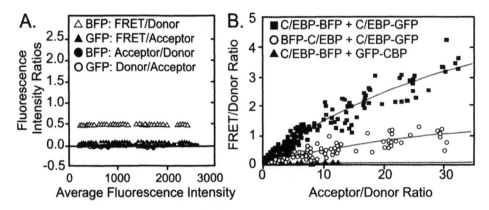

FIGURE 4–7 Ratiometric FRET measurements. FRET between blue fluorescent protein (BFP) and green fluorescent protein (GFP) attached to the carboxy-terminal bZIP domain in full-length CCAAT/enhancer binding protein alpha (C/EBPα) and expressed in GHFT1–5 pituitary progenitor cells. *A:* The spectral bleedthrough (SBT) ratios are plotted as a function of the average fluorescence intensity for individual cells expressing C/EBPα labeled with either BFP or GFP. The contribution of donor cross talk (open triangles) and the acceptor bleedthrough (filled triangles) to the signal in the FRET channel were constant over a wide range of expression levels. The minimal contribution of donor excitation at the acceptor excitation wavelengths (filled circles) and the acceptor emission bleedthrough into the donor channel (open circles) is also shown. Only cells that expressed fluorescent protein (FPs) in this range were processed for FRET determination. *B:* The FRET signal changed when the domains in C/EBPα that were labeled with the fluorochromes were switched. FRET/donor ratios for individual cells expressing the BFP- and GFP-labeled proteins were calculated following background subtraction, and correction for each of the SBT components was made (*A*). For cells co-expressing full-length C/EBPα labeled with the donor (BFP) and acceptor (GFP) at the carboxyl-terminal DNA-binding domain (DBD), the FRET/donor ratio increased proportionally with increasing acceptor relative to donor (filled squares). The maximum FRET/donor level at saturating acceptor/donor amounts was different when the donor flurophore (BFP) was switched to the amino terminus of C/EBPα (compare filled squares with open circles). In contrast, there was no FRET signal detected when C/EBPα-BFP was co-expressed with the co-regulatory protein CBP labeled with GFP (filled triangles). [Adapted from Schaufele et al. (2003).]

channel (open circles, 0.004); these are also shown in Figure 4–7A. Although these later two SBT components are negligible for this donor–acceptor pair (BFP and GFP), they can be significant for other fluorochrome pairs or filter combinations.

For the experimental cells that co-expressed both the GFP- and the BFP-labeled C/EBPα, the affinity of this dimer interaction determines the fraction of proteins that form protein complexes. For this high-affinity dimer interaction, the FRET signal is strongly dependent on the relative concentrations of the donor- and acceptor-labeled proteins. When the donor- and acceptor-labeled proteins are present in a 1:1 ratio, 50% of the labeled proteins will be in donor–acceptor pairs, with the remaining fraction equally distributed as donor–donor and acceptor–acceptor pairs. If there is no energy transfer between the donor–acceptor pairs, then the FRET/donor ratios will be zero after SBT subtraction, and will not change with increasing acceptor/donor ratio. In contrast, when energy transfer occurs, as the acceptor-labeled protein is increased relative to the donor, the fraction of donor–acceptor pairs that can participate in FRET will increase. There will be increased fluorescence in the FRET channel and a decreased signal in the donor channel, resulting in a positive slope for the FRET/donor graph. The amount of acceptor that is required to reach saturation, when all donor-labeled proteins are associated with acceptor-labeled proteins, is determined by the kinetics of protein–protein interaction. The FRET signal obtained at saturation will be a function of the conformation proteins adopted within the complex, which establishes the distance and orientation between the fluorochromes.

In the example shown in Figure 4–7B, donor, acceptor, and FRET images were acquired of many different cells co-expressing BFP- and GFP-labeled C/EBPα. The ratiometric FRET analysis was then used to compare the interactions between the acceptor-labeled C/EBPα and C/EBPα labeled with the donor fluorochrome at different positions. In addition, FRET measurements were also made from cells co-expressing the donor-labeled C/EBPα in combination with acceptor-labeled CREB-binding protein (CBP). In each case, the contributions of SBT determined in the reference cells (Fig. 4–7A) were subtracted from each individual FRET determination (see legend for Fig. 4–7), and the results were expressed as the FRET/donor ratio as a function of increasing acceptor/donor ratio. For cells co-expressing the C/EBPα labeled with BFP or GFP at the carboxyl-terminal DBD, the FRET/donor ratios increased proportionally with the amount of acceptor relative to donor (filled squares in Fig. 4–7B). The FRET/donor ratio varied with the acceptor/donor level with first-order kinetics, indicative of a bimolecular interaction.

Significantly, when BFP was moved to the amino terminus of C/EBPα, and the protein was co-expressed with the carboxyl-terminal acceptor-labeled C/EBPα, the maximum FRET/donor level at saturating acceptor/donor amounts was substantially reduced (compare open circles with filled squares in Fig. 4–7B). This change in the FRET signal obtained when the position of the donor fluorochromes on C/EBPα was switched could reflect different protein conformations that lead to an increased separation of the fluorochromes, or an alteration in their dipole orientations. Importantly, the results in Figure 4–7B also clearly show that FRET was not detected between the donor-labeled C/EBPα and the acceptor-labeled CBP (filled triangles). These two proteins are known to interact either directly or indirectly as part of a larger protein complex (Schaufele et al., 2001), so this result likely reflects the average position of the fluorochromes being beyond the distance that is necessary for efficient FRET.

When this analysis is done on many cells, the results provide a powerful way of contrasting the relative spatial relationships between different proteins in the living cell population (Weatherman et al., 2002; Schaufele et al., 2003; Ross et al., 2004). This population analysis approach assumes that the relationship of FRET to the acceptor/donor ratio is constant for all the cells of the population. Here, the C/EBPα interactions were consistent for the hundreds of cells measured, and the data fit with high confidence to a model describing the dimer interaction. This may not necessarily hold true for other proteins, in which changes in the cell-cycle status or cell physiology may have profound consequences for the protein interactions in individual cells within the population.

6. GENERAL CONSIDERATIONS FOR FRET IMAGING WITH FLUORESCENT PROTEINS

The intermolecular FRET experiments described here involve the expression of two separate proteins labeled with either the donor or acceptor fluorochrome. Success of this approach requires that the interaction of the labeled proteins bring the fluorochromes within the close proximity necessary to be detected by FRET. In this regard, it is important to realize that the detection of FRET provides information about the spatial relationship of the fluorochromes, and not necessarily the proteins they are linked to. As such, FRET results by themselves cannot prove the direct interaction of the labeled proteins. Rather, the FPs serve as surrogates for the relative spatial positions of the specific protein domains to which they are attached.

6.1 Contrasting Intermolecular and Intramolecular FRET Measurements

The most reliable FRET measurements come from the use of biosensor proteins, where donor and acceptor fluorochromes are directly coupled to each other through a bioactive linker peptide. Changes in the conformation of the linker resulting from its modification, or the binding of a substrate, are detected by changes in the FRET signal (reviewed by Zhang et al., 2002). These intramolecular FRET-based indicator proteins have been used to measure diverse intracellular events, including changes in intracellular calcium or protein kinase activity (Nagai et al., 2000; Ting et al., 2001; Zhang et al., 2001). The direct tethering of the donor and acceptor fixes the ratio of the expressed fluorochrome pair at 1:1. Because the donor–acceptor levels are fixed, FRET can be detected as the ratio of acceptor to donor fluorescence. Since the ratio imaging automatically corrects for the SBT background, this approach greatly simplifies the FRET measurements with the biosensor proteins.

In contrast, intermolecular FRET experiments measure the association of separate proteins independently labeled with either the donor or acceptor fluorochrome. Here, the ratio of donor to acceptor is not fixed and will vary among individual cells within the transfected population. The use of a constant amount of input plasmid DNAs in the transfection only establishes the average relative expression levels of donor- and acceptor-labeled proteins within the population. Since the donor–acceptor ratio varies from cell to cell, the SBT background signal will also be different for each transfected cell. Thus, intermolecular FRET measurements must accurately account for the variation in fluorochrome expression level and the corresponding SBT signals in each individual cell. Without those corrections, FRET cannot be measured as ratio of acceptor-to-donor fluorescence.

Further, since intermolecular FRET experiments involve the expression of separate proteins, the formation of mixed complexes involving the endogenous protein partners becomes a concern, and the potential for false negatives is high. Indeed, we showed in Figure 4–6B that two proteins, which are known to interact on the basis of biochemical criteria, did not produce FRET signals when expressed in living cells. Conversely, proteins that do not physically interact with one another but are associated with common protein partners might produce FRET signals because of favorable positioning of the FPs. Therefore, FRET measurements by themselves are not sufficient proof of direct protein interactions, and other biochemical approaches are required to determine whether protein partners are physically interacting with one another. What intermolecular FRET measurements do provide, however, is direct evidence of protein associations within the natural environment of the living cell, revealing information about the structure of protein complexes that form at specific subcellular sites.

6.2 Fluorescent Protein Dimerization

Most of the FPs that have been identified in marine organisms exist either as dimers, tetramers, or higher-order complexes (Ward, 1998). The *Aequorea* GFP is an exception to this rule and was crystallized as a monomer (Ormö et al., 1996; Brejc et al., 1997). However, concerns about the potential of the *Aequorea*-based FPs to form dimers are important considerations for FRET studies using these probes. A hydrophobic patch at the carboxyl terminus of the *Aequorea*-based FPs was identified that allows the β barrels to associate (Zacharias et al., 2002). This association does not appear to be significant when the proteins are free to diffuse within the volume of the cell. For instance, recent fluorescence fluctuation spectroscopy measurements of EGFP expressed in living cells over the concentration range of 50 nM to 10 μM showed no evidence for dimer formation (Chen et al., 2003). This result clearly showed that EGFP was unlikely to form dimers when it was free to diffuse within the cytoplasm of the living cell.

   The potential for FPs to form dimers becomes a significant concern, however, when fusion proteins are expressed at high concentrations in a restricted volume, such as inside a cellular organelle, or in diffusion-limited compartments, such as in the two-dimensional space of biological membranes. Under these conditions, a weak tendency to form dimers could allow even randomly distributed proteins to form anomalous protein–protein interactions that would be detected by FRET (for example, see Chapter 8). Given these considerations, Zacharias and colleagues (2002) showed that substitution of the hydrophobic patch residues in the β barrel with positively charged amino acids prevented dimerization of the FPs. For example, the substitution of alanine[206] with lysine blocked dimer formation without changing any other characteristic of the FPs (Zacharias et al., 2002). Given the concerns for potential artifacts in FRET studies, especially when measuring dynamic protein interactions in restricted volumes inside the living cell, it seems sensible to use the monomeric FP mutants.

6.3 Overexpression Artifacts

It must be recognized that any amount of an expressed (exogenous) protein in a cell is by definition overexpressed relative to the endogenous counterpart. Thus, the ideal approach for the types of studies outlined in this chapter would be to generate "knockin" transgenic animals, in which the FP-protein replaces the endogenous counterpart and is under normal regulatory control (see Brewer et al., 2002, for example). This approach would be unmanageable for most protein interaction studies, however, and the transient expression of fusion proteins from

transfected plasmids provides the greatest flexibility for analyzing protein interactions with FRET. For instance, this approach allows the rapid characterization of
many different targeted mutations in the proteins under study that simply would
not be tenable with stably integrated genetic sequences.

This approach can yield very high levels of fusion-protein expression in the
target cells, especially when strong promoters are used. This can result in improper
protein distribution and dysfunction that could lead to erroneous interpretations
of protein co-localization and interaction. It is critical that the subcellular locations
of the expressed proteins be the same as that of the endogenous protein and that
the function of the expressed FP-fusion proteins be assessed with as many different approaches as possible. Further, it should be recognized that only a small
percentage of the expressed protein detected by fluorescence microscopy might
actually be involved in the cellular function being monitored. Therefore, it is important to work with cells expressing near-physiological ranges of the fusion proteins and to carefully quantify the effect of protein concentration on FRET results.

6.4 Analysis in the Cell Population

Finally, FRET results from single cells are by themselves not sufficient to determine
whether or how proteins interact in living cells. Although FRET measurements,
when collected and quantified properly, are remarkably robust, there is still heterogeneity in the measurements. Further, there may also be substantial cell-to-cell
heterogeneity for some interactions, and it is possible that only a subpopulation of
cells will respond to a particular stimulus. Therefore, data must be collected and
statistically analyzed from multiple cells to prevent the user from reaching false
conclusions from a nonrepresentative measurement.

7. CONCLUSION

Here we reviewed the technique of FRET for studying protein–protein interactions
in the nucleus of the living cell by means of WFM. We outlined some of the important considerations in using the different color variants of the FPs in FRET experiments. The combination of CFP and YFP was used to illustrate how the spectral
overlap between this fluorochrome pair, which is necessary for efficient FRET, creates substantial SBT signals that must be accounted for before quantitative measurements can be made. The continued discovery and modification of FPs holds
great promise for development of new FRET probes with improved characteristics. Until these are available, we must exploit the attributes of the FPs that are
currently in hand. We outlined several methods for measuring FRET with WFM
and discussed some of the limitations of each approach. We used each approach to
demonstrate the acquisition of intermolecular FRET measurements. Because of the
very limited distance over which efficient energy transfer can occur, FRET measurements are prone to false-negative results. Furthermore, positive FRET results
cannot prove the direct interaction of the labeled proteins, and other techniques
are required to substantiate direct protein–protein interactions. For example, the

methods of co-immunoprecipitation, epitope-tagged protein pull-down, and two-hybrid assays provide biochemical evidence of direct protein–protein interactions. However, these approaches are also subject to potential artifacts because of the nonphysiological conditions of protein extraction and analysis. Despite its limitations, live-cell FRET studies provide the most physiologically relevant method for studying protein interactions. Finally, no single approach to the measurement of protein interactions is optimal, and it is only through the combined use of different approaches that robust and consistent results are obtained.

This work was supported by National Institutes of Health grants DK 47301 (RND) and DK 54345 (FS). We thank Cindy Booker for expert assistance and Dr. Ty Voss for critical comments. We also wish to thank our colleagues in the W. M. Keck Center for Cellular Imaging at the University of Virginia for their help.

REFERENCES

Baird, G.S., D.A. Zacharias, and R.Y. Tsien. Biochemistry, mutagenesis, and oligomerization of DsRed, a red fluorescent protein from coral. *Proc. Natl. Acad. Sci. U.S.A.* 97:11984–11989, 2000.

Berney, C. and G. Danuser. FRET or no FRET: a quantitative comparison. *Biophys. J.* 84:3992–4010, 2003.

Bevis, B.J. and B.S. Glick. Rapidly maturing variants of the *Discosoma* red fluorescent protein (DsRed). *Nat. Biotechnol.* 20:83–87, 2002.

Brejc, K., T.K. Sixma, P.A. Kitts, S.R. Kain, R.Y. Tsien, M. Ormo, and S.J. Remington. Structural basis for dual excitation and photoisomerization of the *Aequorea victoria* green fluorescent protein. *Proc. Natl. Acad. Sci. U.S.A.* 94:2306–2311, 1997.

Brewer, J.A., B.P. Sleckman, W. Swat, and L.J. Muglia. Green fluorescent protein–glucocorticoid receptor knockin mice reveal dynamic receptor modulation during thymocyte development. *J. Immunol.* 169:1309–1318, 2002.

Campbell, R.E., O. Tour, A.E. Palmer, P.A. Steinbach, G.S. Baird D.A. Zacharias, and R.Y. Tsien. A monomeric red fluorescent protein. *Proc. Natl. Acad. Sci. U.S.A.* 99:7877–7882, 2002.

Chalfie, M., Y. Tu, G. Euskirchen, W.W. Ward, and D.C. Prasher. Green fluorescent protein as a marker for gene expression. *Science* 263:802–805, 1994.

Chen, Y., L.N. Wei, and J.D. Muller. Probing protein oligomerization in living cells with fluorescence fluctuation spectroscopy. *Proc. Natl. Acad. Sci. U.S.A.* 100:15492–15497, 2003.

Cubitt, A.B., L.A., Woollenweber, and R. Heim. Understanding structure–function relationships in the *Aequorea victoria* green fluorescent protein. *Methods Cell Biol.* 58:19–30, 1999.

Day, R.N., S.K. Nordeen, and Y. Wan. Visualizing nuclear protein interactions in living cells. Minireview. *Mol. Endocrinol.* 13:517–526, 1999.

Day, R.N., A. Periasamy, and F. Schaufele. Fluorescence resonance energy transfer microscopy of localized protein interactions in the living cell nucleus. *Methods* 25:4–18, 2001.

Day, R.N., T.C. Voss, J.F. Enwright, 3rd, C.F. Booker, A. Periasamy, and F. Schaufele. Imaging the localized protein interactions between Pit-1 and the CCAAT/enhancer binding protein alpha in the living pituitary cell nucleus. *Mol. Endocrinol.* 17:333–345, 2003.

Enwright, J.F. 3rd, M.A. Kawecki-Crook, T.C. Voss, F. Schaufele, and R.N. Day. A PIT-1 homeodomain mutant blocks the intranuclear recruitment of the CCAAT/enhancer binding protein alpha required for prolactin gene transcription. *Mol. Endocrinol.* 17:209–222, 2003.

Feng, G., R.H. Mellor, M. Bernstein, C. Keller-Peck, Q.T. Nguyen, M. Wallace, J.M.

Nerbonne, J.W. Lichtman, and J.R. Sanes. Imaging neuronal subsets in transgenic mice expressing multiple spectral variants of GFP. *Neuron* 28:41–51, 2000.

Förster, V.T. Zwischenmolekulare energiewanderung und Fluoreszenz. *Ann. Phys. (Leipzig)* 2:55–75, 1948. Translated in: *Biological Physics*, edited by E.V. Mielczarek, E. Greenbaum, and R.S. Knox, New York: American Institute of Physics, 1993, pp. 148–160.

Gerlich, D., J. Beaudouin, B. Kalbfuss, N. Daigle, R. Eils, and J. Ellenberg. Global chromosome positions are transmitted through mitosis in mammalian cells. *Cell* 112:751–764, 2003.

Hadjantonakis, A.K. and A. Nagy. The color of mice: in the light of GFP-variant reporters. *Histochem. Cell Biol.* 115:49–58, 2001.

Heim, R., D.C. Prasher, and R.Y. Tsien. Wavelength mutations and posttranslational autoxidation of green fluorescent protein. *Proc. Natl. Acad. Sci. U.S.A.* 91:12501–12504, 1994.

Heim, R. and R.Y. Tsien. Engineering green fluorescent protein for improved brightness, longer wavelengths and fluorescence resonance energy transfer. *Curr. Biol.* 6:178–182, 1996.

Inouye, S. and F.I. Tsuji. *Aequorea* green fluorescent protein. Expression of the gene and fluorescence characteristics of the recombinant protein. *FEBS Lett.* 341:277–280, 1994.

Matz, M.V., A.F. Fradkov, Y.A. Labas, A.P. Savitsky, A.G. Zaraisky, M.L. Markelov, and S.A. Lukyanov. Fluorescent proteins from nonbioluminescent *Anthozoa* species. *Nat. Biotechnol.* 17:969–973, 1999.

Matz, M.V., K.A. Lukyanov, and S.A. Lukyanov. Family of the green fluorescent protein: journey to the end of the rainbow. *Bioessays* 24:953–959, 2002.

Miyawaki, A. and R.Y. Tsien. Monitoring protein conformations and interactions by fluorescence resonance energy transfer between mutants of green fluorescent protein. *Methods Enzymol.* 327:472–500, 2000.

Nagai, T., K. Ibata, E.S. Park, M. Kubota, K. Mikoshiba, and A. Miyawaki. A variant of yellow fluorescent protein with fast and efficient maturation for cell-biological applications. *Nat. Biotechnol.* 20:87–90, 2002.

Nagai, Y., M. Miyazaki, R. Aoki, T. Zama, S. Inouye, K. Hirose, M. Iino, and M. Hagiwara. A fluorescent indicator for visualizing cAMP-induced phosphorylation in vivo. *Nat. Biotechnol.* 18:313–316, 2000.

Ormö M., A.B. Cubitt, K. Kallio, L.A. Gross, R.Y. Tsien, and S.J. Remington. Crystal structure of the *Aequorea victoria* green fluorescent protein. *Science* 273:1392–1395, 1996.

Patterson, G., R.N. Day, and D. Piston. Fluorescent protein spectra. *J. Cell Sci.* 114:837–838, 2001.

Patterson, G.H., D.W. Piston, and B.G. Barisas. Förster distances between green fluorescent protein pairs. *Anal. Biochem.* 284:438–440, 2000.

Periasamy, A. and R.N. Day. Visualizing protein interactions in living cells using digitized GFP imaging and FRET microscopy. *Methods Cell Biol.* 58:293–314, 1999.

Perrin, F. Théorie quantique des transferts d'activation entre molécules de même espèce. Cas des solutions fluorescentes. *Ann. Chim. Phys.* 17:283–313, 1932.

Platani, M., I. Goldberg, A.I. Lamond, and J.R. Swedlow. Cajal body dynamics and association with chromatin are ATP-dependent. *Nat. Cell Biol.* 4:502–508, 2002.

Plautz, J.D., R.N. Day, G.M. Dailey, S.B. Welsh, J.C. Hall, S. Halpain, and S.A. Kay. Green fluorescent protein and its derivatives as versatile markers for gene expression in living *Drosophila melanogaster*, plant and mammalian cells. *Gene* 173:83–87, 1996.

Prasher, D.C., V.K. Eckenrode, W.W. Ward, F.G. Prendergast, and M.J. Cormier. Primary structure of the *Aequorea victoria* green-fluorescent protein. *Gene* 111:229–233, 1992.

Riven, I., E. Kalmanzon, L. Segev, and E. Reuveny. Conformational rearrangements associated with the gating of the G protein–coupled potassium channel revealed by FRET microscopy. *Neuron* 38:225–235, 2003.

Rizzo, M.A., G.H. Springer, B. Granada, and D.W. Piston. An improved cyan fluorescent protein variant useful for FRET. *Nat. Biotechnol.* 22:445–449, 2004.

Ross, S.E., H.S. Radomska, B. Wu, P. Zhang, J.N. Winnay, L. Bajnok, W.S. Wright, F. Schaufele,

D.G. Tenen, and O.A. MacDougald. Phosphorylation of C/EBP alpha inhibits granulopoiesis. *Mol. Cell. Biol.* 24:675–686, 2004.

Schaufele, F., J.F. Enwright, 3rd, X. Wang, C. Teoh, R. Srihari, R. Erickson, O.A. MacDougald, and R.N. Day. CCAAT/enhancer binding protein alpha assembles essential cooperating factors in common subnuclear domains. *Mol. Endocrinol.* 15:1665–1676, 2001.

Schaufele, F., X. Wang, X. Liu, and R.N. Day. Conformation of CCAAT/enhancer-binding protein alpha dimers varies with intranuclear location in living cells. *J. Biol. Chem.* 278:10578–10587, 2003.

Selvin, P.R. Fluorescence resonance energy transfer. *Methods Enzymol.* 246:300–334, 1995.

Shimomura, O., F.H. Johnson, and Y. Saiga. Extraction, purification and properties of aequorin, a bioluminescent protein from the luminous hydromedusan, *Aequorea*. *J. Cell. Comp. Physiol.* 59:223–239, 1962.

Spring, K.R. Detectors for fluorescence microscopy. In: *Methods in Cellular Imaging*, edited by A. Periasamy. New York: Oxford University Press, 2001, pp. 40–52.

Stryer, L. Fluorescence energy transfer as a spectroscopic ruler. *Annu. Rev. Biochem.* 47:819–846, 1978.

Swedlow, J.R., K. Hu, P.D. Andrews, D.S. Roos, and J.M. Murray. Measuring tubulin content in *Toxoplasma gondii*: a comparison of laser-scanning confocal and wide-field fluorescence microscopy. *Proc. Natl. Acad. Sci. U.S.A.* 99:2014–2019, 2002.

Tang, Q.Q., and M.D. Lane. Activation and centromeric localization of CCAAT/enhancer-binding proteins during the mitotic clonal expansion of adipocyte differentiation. *Genes Dev.* 13:2231–2241, 1999.

Ting, A.Y., K.H. Kain, R.L. Klemke, and R.Y. Tsien. Genetically encoded fluorescent reporters of protein tyrosine kinase activities in living cells. *Proc. Natl. Acad. Sci. U.S.A.* 98:15003–15008, 2001.

Tramier, M., K. Kemnitz, C. Durieux, and M. Coppey-Moisan. Picosecond time-resolved microspectrofluorometry in live cells exemplified by complex fluorescence dynamics of popular probes ethidium and cyan fluorescent protein. *J. Microsc.* 213:110–118, 2002.

Tsien, R.Y. The green fluorescent protein. *Annu. Rev. Biochem.* 67:509–544, 1998.

van Rheenen, J., M. Langeslag, and K. Jalink. Correcting confocal acquisition to optimize imaging of fluorescence resonance energy transfer by sensitized emission. *Biophys. J.* 86:2517–2529, 2004

van Roessel, P., and A.H. Brand. Imaging into the future: visualizing gene expression and protein interactions with fluorescent proteins. *Nat. Cell Biol.* 4:E15–20, 2002.

van Thor, J.J., and K.J. Hellingwerf. Fluorescence resonance energy transfer (FRET) applications using green fluorescent protein. Energy transfer to the endogenous chromophores of phycobilisome light-harvesting complexes. *Methods Mol. Biol.* 183:101–119, 2002.

Verkhusha, V.V. and K.A. Lukyanov. The molecular properties and applications of *Anthozoa* fluorescent proteins and chromoproteins. *Nat. Biotechnol.* 22:289–296, 2004.

Voss, T.C., I.A. Demarco, C.F. Booker, and R.N. Day. A computer-assisted image analysis protocol that quantitatively measures subnuclear protein organization in cell populations. *Biotechniques* 36:240–247, 2004.

Wallace, W., L.H. Schaefer, and J.R. Swedlow, A workingperson's guide to deconvolution in light microscopy. *Biotechniques* 31:1076–1082, 2001.

Walsh, M.K. and J.W. Lichtman. In vivo time-lapse imaging of synaptic takeover associated with naturally occurring synapse elimination. *Neuron* 37:67–73, 2003.

Ward, W.W. Biochemical and physical properties of green fluorescent protein. In: *Green Fluorescent Protein: Properties, Applications and Protocols*, edited by M. Chalfie, and S. Kain, New York: Wiley-Liss, 1998, pp. 45–75.

Weatherman, R.V., C.Y. Chang, N.J. Clegg, D.C. Carroll, R.N. Day, J.D. Baxter, D.P. McDonnell, T.S. Scanlan, and F. Schaufele. Ligand-selective interactions of ER detected in living cells by fluorescence resonance energy transfer. *Mol. Endocrinol.* 16:487–496, 2002.

Wedel, A. and H.W. Ziegler-Heitbrock. The C/EBP family of transcription factors. *Immunobiology* 193:171–185, 1995.

Zacharias, D.A., J.D., Violin, A.C., Newton, and R.Y. Tsien. Partitioning of lipid-modified monomeric GFPs into membrane microdomains of live cells. *Science* 296:913–916, 2002.

Zhang, J., R.E. Campbell, A.Y. Ting, and R.Y. Tsien. Creating new fluorescent probes for cell biology. *Nat. Rev. Mol. Cell Biol.* 3:906–918, 2002.

Zhang, J., Y. Ma, S.S. Taylor, and R.Y Tsien. Genetically encoded reporters of protein kinase A activity reveal impact of substrate tethering. *Proc. Natl. Acad. Su. U.S.A.* 98:14997–15002, 2001.

Zimmer, M. Green fluorescent protein (GFP): applications, structure, and related photophysical behavior. *Chem. Rev.* 102:759–781, 2002.

5

Confocal FRET Microscopy: Study of Clustered Distribution of Receptor–Ligand Complexes in Endocytic Membranes

HORST WALLRABE AND MARGARIDA BARROSO

1. INTRODUCTION

Confocal microscopy is rapidly gaining acceptance as an important technology because of its capability to produce images free of out-of-focus information. In a conventional epifluorescence microscope (see Chapter 4), the entire object is exposed to excitation light and the emission collected by high numerical-aperture (NA) objectives comes from throughout the specimen, whether above or below the focal plane. This reduces the contrast and sharpness, seriously degrading the image. In confocal microscopy, out-of-focus information (blur) is removed. In addition, confocal microscopy provides a significant improvement in lateral resolution and the capacity for direct, noninvasive serial optical sectioning of intact, thick living specimens.

Introduced in 1957, confocal microscopy has since become more commonly used, particularly after the invention of lasers in the 1960s. In this chapter we describe the various types of confocal imaging, including laser scanning confocal microscopy (LSCM), arc lamp microscopy, and laser-based spinning disk microscopy. We also explain how these microscopes can be configured for FRET imaging. Other related techniques, such as multiphoton microscopy and spectral imaging, are dealt with in Chapters 6 and 10.

2. CONFOCAL MICROSCOPY

2.1 Laser Scanning Confocal Microscopy

Many investigators designed confocal microscopes for use in live specimens to image dynamic events in which a fixed microscope stage is scanned by a laser beam with a rotating disk or mirror galvanometers. Laser scanning confocal microscopes generate a clear, thin image (512 × 512 pixels) within 2 or 3 s, free from out-of-focus information. A single diffraction-limited spot of light is projected on the specimen with a high NA objective lens. The light reflected or fluoresced by the specimen is then collected by the objective and focused on a pinhole aperture, where the signal is detected by a photomultiplier tube (PMT). Light originating from above or below the image plane strikes the walls of the pinhole and is not transmitted to the detector. To generate a two-dimensional image, the laser beam is scanned across the specimen pixel by pixel. To produce an image with LSCM, the laser beam must be moved in a regular two-dimensional raster scan across the specimen. Also, the instantaneous response of the photomultiplier must be displayed with equivalent spatial resolution and relative brightness at all points on the synchronously scanned phosphor screen of a CRT monitor (Fig. 5–1). For a three-dimensional projection of a specimen, one needs to collect a series of images at different z-axis planes. The vertical spatial resolution is approximately 0.5 μm for a 40× 1.3 NA objective; for lenses with higher magnification, the vertical spatial resolution is even smaller. Three-dimensional image reconstruction can be accomplished with many commercially available software systems.

The PMTs used in LSCMs have highly desirable characteristics compared to those of video cameras: (1) stability; (2) low noise; (3) very large dynamic range (>1 million-fold); (4) sensitivity; (5) wide range of spectral response; (6) rapid response; and (7) small physical size. The PMT has a low quantum efficiency (QE) of

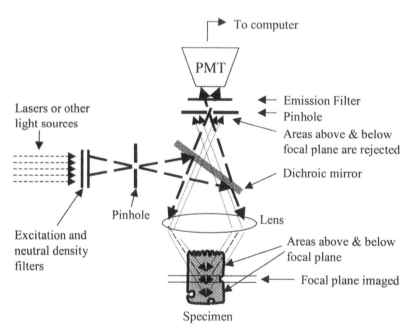

FIGURE 5–1 Schematic representation of a confocal microscope. The light source is closely controlled by various filters, pinholes, and dichroic mirrors before reaching the lens. Signals return from the specimen through the dichroic mirror and those outside the focal plane are rejected by the pinhole. The emission filter restricts the bandwidth in which the signals are received and the photomultiplier (PMT) enhances the signal as necessary.

about 10%–30% and toward the red wavelengths produces very low background noise signal. The alternative, a cooled-(p-doped and n-doped) photodiode, has a QE of 60%–80% but an equivalent noise level of about 100 photons per pixel, so it is not useful for weak signals. The optimum selection of pinhole size is important to achieve a compromise between intensity (brightness) and thickness of the slice observed. Usually, in the default mode, the pinhole size is equivalent to one Airy's disk. For instruments with variable pinholes, an optimum pinhole diameter should be determined empirically to provide the best combination of brightness and slice thickness.

2.2 Spinning Disk Confocal Microscopes

As explained above, LSCMs provide sharper images by rejecting the out-of-focus signals from the specimen. However, time is needed to create an image, as the specimen is scanned by a laser point source. The LSCM is appropriate for many applications, such as monitoring protein associations or co-localization of the protein molecules. For some applications, however, such as in vivo imaging in animals or the study of many rapid biological processes, one may require a faster acquisition mode, such as a spinning disk configuration that can be operated by an arc lamp or laser-based system.

The Yokogawa CSU-10 Nipkow spinning disk confocal microscope (SDCM) was described by Inoue and co-workers (Genka et al., 1999; Maddox et al., 1999). This Nipkow disk contains pinholes fixed in size (50 μm diameter) that are spaced in a constant-pitch helical array. This device can be coupled to any epifluorescence microscope (upright or inverted) with an arc lamp as a light source and a charge-coupled device (CCD) camera as a detector. There are a number of CCD cameras available with high quantum efficiency to obtain good-quality images (Pawley, 1995; Inoue and Spring, 1997; Spring, 2001). The SDCMs reduce the time of acquisition compared to that of the point-scanning detectors such as PMTs. This kind of microscope requires excitation and emission filter wheels as described in Chapter 3 for FRET or any other cellular imaging. Spinning disk confocal microscopes allow the selection of any excitation wavelength, unlike LSCMs, which work with fixed-wavelength lasers.

To attain good-quality images with better illumination, one can use microlens devices rather than pinholes in the spinning disk. However, high-speed data acquisition can be obtained by combining both the microlens device and the Nipkow disk pinholes as described in the literature (Maddox et al., 2003). There are about 800 pinholes in the field of view of the CCD camera at any one time. In this configuration, the microlens disk spins at a constant speed of 1800 rpm and the pinhole array is designed to raster scan 12 image frames in one rotation. At 1800 rpm, the disk scans 360 frames/s (Maddox et al., 2003).

There are several international manufacturers of LSCMs and SDCMs, each with their own proprietary features, advantages and disadvantages, and price ranges. Each can be used for confocal FRET microscopy, but the user must select a system with features important to their particular research and budget. A list of these manufacturers is given in Table 5–1.

TABLE 5–1 Confocal Microscopy Manufacturers

Manufacturer	Web Address
Carl Zeiss	www.zeiss.com
Leica	www.leica.com
Nikon	www.nikonusa.com
Olympus America	www.olympus.com
Perkin-Elmer	www.perkinelmer.com
Univeral Imaging	www.imagl.com

3. LASER SCANNING CONFOCAL FRET MICROSCOPY

Confocal–Förster resonance energy transfer (C-FRET) microscopy has become a mainstream methodology that is ideal for investigating cellular processes. However, there are certain challenges that need to be considered: (1) the limited number of laser wavelengths available limits the number of fluorophores that can be used; (2) suitable FRET pairs need to be used; and (3) a certain level of experience needs to be acquired to apply the FRET technique successfully. Having said that, all three can be managed and are addressed in Chapters 3, 4 and 7 in this book. Table 5–2 lists some of the more frequently used FRET pairs, their excitation and emission wavelengths, and their appropriate filter combinations. Table 5–3 and the following section provide a guide for the application of FRET.

TABLE 5–2 Commonly Used Filter Sets

FRET Fluorophore Pairs		Donor Excitation (nm)			Emission Channels (nm)	
Donor	Acceptor	Laser	Filter	Dichroic	Donor Filter	Acceptor Filter
BFP	GFP	380	365/15	390	460/50	535/50 515/30
BFP	YFP	380	365/15	390	460/50	535/26 528/50
CFP	YFP	457	440/21	455	480/30	535/26 528/50
GFP	Rhod-2	488	488/20	505	535/50	595/60 590/70 590LP
FITC	Rhod-2	488	488/20	505	535/50 535/45 515/30	595/60 590/70 590LP
Cy3	Cy5	543	525/45	560	595/60	695/55
Alexa 488 FITC	Alexa 555 Alexa 546 Cy3	Argon 488	488/20	505	535/50 535/45 515/30	595/60 590/70 590LP

TABLE 5–3 Optimizing Imaging Conditions Before Data Collection

Step	Action	Light Source	Acceptor Channel	Donor Channel
1	Select suitable cell or area of double-label specimen through ocular	Arc lamp		
2a	Switch to acceptor laser only, fine-tune focal plane	Acceptor laser	X	
2b	Check signal level, adjust power, select optimal PMT (low background)		X	
3	Switch to donor laser only, check for acceptable uncorrected FRET signal	Donor laser	X	
3a	Check to avoid wide-spread saturation (limited amount acceptable)			X
4	Final tuning of power, PMT, processing, filters etc to achieve objectives			
5	Record all instrument settings			
6	Select suitable cell or area of single-label acceptor through ocular	Arc lamp		
7a	Switch to acceptor laser only, fine-tune focal plane	Acceptor laser	X	
7b	Check signal level (saturation not expected, signal nonsensitized)		X	
8	Select suitable cell or area of single-label donor through ocular, avoid very bright cells which are likely to be saturated	Arc lamp		
9a	Switch to donor laser only, fine-tune focal plane	Donor laser		X
9b	Check signal level to avoid wide-spread saturation			
9c	If necessary, pick another, less bright cell or area through ocular	Arc lamp		X
10	At final imaging conditions, check unlabeled specimen for background fluorescence	Acceptor laser Donor laser	X X	X X

PMT, photomultiplier tube.

3.1 FRET Microscopy

Fluorescence resonance energy transfer (FRET) offers a significant advance over conventional light microscopy (limit of resolution ~200 nm) in that it can report distances of 1–10 nm between suitable FRET partners. FRET occurs when two fluorophores (donor and acceptor) have a sufficiently large spectral overlap, a favorable dipole–dipole orientation, a proximity of 1–10 nm, and a large enough quantum yield (see Chapter 2 for a more detailed discussion). Upon energy transfer, donor fluorescence is quenched and acceptor fluorescence is increased (sensitized), resulting in a decrease in donor excitation lifetime. Measuring either sensitized acceptor emission, donor quenching, or donor lifetime provides a method for the estimation of energy transfer efficiency ($E\%$).

The spectral overlap between donor and acceptor fluorophores that allows FRET to occur is the cause of FRET signal contamination, i.e., spectral bleedthrough (SBT) spillover. This is due to the overlap of the emission spectra of donor and acceptor (donor SBT) and the part of the acceptor absorption spectrum that is excited by the donor wavelength (acceptor SBT).

There are a number of methods to avoid, minimize, or correct SBT contamination, each having specific limitations depending on the level of sensitivity desired (see Chapter 7). The method used in our experiments is based on an algorithm that corrects the SBT spillover pixel by pixel. It is favorable to obtain highly sensitive corrected FRET signals and the use of single-labeled reference specimens is required. The algorithm-based method represents a significant improvement when using cell-based FRET assays because it does not assume constant cross-talk ratios. Instead, the cross-talk ratios are determined at different fluorescence intensities through generation of an intensity-dependent look-up table, which is then used in the final calculation (Berney and Danuser, 2003; Elangovan et al., 2003).

Widely used methods that avoid the need for SBT correction are acceptor photobleaching and the less-used donor photobleaching (see Chapter 8). Both of these may cause unforeseen changes in the specimen as a result of significant laser power exposure, which may or may not be of importance to a particular experiment. This methodology has significant disadvantages for live-cell imaging, as photobleaching takes several minutes. During this time cellular processes proceed, including recovery and replacement of bleached acceptors with intact acceptors from outside the bleaching area. Furthermore, photobleaching results in the destruction of the fluorophore, thus preventing its use for time-series FRET analysis in live cells. To overcome this limitation (and avoid the use of an SBT spillover-correction algorithm), two different approaches have recently been developed:

1. Partial acceptor photobleaching within a much shorter time frame, with extrapolation of the data predicting full-bleaching outcomes (Nashmi et al., 2003)
2. Spectral imaging, as mentioned above and discussed in Chapter 10.

The energy transferred from the donor to the acceptor ($E\%$) is expressed as a percentage of total unquenched donor fluorescence (uD)—for example, the donor fluorescence after acceptor photobleaching or, in the case of precise FRET (PFRET) methodology, the product of quenched donor plus energy transfer. $E\%$ should correctly be called "apparent $E\%$," because the computation is based on total donor fluorescence, including those donors not participating in energy transfer. $E\%$ can be used as a "spectroscopic ruler," measuring average distance between fluorophores. This requires the use of R_0, the distance at which $E\%$ is 50% (Förster distance; Lakowicz et al., 1999; Chapter 2). Energy transfer efficiency decreases rapidly with increasing distance between its two fluorophores. For situations in which there is a consistent spatial arrangement between fluorophores (fusion proteins, stable macromolecular assemblies), the distance measurement is likely to be more accurate than for co-localizations of components with varying distances (e.g., clusters). Additional computations relate $E\%$ to acceptor (A) intensity levels, uD levels, and uD:A ratios, yielding important information about the random vs. clustered distribution of membrane-bound proteins. Furthermore, in live-cell FRET microscopy, the analysis can be extended to fluorescence correlation spectroscopy (FCS) (Berland, 2004), image correlation spectroscopy (ICS), and image cross-correlation spectroscopy (ICCS) (Wiseman et al., 2000) to investigate the dynamic movement of cellular components and the measurement of cluster densities (see Chapters 15 and 16).

3.2 Use of Conventional Fluorophores for FRET Imaging

In addition to the GFP-based fluorophores (see Chapter 4), there are a number of other fluorophore dyes that can be good FRET partners when conjugated to proteins. The most frequently used pairings include donors excited by the 488 nm laser line (e.g., FITC, Alexa 488), combined with acceptor dyes excited by the 543 nm laser line (e.g., Alexa 546/555, Cy3). Another popular FRET pair is Cy3 as donor and Cy5 as acceptor, using the 543 nm and 615 nm laser lines, respectively.

It is important to carry out a spectrofluorometry analysis of the actual proteins–dye conjugate, as a small shift in the emission spectra is common compared to the published fluorophore dye spectrum. By way of example, we have compared the published Cy3 spectrum with that of our [Fab]₂–Cy3 conjugate (Fig. 5–2). A clear shift was seen between the Cy3 fluorophore and the Cy3-labeled protein conjugate emission spectra (Fig. 5–2). Establishing the emission spectrum of the particular protein–dye conjugate may facilitate fine-tuning the emission filter. When working with a particular FRET pair regularly, it is also important to establish the degree of labeling (average dye/protein ratio) of the conjugate each time a new batch

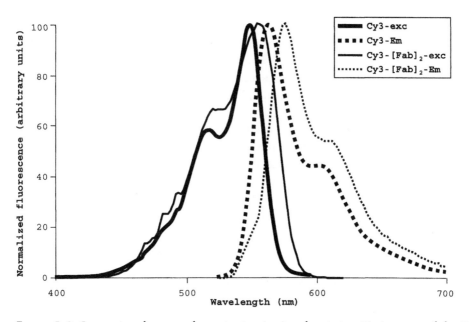

FIGURE 5–2 Comparison between the excitation (exc) and emission (Em) spectra of the Cy3 fluorophore and of the Cy3-labeled polymeric IgA receptor–ligand complexes.

is produced; the fluorophore manufacturers provide the methodology to do just that. This ensures that the conjugates are consistently labeled from batch to batch. In summary, conventional fluorophores provide an excellent opportunity for tracking cellular trafficking of receptor–ligand complexes by prior labeling of the ligand (Barroso and Sztul, 1994; Wallrabe et al., 2003a, 2003b).

3.3 Configuration of Confocal FRET Imaging

Any confocal setup can be used for FRET imaging. Most LSCMs have two or three of the standard laser lines (457, 488, 514, 543, 594, 633 nm), and this standardization limits the choice of compatible FRET pairs (Table 5–2). In the case of the arc lamp–based spinning disc system, the appropriate filter wheels are usually available. Most systems allow the user to vary the power level, the PMT setting (manufacturers usually recommend an optimum or maximum level not to be exceeded to avoid excessive background noise), the zoom level, the size of the pinhole (iris), background control, size of the image (which influences resolution), and various forms of processing (e.g., accumulating, averaging, binning). Moreover, in spite of effective SBT correction methodologies, it is wise to optimize emission filter configurations, which maximize the FRET signal and minimize SBT.

Before embarking on a series of FRET experiments, a great deal of homework needs to be completed (Table 5–3) to achieve the following objectives: maximize the FRET signal, minimize SBT, avoid widespread saturation, avoid unintended photobleaching, and minimize background noise. Unfortunately, there is no hard-and-fast rule of the best combination of these factors, since they are influenced by the particular system, specimens, and fluorophores. As practical first steps, any photobleaching and high levels of saturation need to be controlled by reduction of laser power at the specimen plane and PMT settings.

A combination of lower-power PMT and some processing (e.g., accumulation) may achieve the objective. The next steps involve attaining a viable uncorrected FRET signal with a good distribution within several potential regions of interest (ROIs) in the range of, for example, >20 units (as an example of an 8-bit gray-level range of 0–255, a predictive SBT level of 40%–60%; Fig. 5–3G). This avoids excessive SBT as measured by single-labeled control samples (see below). By just checking levels outside the immediate cell, one can easily establish background fluorescence for single-cell images. With confluent cell layers it is advisable to use an unlabeled sample with cells only when imaged under identical conditions. In summary, by optimizing all of the conditions, the researcher can have confidence in the results of a series of experiments conducted under the same conditions. This is particularly important when biological treatments are introduced that affect intracellular processes, such as membrane trafficking and cytoskeleton and cell migration.

The following objectives should be achieved prior to data collection:

1. Establish the optimal laser power level at the double-label specimen plane and PMT setting. The right level and setting help prevent photobleaching and saturation. Ideally, the PMT setting should be the same or similar for both channels and once established should be kept the

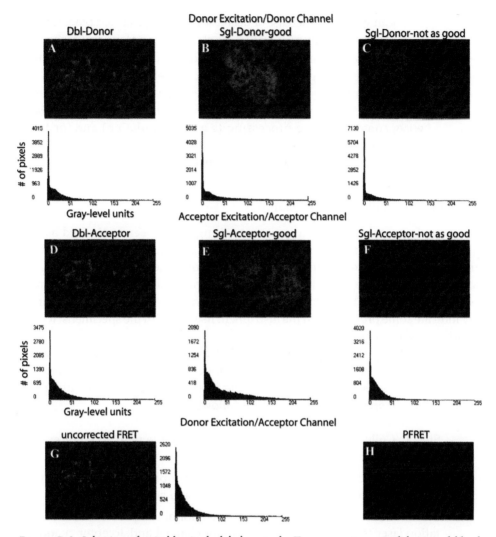

FIGURE 5–3 Selection of suitable single-label controls. To carry out successful spectral bleed-through (SBT) correction using the PFRET software, single (5gl)-label control images (*B, C, E–F*) should match or exceed the gray-level ranges of the double (Dbl)-label range (*A, D*). The suitablity of any particular single label can be checked by the histograms of several regions of interest while imaging. Two to three different control single-labeled images should be taken to find a suitable match for SBT correction. There are clearly choices to make, as the above images demonstrate. Matching *A* finds *B* more suitable than *C*; to match *D*, image *E* provides a wider range than *F*. Having completed the SBT correction program with the *B* and *E* single-labeled images, the uncorrected FRET image *G* emerges as PFRET (*H*), representing the actual amount of energy transfered from donor to acceptor.

same between each different experiment. The following steps can be performed to achieve this objective:

Set recommended filters (see Table 5–2). Set at the manufacturer's optimal PMT level (e.g., 30%), start with low laser power (e.g., Argon 488 at 2% and HeNe 543 at 10%) and no processing, and fast scan with HeNe alone to find the desired focal plane. Set normal scan speed and increase or decrease power until an acceptable image appears, avoiding saturation. Switch to argon alone and repeat the process, avoiding saturation in the donor channel and achieving a viable FRET signal in the acceptor channel.

2. Ensure that these settings produce a viable uncorrected FRET level (e.g., 60–100 gray-level units; see discussion above and Fig. 5–3). Otherwise, consider image processing (e.g., accumulation) and/or finetuning emission filters.

3. Establish level background fluorescence for later subtraction.

3.4 FRET Data Collection and Analysis

Once sample preparation and instrument settings have been optimized as per Table 5–3 (which is most critical to successful FRET microscopy), data collection can begin. The optimized settings should be good for successive experiments, provided that fluorophore concentrations or other parameters have not dramatically been changed. In fact, the need for significant changes to the instrument settings may indicate some problem with the instrument itself or with the specimens, which would then have to be addressed.

To be able to use the SBT correction PFRET algorithm, double-labeled specimens and one specimen each of a single label donor and acceptor are required. Of these, five images need to be taken, resulting in seven images after splitting the image into donor and acceptor channels as follows:

Step 1. Collect image from double-labeled sample with acceptor excitation laser (one image: acceptor channel).

Step 2. Collect image from double-labeled sample with donor excitation laser (split into two images: donor channel and acceptor channel).

Step 3. Collect image from single-labeled donor sample with donor excitation laser (split into two images: donor channel and acceptor channel).

Step 4. Collect image from single-labeled acceptor sample with acceptor excitation laser (one image: acceptor channel).

Step 5. Collect image from single-labeled acceptor sample with donor excitation laser (one image: acceptor channel).

The use of multitracking with line scanning or of software macros to automatically switch between donor and acceptor excitation lasers enables quasisimultaneous acquisition of confocal images by separate laser lines. This is preferable to manually switching between lasers and can be critical for live-cell experiments. The image collection sequence is virtually identical to that of Table 5–3, always starting with the acceptor laser to avoid potential donor bleaching during precollection finetuning of the focal plane and imaging conditions. Throughout data collection, one must also be careful to avoid widespread saturation (<5%). Selection of cells with a wide range of fluorescence levels is essential for a detailed $E\%$ analysis when plotting of $E\%$ vs. acceptor- and donor-based parameters.

Proceeding as per Table 5–3, save each image set (automatically in two channels) with a suitable reference denoting the laser used. Separate the images into their respective channels after data collection with the instrument software, ImageJ, or PFRET.

To minimize the time on the confocal microscope and to choose the best single-label image with the widest fluorescence range for SBT correction at the analysis stage, we take three different single-labeled images. These images must show similar fluorescence ranges to that of the double label; when in doubt, they are best checked by means of histogram analysis during imaging (see Fig. 5–3). Depending on the type of biological model (individual cells, confluent or polarized cells, axons, tissue, membranes, etc.) and the aim of the experiment, a series of double-labeled images would provide a large number of ROIs and increase the statistical power of subsequent quantitative analysis (see below).

After the collection of images, two image correction steps are necessary, followed by quantitative analysis. The following remarks apply to the use of PFRET software (www.circusoft.com) and a proprietary expanded version that provides some additional features. First, all images need to be background subtracted, either by selecting an area outside of a cell as a reference point or by establishing the level of background in an unlabeled specimen. Second, SBT needs to be removed from the uncorrected FRET image; the PFRET software will achieve that. It is important to spend some time examining single-label images for the best match with the double-label specimen, as it will provide the best basis for the SBT correction program.

For data analysis, after loading the required double-label images into the PFRET analysis program (see Chapter 7), ROIs are selected and the average $E\%$ and distance (where appropriate) is calculated. The expanded software version provides details for each ROI of the average quenched donor fluorescence intensity, number of pixels before and after removal of saturated donor pixels, average acceptor fluorescence intensity, and average uncorrected FRET and PFRET values. These data allow the calculation of the uD and A levels, uD:A ratios, and percent of correction level (see Fig. 5–4). We have found these parameters to be of importance for extensive data analysis and for tracking specimen and instrument consistency over several experiments.

4. APPLICATION OF FRET TECHNIQUES TO BIOLOGIOCAL QUESTIONS

4.1 Clustered Distribution of Receptor–Ligand Complexes in Endocytic Membranes

There are many biological questions that can be answered with FRET microscopy, one being the investigation of receptor distribution in endocytic membranes of polarized cells.

Membrane trafficking is essential to many cellular functions—e.g., cell polarity, receptor function, and signal transduction. We have taken advantage of FRET

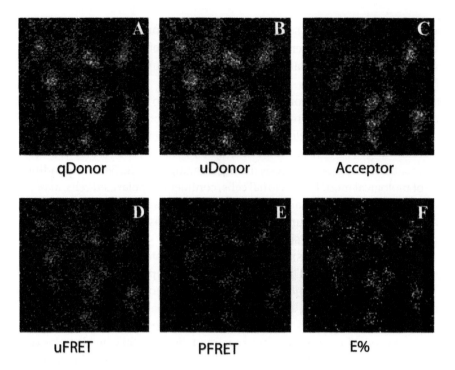

FIGURE 5–4 Major components for data analysis based on double-label images. Quenched donor (qDonor): donor excitation/donor channel; unquenched donor (uDonor): qDonor + PFRET; Acceptor: acceptor excitation/acceptor channel; uncorrected FRET (uFRET): donor excitation/acceptor channel; PFRET-processed uFRET (PFRET): uFRET – SBT correction; E%: energy transfer efficiency.

to follow the endocytic and transcytotic transport of membrane-bound receptors internalized from opposite membranes in epithelial polarized MDCK cells.

MDCK cells provide an ideal biological system in which to study protein transport and the sorting and signaling in polarized epithelial cells. These cells have two distinct plasma membrane (PM) domains: apical and basolateral, which are separated by tight junctions (TJ) (Fig. 5–5). Polarized epithelial MDCK cells stably transfected with polymeric immunoglobulin A receptor (pIgA-R) are a well-known model for the study of endosomal trafficking and sorting (Brown et al., 2000; Wang et al., 2000). Polymeric IgA-R uses different endocytic pathways throughout its intracellular protein transport. In this process, pIgA-R is transported across the cell (transcytosed) from the basolateral to the apical PM domains via the basolateral early endosome, the common endosome, and the apical recycling endosome. It is recycled back to the apical or basolateral PM via the apical early, common and apical recycling endosomes or basolareral early and common endosomes, respectively.

In our FRET-based assay of receptor distribution and sorting, pIgA-R ligands, labeled with different fluorophores, are internalized from opposite PMs in MDCK cells, which are stably transfected with rabbit pIgA-R. Upon binding, these receptor–ligand complexes are transcytosed from the basolateral PM and endocytosed from the apical PM to eventually co-localize in the apical recycling endosomal compartment(Barroso and Sztul, 1994; Wallrabe et al., 2003a, 2003b; Fig. 5–5). Then

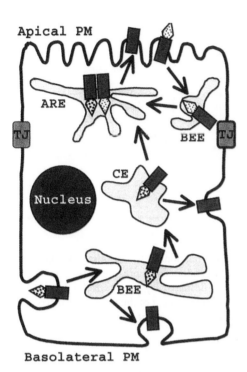

FIGURE 5–5 FRET is used to assay the distribution and sorting of receptor–ligand complexes in endocytic membranes, upon internalization from basolateral and/or apical plasma membrane (PM) domains. The triangles with light spots represent the basolaterally internalized Cy3-polymeric IgA receptor (pIgA-R)–ligand complexes (acceptor). The triangles with dark spots are the apically internalized Alexa 488–pIgA-R–ligand complexes (donor). In this example, both complexes accumulate and meet at the apical recycling endosome (ARE), where FRET may occur. The dark rectangles represent pIgA-R. AEE, apical early endosome; BEE, basolateral early endosome; CE, common endosome; TJ, tight junctions.

the cells are subjected to LSCM at the level of the apical recycling endosome; subsequently, the images are processed by the PFRET correction algorithm to detect FRET between the Alexa 488– (donor) and Cy3- (acceptor) labeled pIgA-R–ligand complexes (Wallrabe et al., 2003a, 2003b; Fig. 5–5).

Although FRET is often used to establish co-localization or measure distance, we have used the data for more extensive quantitative analysis. The dependence of $E\%$ on acceptor levels and uD:A ratios is calculated to distinguish a clustered arrangement from a random distribution of labeled receptor–ligand complexes, e.g., pIgA-R (Wallrabe et al., 2003a, 2003b). In summary, our receptor distribution and sorting FRET-based assay provides a snapshot of receptor distribution and sorting throughout the endocytic pathway of polarized cells.

Theories of ways to distinguish between clustered and random distribution of membrane-bound receptors labeled with donor or acceptor fluorophores have been described in the literature (Kenworthy and Edidin, 1998; Kenworthy et al., 2000; Kenworthy, 2001). In the random situation, the likelihood of an acceptor co-localizing with a given, equally random, donor population increases with increasing acceptor fluorescence and leads to an increase in $E\%$ (Fig. 5–6A). In contrast,

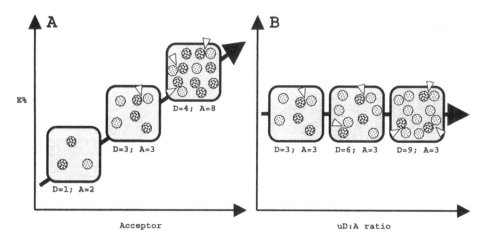

Figure 5–6 Two parameters to determine a random organization. *A:* Increasing the acceptor (A) density leads to an increase in $E\%$, as the chances of a random encounter increase—i.e., $E\%$ is dependent on acceptor levels for arbitrary uD:A ratios. *B:* Increasing the donor (D) density has no effect on $E\%$ at a fixed acceptor density—i.e., $E\%$ is independent of the uD:A ratio if acceptor density is kept fixed. Lighter dots represent donor, darker dots represent acceptor. Donor–acceptor pairs are involved in FRET (arrowheads).

increasing acceptor fluorescence does not increase $E\%$ in clusters that by definition have molecules in proximity. $E\%$ is therefore independent of the acceptor, which has been used as the main indicator for a clustered distribution pattern. Our results demonstrate that pIgA-R–ligand complexes are organized in clusters within apical endocytic membranes of polarized MDCK cells, since $E\%$ levels are independent of acceptor fluorescence (a standard parameter to confirm clustered distribution).

Our experimental results have led us to a second indicator to determine clusters, signified by the phenomenon of a decreasing $E\%$ with an increase in uD levels and uD:A ratios. An increasing number of donors with respect to acceptors within a cluster that prevent other donors from being in FRET distance of an acceptor causing $E\%$ to decrease is referred to as "donor geometric exclusion." (Fig. 5–7B). We have not excluded donor–donor competition or donor–donor energy transfer as contributory factors, which, while not appearing in the FRET channel, may nevertheless impact $E\%$. In summary, our results showing that $E\%$ is independent of acceptor levels and negatively dependent on the uD:A ratio led us to establish a clustered distribution for pIgA-R–ligand complexes in apical endocytic membranes of epithelial polarized MDCK cells (Wallrabe et al., 2003a, 2003b).

To follow up our analysis of receptor distribution and sorting throughout the endocytic pathways, we have hypothesized that there are differences in the configuration and density of clusters of receptor–ligand complexes. We used the dependency of cluster density on the total protein content as expressed by total uD+A fluorescence to account for differences between ROIs with the same uD:A ratios but different $E\%$. This also addresses the mechanisms of receptor cluster assembly.

Because pIgA-R–ligand complexes occupy a cylindrical space with a diameter of some 8 nm, facing the lumen of the endosome, we imagined a disk-shaped ref-

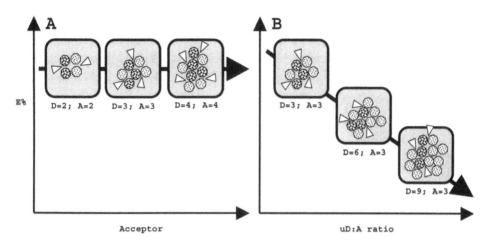

FIGURE 5–7 Two parameters to determine a clustered organization (where donor [D] and acceptor [A] by definition are in proximity). *A:* Increasing the acceptor density does not increase energy transfer, as one donor can only interact with one acceptor at a time—i.e., *E%* is independent of the acceptor at a fixed uD:A ratio. *B:* Increasing the donor density has the effect of decreasing *E%*—i.e., *E%* is negatively dependent on uD:A ratios. We attribute this to donor geometric exclusion, in which one donor prevents another from interacting with an acceptor. Lighter dots represent donor, darker dots represent acceptor. Donor–acceptor pairs are involved in FRET (arrowheads).

erence donor surrounded by a maximum of six other disk-shaped labeled receptor–ligand complexes (either uD or A), unoccupied disks, or some unknown proteins. A perfect cluster would have all positions around a reference donor occupied by labeled complexes at a low uD:A ratio, resulting in a high *E%* and indicating a high level of density. At the other extreme, fewer positions would be occupied by labeled complexes, with a high uD:A ratio resulting in a low *E%*. The probability of the space around a reference donor being occupied by other receptor–ligand complexes is *s*, such that $0 < s < 1$. Applying our computational method with the adjustable parameter *s* to a large data set, we calculated a value of $s = 0.447$, i.e., some 45% of the space in the neighborhood of a reference donor is occupied by another receptor–ligand complexes (Wallrabe et al., 2003a, 2003b).

Upon further analysis, we observed that by dividing that same data set into two equal groups based on total uD+A fluorescence, which we equate with total protein available, different levels of local density *s* result. The unexpected finding was the association of lower average total fluorescence (uD+A) with higher *E%* values and viceversa (Wallrabe et al., 2003a).

To address the possible biological implications, we have explored several scenarios for the distribution of our pIgA-R–ligand complexes on apical endocytic membranes. On the basis of the data above, *E%* values of the two groups, sorted by total uD+A, are statistically different populations. Here, higher total fluorescence coincides with lower local density and vice versa. This implies a saturable spot in the membrane with a finite capacity. Once this is reached, incoming receptor–ligand complexes cannot be accommodated and have to find another spot. This spot may be at discrete membrane locations and/or require a specific protein that can be saturated. At the moment of cell fixation, unorganized complexes outside the cluster spot have the effect of lowering overall density. It is likely that the data represent a snapshot of a continuum of sorting throughout endocytic trafficking. The above work demonstrates that quantitative analyses based on FRET microscopy and $E\%$ offer powerful tools to elucidate intracellular processes, such as protein transport and sorting.

5. EXPERIMENTAL VERIFICATION USING DIFFERENT FRET MICROSCOPY TECHNIQUES

5.1 Acceptor Photobleaching

A direct validation of the PFRET algorithm is a comparison with the bleaching-the-acceptor method. This was done within the same experiment and the results fell within the same statistical ranges for both methods (Wallrabe et al., 2003a).

5.2 Donor Photobleaching

The phenomenon of donor geometric exclusion (see Fig. 5–7B) and our earlier definition of apparent $E\%$ imply the existence of donors not participating in energy transfer. Because these non-FRET donors spend a longer time in the excited state (FRET donors spend less time there, as energy transfer provides an additional path for deexcitation), we hypothesized that bleaching the donor would bleach non-FRET donors first and would lead to an increase in $E\%$. This indeed was the case, thus our hypothesis and the utility of the PFRET software were supported (Wallrabe et al., 2003a).

5.3 One-Photon vs. Two-Photon FRET Microscopy

When relating uD levels to $E\%$, both one- and two-photon microscopy results collected from the same specimens showed the similar trends explained above—increasing uD values and $E\%$ markedly decreasing (Wallrabe et al., 2003b).

REFERENCES

Barroso, M. and E.S. Sztul. Basolateral to apical transcytosis in polarized cells is indirect and involves BFA and trimeric G protein sensitive passage through the apical endosome. *J. Cell Biol.* 124:83–100, 1994.

Berland, K.M. Fluorescence correlation spectroscopy: a new tool for quantification of molecular interactions. *Methods Mol. Biol.* 261:383–398, 2004.

Berney, C. and G. Danuser. FRET or no FRET: a quantitative comparison. *Biophys. J.* 84:3992–4010, 2003.

Brown, P.S., E. Wang, B. Aroeti, S.J. Chapin, K.E. Mostov, and K.W. Dunn. Definition of distinct compartments in polarized Madin-Darby canine kidney (MDCK) cells for membrane-volume sorting, polarized sorting and apical recycling. *Traffic* 1:124–140, 2000.

Elangovan, M., H. Wallrabe, Y. Chen, R.N. Day, M. Barroso, and A. Periasamy. Characterization of one- and two-photon excitation fluorescence resonance energy transfer microscopy. *Methods* 29:58–73, 2003.

Genka, C., H. Ishida, K. Ichimori, Y. Hirota, T. Tanaami, and H. Nakazawa. Visualization of biphasic Ca^{2+} diffusion from cytosol to nucleus in contracting adult rat cardiac myocytes with an ultra-fast confocal imaging system. *Cell Calcium* 25:199–208, 1999.

Inoué, S. and K.R. Spring. *Video Microscopy*, 2nd edition. New York: Plenum Press, 1997.

Kenworthy, A.K. Imaging protein-protein interactions using fluorescence resonance energy transfer microscopy. *Methods* 24:289–296, 2001.

Kenworthy, A.K. and M. Edidin. Distribution of a glycosylphosphatidylinositol-anchored protein at the apical surface of MDCK cells examined at a resolution of <100 A using imaging fluorescence resonance energy transfer. *J. Cell Biol.* 142:69–84, 1998.

Kenworthy, A.K., N. Petranova, and M. Edidin. High-resolution FRET microscopy of cholera toxin B-subunit and GPI-anchored proteins in cell plasma membranes. *Mol. Biol. Cell* 11:1645–1655, 2000.

Lakowicz, J.R., I. Gryczynski, Z. Gryczynski, and J.D. Dattelbaum. Anisotropy-based sensing with reference fluorophores. *Anal. Biochem.* 267:397–405, 1999.

Maddox, P., A. Desai, E.D. Salmon, T.J. Mitchison, K. Oogema, T. Kapoor, B. Matsumoto, and S. Inoué. Dynamic confocal imaging of mitochondria in swimming *Tetrahymena* and of microtubule poleward flux in *Xenopus* extract spindles. *Biol. Bull.* 197:263–265, 1999.

Maddox, P.S., B. Moree, J.C. Canman, and E.D. Salmon. Spinning disk confocal microscope system for rapid high-resolution, multimode, fluorescence speckle microscopy and green fluorescent protein imaging in living cells. *Methods Enzymol.* 360:597–617, 2003.

Nashmi, R., M.E. Dickinson, S. McKinney, M. Jareb, C. Labarca, S.E. Fraser, and H.A. Lester. Assembly of α4β2 nicotinic acetylcholine receptors assessed with functional fluorescently labeled subunits: effects of localization, trafficking, and nicotine-induced upregulation in clonal mammalian cells and in cultured midbrain neurons. *J. Neurosci.* 23:11554–11567, 2003.

Pawley, J.B. *Handbook of Biological Confocal Microscopy*, 2nd edition. New York: Plenum Press, 1995.

Spring, K.R. Detectors for fluorescence microscopy. In: *Methods in Cellular Imaging*, edited by A. Periasamy. New York: Plenum Press, 2001, pp. 40–52.

Wallrabe, H., M. Elangovan, A. Burchard, A. Periasamy, and M. Barroso. Confocal FRET microscopy to measure clustering of ligand–receptor complexes in endocytic membranes. *Biophys. J.* 85:559–571, 2003a.

Wallrabe, H., M. Stanley, A. Periasamy, and M. Barroso. One- and two-photon fluorescence resonance energy transfer microscopy to establish a clustered distribution of receptor–ligand complexes in endocytic membranes. *J. Biomed. Opt.* 8:339–346, 2003b.

Wang, E., P.S. Brown, B. Aroeti, S.J. Chapin, K.E. Mostov, and K.W. Dunn. Apical and basolateral endocytic pathways of MDCK cells meet in acidic common endosomes distinct from a nearly-neutral apical recycling endosome. *Traffic* 1:480–493, 2000.

Wiseman, P.W., J.A. Squier, M. H. Ellisman, and K.R. Wilson. Two-photon image correlation spectroscopy and image cross-correlation spectroscopy. *J. Microsc.* 200:14–25, 2000.

6

Multiphoton FRET Microscopy for Protein Localization in Tissue

James D. Mills, James R. Stone, David O. Okonkwo,
Ammasi Periasamy and Gregory A. Helm

1. Introduction

The physics and chemistry of Förster resonant energy transfer (FRET) have been studied theoretically and experimentally for many years (Förster, 1965; Stryer, 1978; Dale et al., 1979; Jovin and Arndt-Jovin, 1989; Wu and Brand, 1994). Only with recent technical advances in fluorophores and detectors has it become feasible to apply FRET in biomedical research (Clegg, 1996; Day, 1998). FRET microscopy is a better method for studying the localization of proteins in live specimens than X-ray diffraction, nuclear magnetic resonance, or electron microscopy (Guo et al., 1995), all of which use fixed specimens. Light microscopy FRET provides spatial and temporal distribution of protein associations in a single living cell. Digitized video FRET (DVFRET) imaging shows the two-dimensional (2-D) spatial distribution of steady-state protein–protein interactions (Jurgens et al., 1996; Mahajan et al., 1998; Periasamy and Day, 1999; Chapter 4). The disadvantage of DVFRET is that it includes out-of-focus information in the FRET signal of the protein under investigation. This out-of-focus signal is discriminated in confocal FRET (C-FRET) and in multiphoton excitation FRET (mp-FRET) microscopy (Periasamy, 2001; Periasamy et al., 2001; Chapter 5). In mp-FRET microscopy, infrared (IR) laser light is used as an excitation wavelength and has the ability to excite most of the selected fluorophore pairs, unlike C-FRET microscopy, in which few visible laser lines are used as an excitation wavelength. It is important that one choose an mp-FRET fluorophore pair with different mp absorption cross sections for the fluorophores (Xu and Webb, 1996; Chen and Periasamy, 2004). The conventional one-photon (1p) FRET pair, fluorescein and rhodamine, may not be a good mp-FRET fluorphore pair because both are excitable around 790 nm. But fluorescein (or Alexa488) and Cy3 are an ideal pair for mp-FRET imaging (Wallrabe et al., 2003). The use of IR laser light excitation instead of ultraviolet (UV) laser light also reduces phototoxicity in the living cell. It has been reported that the extent of photobleaching with two-photon (2p) excitation is greater than 1p excitation (Patterson and Piston, 2000), but in general the photobleaching is considerably less in 2p excitation than in 1p excitation (Denk et al., 1990, 1995). Although both statements are true under varying measurement conditions, it is possible that one could minimize the photobleaching in mp-FRET microscopy by reducing the average power of the excitation IR laser (see Chapter 13).

In this chapter we will describe the technique of mp–FRET microscopy and its use in localizing the proteins in tissue following traumatic brain injury.

2. Basics of Multiphoton Microscopy

The theory behind 2p absorption was described by Göppert-Mayer in 1931, and it was experimentally observed for the first time in 1961 with use of a ruby laser as the light source (Kaiser and Garrett, 1961). Denk et al. (1990) demonstrated laser scanning 2p fluorescence microscopy for biological applications for the first time using a mode-locked dye laser producing a 100 fs stream of pulses at 80 MHz and a Biorad MRC500 confocal scan head. The probability of 2p absorption depends on the co-localization of two photons within the absorption cross section of the fluorophore. In addition, the rate of excitation is proportional to the square of the instantaneous intensity. Two-photon excitation is made possible by the extremely high local instantaneous intensity. This is provided by the combination of diffraction-limited focusing of the laser beam and the temporal concentration of the incident light energy into femtosecond pulses by a mode-locked laser, which is typically on the order of 10^{-50} to 10^{-49} cm^4 s/photon/molecule (Denk et al., 1995). For example, a fluorophore might absorb one photon at 365 nm and fluoresce at a blue wavelength around 460 nm. In 2p excitation the fluorophore is excited at approximately twice their 1p absorption wavelength, and emission is the same as that of 1p emission. The emission wavelength is widely separated from excitation, unlike that in 1p excitation (Stokes shift <50 nm). Three- or four-photon excitation (multiphoton) is the extension of 2p excitation.

In a confocal microscope, 1p illumination occurs throughout the excitation beam path, as shown in Figure 6–1A, in an hourglass-shaped volume. This results in absorption within the entire volume, giving rise to substantial fluorescence emission both above and below the focal plane (see Fig. 6–2; Periasamy et al., 1999). This excitation out of the focal plane contributes to photobleaching and photodamage in specimen planes not excited in 2p-FRET imaging. In the case of 2p illumination, the IR light beam is localized in the focal plane and does not excite outside the focal plane (see Fig. 6–1B). The IR illumination in 2p-FRET imaging introduces considerably less autofluorescence and penetrates deeper into living specimens than visible light excitation.

3. Configuration of Multiphoton FRET Microscopy

In mp-FRET microscopy, the emission process is the same as in 1p-FRET imaging. To localize proteins one needs two fluorophore molecules to be attached to each protein molecule under investigation (see Fig. 6–3). Conventional (Alexa 488 and Alexa 555 or FITC and Cy3) or different mutant forms of green fluorescent proteins

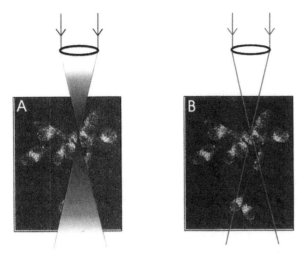

FIGURE 6–1 Illustration of illumination of excitation light in one- and two-photon microscopy.

(GFP) (Heim and Tsien, 1996; Periasamy et al., 2001) could be used as a FRET pair. It is important that one have good expression (or labeling) level of the proteins with the selected fluorophores. It is also important that one use the best optics, a high–quantum efficiency charge-coupled device (CCD) camera (for real-time 2p-FRET), or photomultiplier tubes (PMT) to acquire the FRET images.

The spectral emission of donor should have at least a 30% overlap with the absorption spectrum of acceptor. FRET efficiency increases with an increase in the percentage of overlap and, correspondingly, the bleedthrough signal increases in the FRET channel. The efficiency of the FRET could also be improved by optimizing the concentration of the fluorophore used for protein labeling. One should reduce or optimize the fluorophore concentration to reduce self-interactions, i.e., donor–donor (D-D) or acceptor–acceptor (A-A).

3.1 Instrumentation

Commercially available multiphoton microscopes have recently become available from Carl Zeiss (www.zeiss.de) and Leica (www.leica-microsystems.com). Many other laboratories and imaging centers have built their own multiphoton microscopes, often incorporating the appropriate Ti:sapphire excitation laser into existing confocal laser microscopes. Here we describe a multiphoton microscope system using a Coherent Ti:sapphire tunable laser incorporated into a Bio-Rad Radiance 2100 confocal microscope system (see Fig. 6–4; Mills et al., 2003; Chen and Periasamy, 2004). The system consists of a Nikon TE300 epifluorescent microscope with a 100W Hg arc lamp. The TE300 is coupled to a Biorad Radiance 2100 confocal/multiphoton system (www.zeiss.de/cellscience). A 10W Verdi pumped, tunable (model 900 Mira, www.coherentinc.com), mode-locked ultrafast (78 MHz) pulsed (<150 fs) laser is coupled to the laser port of a Radiance 2100. This laser is equipped with x-wave optics for easy tunable range of the entire wavelength (700 to 1000 nm). The system is equipped with a laser spectrum analyzer (Model

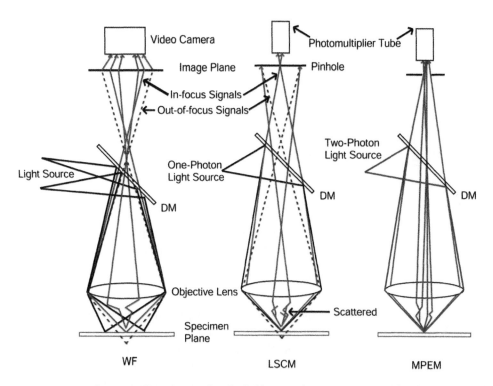

FIGURE 6–2 Schematic illustration of wide-field (WF), laser scanning confocal (LSCM), and multiphoton excitation (MPEM) microscopy. DM *Dichroic mirror*. [Adapted from Periasamy et al. (1999).]

E201; www.istcorp.com) to monitor the excitation wavelength and a power meter to measure the laser power at the specimen plane (Model SSIM-VIS and IR; www. coherentinc.com). The Radiance system is equipped with external detectors and four internal detectors for fluorescence imaging. In mp-FRET imaging, we select the appropriate filters and high-sensitivity PMTs to acquire donor and acceptor images. The system is equipped with bialkali (high sensitivity in the blue spectrum) and multialkali (high sensitivity in the visible spectrum). For the data presented here, we used LaserSharp 2000 software to acquire both laser scanning confocal microscopy (LSCM) and multiphoton images with the internal detectors.

An argon laser emitting at 488 nm was used to excite the donor fluorophore, an HeNe green laser emitting at 543 nm was used to excite the acceptor fluorophore, and a red diode laser emitting at 637 nm was used to excite separately the triple-label fluorophore. Emissions from donor and acceptor fluorophores were split using a 560 nm dichroic mirror and filtered with an HQ528/30 nm filter for the donor emission channel and an HQ590/70 nm filter for the acceptor channel; a 660LP filter was used for the triple-label channel (www.chromatech.com). For mp-FRET microscopy, following confocal laser scanning image acquisition, an inline tunable Ti:sapphire laser was used to excite the donor fluorophore at 790 nm and at 730 nm for the acceptor fluorophore, and emissions were captured with the same filters as those for confocal microscopy. A key benefit of mp-FRET microscopy is the

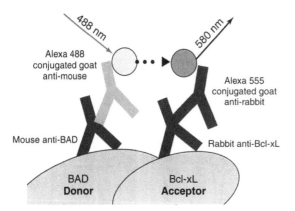

FIGURE 6–3 The heterodimerization of BAD labeled with Alexa 488 and Bcl-xL labeled with Alexa 555 approximates the fluorophores within 10 nm of each other, enabling the sensitized emission. [Adapted from Mills et al. (2003).]

ability to acquire images from tissue specimens > 40 μm thick, the normal limits of C-FRET microscopy. Immunohistochemical labeling methods, however, have a disadvantage in labeling thick tissue, thus adequate diffusion of antibodies is generally limited to the outer 10 μm of tissue. To determine if images acquired with mp-FRET microscopy in thicker tissue specimens were appropriate for FRET image analysis, we used the z-plane focus motor of the microscope to focus on the distal surface of the specimen.

3.2 Selection of Donor and Acceptor Excitation Wavelength

If the 2p absorption cross sections are known for the selected fluorophore, then that wavelength should be selected for excitation of the donor and acceptor molecule (Xu and Webb, 1996; Chen and Periasamy, 2004). If the excitation wavelength is not known, the following steps should be used to identify the wavelength.

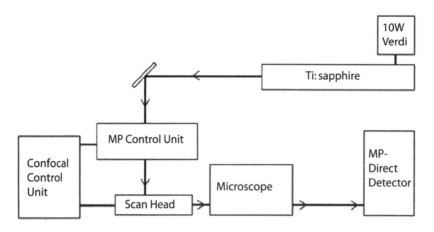

FIGURE 6–4 Schematic illustration of multiphoton (MP) FRET microscopy system.

For mp-FRET, the ti:sapphire laser is tuned to detect the maximum and minimum signal for the donor- and acceptor-alone expressed proteins. The wavelength corresponding to maximum donor signal and minimum acceptor signal, λ_D, is used as the donor excitation wavelength. The wavelength corresponding to minimum donor signal and maximum acceptor signal, λ_A, is used as acceptor excitation wavelength. An excitation wavelength of λ_D is used to acquire donor and acceptor images from doubly expressed cells. This method of selecting donor and acceptor excitation wavelengths can be used for any possible mp-FRET fluorophore pairs.

It is important to note that by adjusting appropriate average power to reduce photobleaching (see Chapter 13), the donor excitation wavelength should be chosen so that it provides maximum donor signal and minimum or no acceptor signal. Since any selected wavelength could excite any fluorophore molecule (both donor and acceptor), the technique requires correction to remove unwanted fluorescence signal in the FRET image (Elangovan et al., 2003; Chapter 7). The filter configurations for various FRET pairs are shown in Table 6–1.

4. IMAGING APOPTOTIC PROTEIN INTERACTIONS FOLLOWING TRAUMATIC BRAIN INJURY

4.1 Traumatic Brain Injury

Traumatic brain injury (TBI) remains the most common cause of death in persons under age 45 in the Western world. The societal impact is profound, with 2 million cases, 220,000 hospitalizations, and 52,000 deaths from head trauma occurring each year in the United States alone (Sosin et al., 1995). Another 80,000 to 90,000 persons each year suffer permanent debilitation. While multiple tissue and animal studies have proposed various pharmacological or physiological interventions to reduce

TABLE 6–1 Filter Configurations for Multiphoton FRET Image Acquisition for Selected Fluorophore Pairs

Fluorophore	Excitation (nm)	Emission (nm)
Alexa 488	790	515/30
Cy3	735	590/70
Alexa 488	790	528/30
Alexa 555	730	590/70
BFP	740	450/80
EGFP	880	515/30
CFP	820	485/30
YFP	920	528/30
CFP	820	485/30
dsRED1	780	590/70
BFP	740	450/80
S65T (GFP)	770	515/30

Adapted from Sekar and Periasamy (2003).

axonal injury, to date, no phase III clinical study has shown significant effect in reducing the morbidity and mortality of TBI (Doppenberg et al., 1997). However, development of new methods for detecting protein interactions and elucidation of critical mechanisms of injury in animals and humans hold the potential to greatly enhance the transition of promising treatments from benchtop to bedside.

One of the principal determinants of morbidity and mortality following TBI is traumatic axonal injury (TAI). Initially, TAI was thought to involve immediate axonal tearing through the direct action of forces associated with the traumatic insult. More recently, experiments employing anterograde tracers have revealed that TAI is a progressive event involving a focal impairment of axonal transport leading to axonal swelling and ultimate disconnection in the hours to days following TBI. Recent evidence has demonstrated that mitochondrial damage may be a key element of TAI pathogenesis. Specifically, mitochondrial damage with cytochrome *c* release has been demonstrated in relation to caspase-3-mediated proteolytic cleavage of cytoskeletal substrates (Buki et al., 2000), impaired axoplasmic transport, and axonal disconnection in TAI (Eldadah and Faden, 2000).

4.2 BAD, Bcl-xL, and Mitochondrial Disruption

Recent in vitro work has described a calcium-sensitive pro-apoptotic cascade known as the 14-3-3/BAD/Bcl-xL pathway (Wang et al., 1999). In this process, an increase in free Ca^{2+} within cells results in the activation of calcineurin, a serine–threonine phosphatase. Activated calcineurin has multiple targets within neurons including the protein BAD, a pro-apoptotic member of the bcl-2-family of proteins. BAD exists in its native phosphorylated state in complex with protein 14-3-3. BAD is activated upon dephosphorylation by calcineurin and dissociates from protein 14-3-3. Active BAD then translocates to mitochondria where it binds Bcl-xL. BAD heterodimerization with Bcl-xL negates the anti-apoptotic effect of Bcl-xL, allowing for the release of cytochrome *c* into the cytosol through aberrant gating of the mitochondrial voltage-dependent anion channel (VDAC). This release of cytochrome *c* into the cytosol precipitates formation of an apoptosome, a molecular complex comprised of cytochrome *c*, ATP, Apaf-1, and caspase-9. Activation of caspase-9 results in the cleavage of downstream effectors, including caspase-3, responsible for proteolytic destruction of the axon. The dissociation of BAD from protein 14-3-3 and subsequent heterodimerization with Bcl-xL is a critical but potentially reversible step in this pathway. Because both cytochrome *c* release and caspase-3 activation have been demonstrated in relation to TAI, we assessed whether BAD/Bcl-xL interaction may occur within traumatically injured axons (described below).

4.3 Animal Surgery

Adult male Sprague-Dawley rats were subjected to an impact acceleration injury as described previously (Marmarou et al., 1994). Specifically, rats weighing between 350 and 400 received induction anesthesia with 5% isoflurane in a bell jar for 8 to 10 min. Following anesthesia induction, animals were endotracheally intubated and maintained on 3% isoflurane in 30% O_2, 70% N_2O by means of a modified medical anesthesia machine. The animals were then shaved and prepared in sterile fashion for surgery, followed by subcutaneous injection of 1 ml of 0.25% bupivicane into the planned incision site. A 3 cm midline incision in the scalp was then made, and periostial membranes were separated, exposing bregma and lambda. A metal disc 10 mm in diameter and 3 mm thick was attached to the skull with cyanoacrylate and centered between bregma and lambda. Isoflurane was stopped until animals regained respiratory and hindleg reflexes. The animal was then placed prone on a foam bed with the metal disk directly under a plexiglass tube. A 450 g brass weight was dropped through the tube from a height of 2 meters, striking the disc. The animal was then ventilated on 100% O_2 while the skull was inspected for fracture, which would eliminate the animal from the study, and the incision was sutured with 3-0 Vicryl sutures. When the animal recovered spontaneous respirations, the endotracheal tube was removed and the animal was returned to its cage for post-operative observation. Sham surgery animals underwent preparation for injury, but just prior to impact acceleration, the steel disc was removed and the scalp was sutured with 3-0 Vicryl sutures.

All procedures involving live animals were approved by the Institutional Animal Care and Use Committee of the University of Virginia, and were performed according to the principles of the Guide for the Care and Use of Laboratory Animals, published by the Institute of Laboratory Resources, National Research Council (NIH publication 85-23-2985).

4.4 Tissue Preparation

Animals were divided into two groups and euthanized with a lethal-dose injection of 0.5 ml ketamine and 0.5 ml xylazine at 6 h following injury or 1 h following sham injury. The animals were then immediately perfused transcardially with 200 ml cold 0.9% saline. This was followed by 4% paraformaldehyde in Millonigs buffer, initially at 100 ml/min for 2 min, then at 20 ml/min for 40 min. The entire brain, brainstem, and rostral spinal cord were removed and immediately placed in 4% paraformaldehyde for 24 h.

Following 24 h fixation, the brain was blocked by cutting the brainstem above the pons, cutting the cerebellar peduncles, and then making sagittal cuts lateral to the pyramids. The resulting tissue containing the corticospinal tracts and the medial lemnisci was sagitally cut on a Vibratome into 40, 100, or 200 μm–thick sections. These sections were serially placed in phosphate-buffered saline (PBS) in 12-well plates. To eliminate endogenous peroxidase activity, the tissue was incubated for 1 h in 1% hydrogen peroxide in PBS, followed by additional washes in PBS. The tissue underwent temperature-controlled microwave antigen retrieval with our previously described technique (Stone et al., 1999). The tissue was preincubated in a solution containing 10% normal goat serum (NGS) and 0.2% Triton X in PBS for 40 min.

4.5 Immunohistochemistry

Following preincubation in 10% NGS in PBS, the tissue was incubated overnight at 4°C in polyclonal antibody raised in rabbit against Bcl-xL amino acids 126–188 (sc-7195, Santa Cruz Biotechnology) at a dilution of 1:200 and in monoclonal anti-BAD amino acids 1–168 raised in mouse (sc-8044, Santa Cruz Biotechnology) at a dilution of 1:50 in 1% NGS in PBS overnight at 4°C. Following incubation in the primary antibodies, the issue was washed three times in 1% NGS in PBS, then incubated in a secondary anti-rabbit IgG antibody conjugated with Alexa555 fluorophore (A-21428, Molecular Probes) and anti-mouse IgG antibody conjugated with Alexa 488 (A-11001, Molecular Probes) for 2 h at room temperature. The tissue underwent a final wash in 0.1 M phosphate buffer and then was mounted using Pro-long antifade agent (Molecular Probes) and coverslipped. The slides were sealed with acrylic and stored in the dark in a laboratory refrigerator.

For labeling of cytosolic cytochrome c and active caspase-3-positive axons, the tissue underwent a fluorescent tyramide signal amplification (TSA) staining process prior to BAD and Bcl-xL labeling. As described previously (Wang et al., 1999), this process enables the use of two primary antibodies raised in the same species to be distinctly differentiated within a given tissue section. Specifically, the tissue was incubated overnight at 4°C in polyclonal antibody raised in rabbit against cytochrome c amino acids 1–104 (sc-7159, Santa Cruz Biotechnology) at a dilution of 1:500 in 1% NGS/PBS or active caspase-3 p17 subunit amino acids 163–173 (AF-835, R&D Systems) at a dilution of 1:1000 in 1% NGS/PBS. Following incubation and rinsing in 1% NGS/PBS, the primary antibodies were labeled with biotinylated anti-rabbit antibody at a dilution of 1:200 in 1% NGS/PBS (BA-1000, Vector Labs) for 2 h at room temperature. The tissue then underwent processing to conjugate horseradish peroxidase to the biotin with an ABC kit (PK-6100, Vector Labs). Fluorescent labeling of the proteins was performed with an Alexa647 dye TSA kit (T-20916, Molecular Probes) with a dye dilution of 1:300 for 12 min.

Precision FRET (PFRET) analysis requires the acquisition of images from tissue labeled with either donor or acceptor fluorophore alone. Simultaneous with the triple labeling of tissue, the double-labeled donor-alone and acceptor-alone specimens were processed. Following labeling for cytochrome c or caspase-3, the tissue underwent processing for either BAD (donor) or Bcl-xL (acceptor). For the donor-only specimen, both the anti-Bcl-xL primary antibody and the anti-rabbit IgG Alexa555 secondary antibody were replaced with 1% NGS/PBS in the process described above. Likewise, for the acceptor-only specimen, both the anti-BAD primary antibody and the anti-mouse IgG Alexa 488 were replaced with 1% NGS/PBS.

4.6 IMAGE ACQUISITION AND FRET ANALYSIS

Axons of the medial lemniscus were imaged from each animal. In the sham groups, which did not have cytochrome c or caspase-3-positive axons, double-labeled axons were selected. In the 6-h postinjury axons, cytochrome c and caspase-3-positive axons were selected for imaging. Eight images were acquired using the appropriate filters as described and processed by means of PFRET 2.0 software (www.circusoft.com). The data analysis algorithm works under the assumption that the triple-labeled specimens and the double-labeled donor and acceptor specimens, imaged under the same conditions, provide the same spectral bleedthrough (SBT) dynamics

(Elangovan et al., 2003; Wallrabe et al., 2003; Chapter 7). The algorithm follows fluorescence levels pixel by pixel to establish the level of SBT in the donor- or acceptor-labeled specimens and then applies these values as a correction factor to the appropriate matching pixels of the triple-labeled specimen. Precision FRET is then

$$PFRET = uFRET - DSBT - ASBT \tag{1}$$

where uFRET is the uncorrected FRET image, ASBT is the acceptor spectral bleedthrough signal, and DSBT is the donor spectral bleedthrough signal, established in the corresponding double-labeled specimens (Chapter 7).

This software also allows pixel-by-pixel estimation of the energy transfer efficiency (E_n) and the distance (r_n) between donor and acceptor molecules, and the creation of efficiency and distance images using the equations listed below. Regions of interest (ROI) within the labeled axons were drawn using the software; the calculated mean FRET efficiency for each ROI was calculated and exported to an Excel spreadsheet (Microsoft Corp) for ANOVA data analysis (Elangovan et al., 2003; Mills et al., 2003; Chapter 7).

$$E_n = 1 - [I_{DA} / I_{DA} + PFRET] \tag{2}$$

$$r_n = R_0 \{(1/E_n) - 1\}^{1/6} \tag{3}$$

4.7 Confocal and Multiphoton FRET

To evaluate heterodimerization between the pro-apoptotic protein BAD and the anti-apoptotic protein Bcl-xL following TAI, we developed a novel tissue FRET technique. Specifically, following an impact acceleration injury, animals were allowed to survive for varying periods of time, and then tissue collection and immunohistochemical processing were performed. Tissue sections from the pons and medulla were first labeled for BAD and Bcl-xL, then double-labeled with fluorescently labeled antibodies. The tissue was examined with a LSCM system using the image analysis method described above. Following collection of 1p images, images of the identical axon were collected using a 2p excitation methodology. FRET analysis, including correction of the FRET signal for both donor and acceptor spectral bleedthrough and pixel-by-pixel efficiency analysis, was then performed (Fig. 6–5).

BAD and Bcl-xL ubiquitously labeled axons of the medial lemniscus in both sham-injured and impact acceleration–injured animals. FRET analysis of images acquired by means of the confocal laser scanning system from animals at 6 h postinjury demonstrated FRET efficiencies >20%, consistent with BAD and Bcl-xL heterodimerization. In contrast, axons imaged from sham-injured animals demonstrated low FRET efficiency, consistent with separation between the cytosolic location of BAD and the mitochondrial location of Bcl-xL in the noninjured state (Mills et al., 2003). Images of identical axons collected with the 2p methodology likewise demonstrated similar results. Swollen axons from animals 6 h postinjury

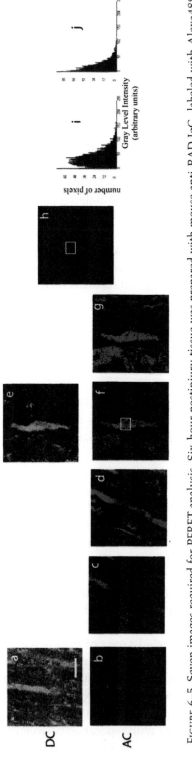

FIGURE 6–5 Seven images required for PFRET analysis. Six-hour postinjury tissue was prepared with mouse anti-BAD IgG₁ labeled with Alexa488 anti-mouse antibody (donor) and/or rabbit anti-Bcl-xL IgG labeled with Alexa555 anti-rabbit IgG (acceptor). *a:* Donor labeled, donor excitation, donor channel (DC); *b:* donor labeled, donor excitation, acceptor channel (AC); *c:* acceptor labeled, donor excitation, AC; *d:* acceptor labeled, acceptor excitation, AC; *e:* double labeled, donor excitation, DC; *f:* double labeled, donor excitation, AC; *g:* double labeled, acceptor excitation, AC; *h:* PFRET image following processing as described above. Calculated mean energy transfer efficiency was 22.4 +/− 2.2 and estimated distance was 83.0 +/− 1.7 Å. [Adapted from Mills et al. (2003).]

demonstrated FRET efficiencies >20%, whereas axons from sham-injured animals demonstrated low FRET efficiency. These results demonstrate that heterodimerization between BAD and Bcl-xL occurs in the hours following traumatic axonal injury. Further, both LSCM and multiphoton excitation microscopy can be used to acquire images for FRET analysis with similar results.

4.8 Thick-Tissue FRET

An advantage of 2p-FRET microscopy is the ability to penetrate tissue specimens and visualize deep focal planes. To determine if we could image axons in tissue sections >40 μm, we prepared tissue specimens 100 and 200 μm thick with an adjustable vibratome. After immunohistochemcial labeling of the specimens, we imaged the proximal surface of the specimens with C-FRET microscopy. Then, using a calibrated z-focus motor, we focused on the distal surface of the specimen by changing the z-plane focus 40–200 μm, respectively (Fig. 6–6).

Using the tunable Ti:sapphire laser, the images required for FRET analysis were acquired. Axons ubiquitously labeled for both BAD and Bcl-xL were visualized in all tissue specimens. However, the relative number of axons seen decreased as the thickness increased. Likewise, the amount of background fluorescence decreased as the thickness increased. This suggests that as the thickness of the tissue increased, the emission signal decreased, and we therefore were only able to visualize the most intensely labeled axons. In 6-h postinjury tissue, swollen axons were imaged for PFRET analysis (Fig. 6–6). At 100 and 200 μm, we obtained FRET efficiencies >20%, consistent with both C-FRET and 2p-FRET data from the proximal surface of 40 μm tissue. These results suggest that multiphoton microscopy can be used to image thick tissue sections with minimal loss of FRET signal.

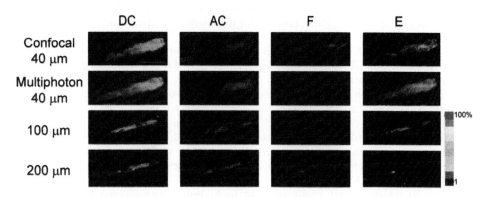

FIGURE 6–6 Deep-tissue FRET imaging. The same tissue was used for confocal FRET and multiphoton FRET imaging using Biorad Radiance 2100 confocal/multiphoton microscopy. The confocal PFRET image was obtained at 40 μm depth and did not have decent signal beyond that depth. By contrast, multiphoton PFRET images (F) up to 200 μm depth were obtained. As shown in the figure, the signal appears to be less when we moved from 40 to 200 μm. This can be attributed to the concentration of the fluorophore deep inside the tissue and to the visible FRET signal lost in the tissue before reaching the detector. Bar = 10 μm; 20 × MIMM numerical aperature 0.75; AC, acceptor channel; DC, donor channel. [Adapted from Chen et al. (2003).]

We wish to thank the University of Virginia and the Commonwealth Neurotrauma Initiative for their support. We also thank Ms. Ye Chen and Erica Caruso for help in the preparation of this chapter.

References

Buki, A., D.O. Okonkwo, K.K. Wang, and J.T. Povlishock. Cytochrome *c* release and caspase activation in traumatic axonal injury. *J. Neurosci.* 20:2825–2834, 2000.

Chen, Y., J.D. Mills, and A. Periasamy. Protein interactions in cells and tissues using FLIM and FRET. *Differentiation* 71:528–541, 2003.

Chen, Y. and A. Periasamy. Characterization of two-photon excitation fluorescence lifetime imaging microscopy for protein localization. *Microsc. Res. Tech.* 63:72–80, 2004.

Clegg, R.M. Fluorescence resonance energy transfer. In: *Fluorescence Imaging Spectroscopy and Microscopy*, edited by X.F. Wang and B. Herman. Chemical Analysis Series Vol. 137. New York: John Wiley & Sons, pp. 179–251, 1996.

Dale, R.E., J. Eisinger, and W.E. Blumberg. The orientational freedom of molecular probes. The orientation factor in intramolecular energy transfer. *Biophys. J.* 26:161–194, 1979.

Day, R.N. Visualization of Pit-1 transcription factor interactions in the living cell nucleus by fluorescence resonance energy transfer microscopy. *Mol. Endosc.* 12:1410–1419, 1998.

Denk, W., D.W. Piston, and W.W. Webb. Two-photon molecular excitation in laser-scanning microscopy. In: *Handbook of Biological Confocal Microscopy*, edited by J.B. Pawley. New York: Plenum Press, pp. 445–458, 1995.

Denk, W., J.H. Strickler, and W.W. Webb. Two-photon laser scanning fluorescence microscopy. *Science* 248:73–75, 1990.

Doppenberg, E.M., S.C. Choi, and R. Bullock. Clinical trials in traumatic brain injury. What can we learn from previous studies? *Ann. N Y Acad. Sci.* 825:305–322, 1997.

Elangovan, M., W. Horst, Y. Chen, R.N. Day, M. Barroso, and A. Periasamy. Characterization of one- and two-photon excitation energy transfer microscopy. *Methods* 29:58–73, 2003.

Eldadah, B.A. and A.I. Faden. Caspase pathways, neuronal apoptosis, and CNS injury. *J. Neurotrauma* 17:811–829, 2000.

Förster, T. Delocalized excitation and excitation transfer. In: *Modern Quantum Chemistry*, Vol. 3, edited by O. Sinanoglu, New York: Academic Press, pp. 93–137, 1965.

Göppert-Mayer, M. Ueber Elementarakte mit Quantenspreungen. *Ann. Phys.* 9:273–295, 1931.

Guo, C., S.K. Dower, D. Holowka, and B. Baird. Fluorescence resonance energy transfer reveals interleukin (IL)-1-dependent aggregation of IL-1 type I receptors that correlates with receptor activation. *J. Biol. Chem.* 270:27562–27568, 1995.

Heim, R. and R.Y. Tsien. Engineering green fluorescent protein for improved brightness, longer wavelengths and fluorescence resonance energy transfer. *Curr. Biol.* 6:178–182, 1996.

Jovin, T.M. and D. Arndt-Jovin. FRET microscopy: Digital imaging of fluorescence resonance energy transfer: application in cell biology. In: *Cell Structure and Function by Microspectrofluorometry*, edited by E. Kohen and J.G. Hirschberg. San Diego: Academic Press, 1989, pp. 99–117.

Jurgens, L., D. Arndt-Jovin, I. Pecht, and T.M. Jovin. Proximity relationships between the type I receptor for Fc epsilon (Fc epsilon RI) and the mast cell function-associated antigen (MAFA) studied by donor photobleaching fluorescence resonance energy transfer microscopy. *Eur. J. Immunol.* 26:84–91, 1996.

Kaiser, W. and C.G.B. Garrett. Two-photon excitation in $CaF_2:Eu^{2+}$. *Phys. Rev. Lett.* 7:229–231, 1961.

Mahajan, N.P., K. Linder, G. Berry, G.W. Gordon, R. Heim, and B. Herman. Bcl-2 and Bax interactions in mitochondria probed with green fluorescent protein and fluorescence resonance energy transfer. *Nat. Biotechnol.* 16:547–552, 1998.

Marmarou, A., M.A. Foda, B.W. van Den, J.K.H. Campbell, and K. Demetriadou. A new model of diffuse brain injury in rats. Part I: Pathophysiology and biomechanics. *J. Neurosurg.* 80:291–300, 1994.

Mills, J.D., J.R. Stone, J.D. Rubin, D.E. Melon, D.O. Okonkwo, A. Periasamy, and G.A. Helm. Illuminating protein interactions in tissue using confocal and two-photon excitation fluorescent resonance energy transfer (FRET) microscopy. *J. Biomed. Opt.* 8:347–356, 2003.

Patterson, G.H. and D.W. Piston. Photobleaching in two-photon excitation microscopy. *Biophys. J.* 78:2159 2162, 2000.

Periasamy, A., Ed. (2001). *Methods in Cellular Imaging*. New York, Oxford University Press.

Periasamy, A. and R.N. Day. Visualizing protein interactions in living cells using digitized GFP imaging and FRET microscopy. *Methods Cell Biol.* 58:293–314, 1999.

Periasamy, A., M. Elangovan, W. Horst, M. Barroso, J.N. Demas, D.L. Brautigan, and R.N. Day. Wide-field, confocal, two-photon and lifetime energy transfer imaging microscopy. In: *Methods in Cellular Imaging*, edited by A. Periasamy, New York: Oxford University, 2001, pp. 295–308.

Periasamy, A., P. Skoglund, C. Noakes, and R. Keller. An evaluation of two-photon excitation versus confocal and digital deconvolution fluorescence microscopy imaging in Xenopus morphogenesis. *Micros. Res. Tech.* 47:172–181, 1999.

Sekar, R.B. and A. Periasamy. Fluorescence resonance energy transfer microscopy imaging of live cell protein localizations. *J. Cell Biol.* 160:629–633, 2003.

Sosin, D.M., J.E. Sniezek, and R.J. Waxweiler. Trends in death associated with traumatic brain injury, 1979 through 1992. Success and failure. *Journal of American Medical Association* 273:1778–1780, 1995.

Stone, J.R., S.A. Walker, and J.T. Povlishock. The visualization of a new class of traumatically injured axons through the use of a modified method of microwave antigen retrieval. *Acta Neuropathol. (Berl)* 97:335–345, 1999.

Stryer, L. Fluorescence energy transfer as a spectroscopic ruler. *Annu. Rev. Biochem.* 47:819–846, 1978.

Wallrabe, H., M. Elangovan, A. Burchard, A. Periasamy, and M. Barroso. Confocal FRET microscopy to measure clustering of receptor–ligand complexes in endocytic membranes. *Biophys. J.* 85:559–571, 2003.

Wang, H.G., N. Pathan, I.M. Ethell, S. Krajewski, Y. Yamaguchi, F. Shibasaki, F. McKeon, T. Bobo, T.F. Franke, and J.C. Reed. Ca^{2+}-induced apoptosis through calcineurin dephosphorylation of BAD. *Science* 284:339–343, 1999.

Wu, P. and L. Brand. Review–resonance energy transfer: methods and applications. *Anal. Biochem.* 218:1–13, 1994.

Xu, C. and W.W. Webb. Measurement of two-photon excitation cross-sections of molecular fluorophores with data from 690 nm to 1050 nm. *J. Opt. Soc. Am.* B13:481–491, 1996.

7

FRET Data Analysis: The Algorithm

YE CHEN, MASILAMANI ELANGOVAN, AND AMMASI PERIASAMY

1. INTRODUCTION

While light microscopy initiated our understanding of cellular structure and function, molecular biological studies over the past few decades have shown that cellular events, such as signal transduction and gene transcription, require the assembly of proteins into specific macromolecular complexes. Traditional biophysical or biochemical methods, however, do not provide direct access to the interactions of these protein partners in their natural environment. Intensity-based imaging techniques using FRET (Förster resonance energy transfer) microscopy were developed for the analysis of these interactions inside intact living cells (Jovin et al., 1989; Cubitt et al., 1995; Miyawaki et al., 1997; Day, 1998; Chen et al., 2003; Sekar and Periasamy, 2003). New imaging technologies, coupled with the development of new genetically encoded fluorescent labels and sensors and the increasing capability of computer software for image acquisition and analysis, have permitted increasingly sophisticated studies of protein functions and processes ranging from gene expression to second-messenger cascades and intracellular signaling (Roessel and Brand, 2002).

Several different FRET microscopy techniques exist, and each has its own advantages and disadvantages. Wide-field FRET (W-FRET) microscopy is the simplest and most widely used technique (Day et al., 2003; see Chapter 4). It is used for quantitative comparisons of cellular compartments and time-lapse studies of cell motility, intracellular mechanics, and molecular movement. Laser scanning confocal and multiphoton FRET microscopy have the advantage of rejecting out-of-focus information (see Chapters 5 and 6). They also allow associations occurring inside the cell to be localized in three dimensions. Confocal FRET (C-FRET) microscopy, however, is limited to standard laser lines of defined wavelengths. Multiphoton or two-photon FRET microscopy overcomes this limitation by using a tunable laser (range 700–1000 nm) that allows excitation of a wide variety of fluorophores with higher axial resolution, greater sample penetration, reduced photobleaching of marker dyes outside of the focal plane, and increased cell viability. These advantages enable investigations on thick living tissue specimens that would not otherwise be possible with conventional techniques (Mills et al., 2003; Chapter 6). However, each of these intensity-based FRET techniques requires post-image processing methods to remove the unwanted bleedthrough components in the sensitized signal (FRET image).

2. Review of FRET Data Analysis Methods

One of the important conditions for FRET to occur is the overlap of the emission spectrum of the donor (D) with the absorption spectrum of the acceptor (A). Because of spectral overlap, the FRET signal is contaminated by donor emission into the acceptor channel and by the excitation of acceptor molecules by the donor excitation wavelength (Fig. 7–1). The combination of these signals is called the *spectral bleedthrough* (SBT) signal into the acceptor (or FRET) channel. In principle, the SBT signal is same for one-photon or two-photon FRET microscopy (Elangovan et al., 2003). In addition to SBT, other sources of noise contaminate the FRET signals, including spectral sensitivity variations in donor and acceptor channels, auto-fluorescence, and detector and optical noise.

There are various methods to assess SBT contamination in the FRET image. The donor bleedthrough can be calculated and corrected using the percentage of the spectral area of the donor emission spill over into the acceptor emission spectrum or FRET channel (Fig. 7–1). But it is difficult to determine the acceptor bleedthrough because one does not know the percentage of donor excitation wavelength that may

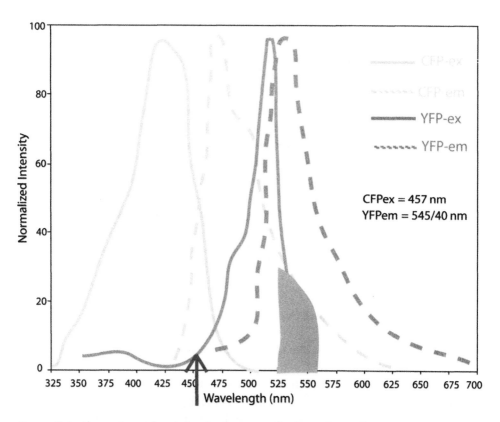

FIGURE 7–1 Absorption and emission (em) spectra of enhanced cyan fluorescent protein (CFP) and enhanced yellow fluorescent protein (YFP). The green area in the enhanced YFP emission channel represents the donor spectral bleedthrough in the FRET channel and the red arrow shows the donor excitation (ex) wavelength exciting acceptor molecules to cause acceptor bleedthrough.

excite the acceptor absorption spectrum. Moreover, this does not correct for variations in expression or concentration of the fluorophore-labeled cells.

One way to measure FRET that overcomes some of the problems associated with SBT is acceptor photobleaching FRET. In this method the difference between quenched donor (donor fluorescence in the presence of acceptor) and unquenched donor (donor fluorescence after the acceptor has been bleached) is used to calculate the energy transfer efficiency (see Chapters 4 and 8; Wouters et al., 1998; Bastiaens and Jovin, 1996; Day et al., 2001; Zal and Gascoigne, 2004). While this approach has the advantage of using a single double-label specimen, there are a number of limitations with this method: it cannot be used to examine highly mobile protein complexes (Day et al., 2001), and it is not known whether the exposure of the specimen to extended laser energy has any ill effects for live-cell imaging, which could introduce artifacts to the data analysis.

An alternative method is to use computer-based algorithms to measure and then remove the contaminating signals from the FRET image based on single-labeled reference samples. Depending on the level of sensitivity desired, other methods use two or three filter sets and normalize the FRET signal for the donor and acceptor (Gordon et al., 1998; Xia and Liu, 2001) bleedthrough signal. There are other approaches that correct after image acquisition with single-label reference specimens (Chamberlain et al., 2000). These methods assume a linear relationship between single- and double-labeled specimens imaged under the same conditions and use that for the correction of donor and acceptor bleedthrough signal. The relationship, however, is seldom linear, but for a certain level of sensitivity this approach may be perfectly adequate.

We believe that our new algorithm-based image acquisition and processing approach to correct SBT represents an advance over the other methods previously described, particularly if the algorithm is used for estimation of rate of energy transfer efficiency ($E\%$) or distance between donor and acceptor molecules or for higher-level analysis (see Chapter 5). In our algorithm the data processing is considered in various intensity ranges that would normalize the expression level variation in the cellular images. Moreover, this algorithm allows estimation of the spectral bleedthrough correction for linear and nonlinear intensity distributions. This method can be used for any intensity-based FRET imaging techniques, including wide-field, total internal reflection (TIRF), confocal, and multiphoton microscopy FRET data analysis.

3. PRECISE FRET DATA ANALYSIS

Here we describe a computer algorithm designed to remove both the donor and acceptor SBT signals and correct for variation in fluorophore expression level (FEL) associated with FRET imaging. These corrections then allow determination of the energy transfer efficiency ($E\%$) and provide an estimate of the distance between donor and acceptor molecules with the same cell used for FRET imaging. Both the SBT and FEL are incorporated together in our mathematical calculations (see below). Seven images are acquired with appropriate filters as shown in Table 7–1 for PFRET data analysis. Our approach works on the assumption that the double-labeled cells and

TABLE 7–1 Control and FRET Images Acquired for FRET Data Analysis Using the Algorithm

Symbol	Fluorophone or Sample	Excitation Wavelength	Emission Wavelength	Image
a	Donor only	Donor	Donor	
b	Donor only	Donor	Acceptor	
c	Acceptor only	Donor	Acceptor	
d	Acceptor only	Acceptor	Acceptor	
e	Donor and acceptor	Donor	Donor	
f	Donor and acceptor	Donor	Acceptor	
g	Donor and acceptor	Acceptor	Acceptor	

single-labeled donor and acceptor cells, imaged under the same conditions, exhibit the same SBT dynamics. The hurdle we had to overcome is having three (D, A, and D+A) different cells or specimens, where individual pixel locations cannot be compared. What can be compared, however, are pixels with matching fluorescence levels. Our algorithm follows fluorescence levels pixel by pixel and establishes the level of SBT in the single-labeled cells, and applies these values as a correction factor to the appropriate matching pixels of the double-labeled cell. PFRET then is

$$PFRET = uFRET - ASBT - DSBT \qquad (1)$$

where uFRET is uncorrected FRET (uFRET, Table 7–1f), ASBT is the acceptor spectral bleedthrough signal, and DSBT is the donor spectral bleedthrough signal. The DSBT and ASBT are established by using the corresponding single-labeled cells described below.

3.1 Donor Spectral Bleedthrough Signals

To correct the DSBT, three images are required (one double-labeled *e* and two single-labeled donor images *a* and *b*), referenced in Table 7–1. The gray-level intensity values of the three images are assembled as matrix elements to implement the algorithm. Image b is the bleedthrough signal due to the donor molecules. Since b and f are different cells, it is not appropriate to subtract b from f. The algorithm corrects the DSBT signal as follows. Figure 7–2 shows a typical example of DSBT removal. A certain range of intensity values was chosen in the intensity matrix e (see Fig. 7–2 (p1)). The same range of selected pixel intensity values in e were marked in a. Then the corresponding "coordinates" of marked intensity values in a were selected in b. To obtain the DSBT values, we need to average the ratio of all the elements in b and a for different ranges to get donor bleedthrough ratio rd (see Fig. 7–2 (p3)). Finally, we apply those ratios of different ranges one by one by multiplying them with e to get donor bleedthrough DSBT ($e * (b/a)$) image (see Fig. 7–2 (p4)). The process can be represented using the equation given below:

$$rd_{(j)} = \frac{\sum_{i=1}^{i=m}(b_i/a_i)}{m} \tag{2}$$

$$DSBT_{(j)} = \sum_{p=1}^{p=n}\left(e_p * rd_{(j)}\right) \tag{3}$$

$$DSBT = \sum_{j=1}^{j=k} DSBT_{(j)} \tag{4}$$

where j is the jth range of intensity, $rd_{(j)}$ is the donor bleedthrough ratio for the jth intensity range, m is the number of pixels in a for the jth range, $a_{(i)}$ is the intensity of pixel i, $DSBT_{(j)}$ is the donor bleedthrough factor for the range j, n is the number of pixels in e for the jth range, e_p is the intensity of pixel (p), k is the number of range, and DSBT is the total donor bleedthrough.

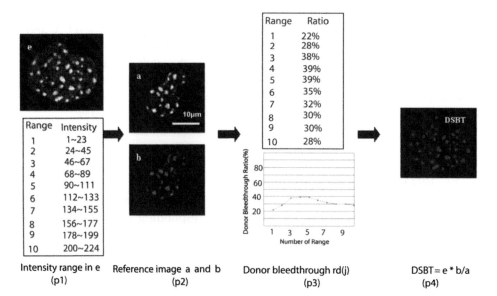

Intensity range in e (p1) Reference image a and b (p2) Donor bleedthrough rd(j) (p3) DSBT = e * b/a (p4)

FIGURE 7–2 Demonstration of donor spectral bleedthrough signal (DSBT) correction. As explained in the text, the b/a ratio (p3) is not linear for different intensity ranges (p1) (see also Fig. 7–6).

3.2 Acceptor Spectral Bleedthrough Signal

The ASBT correction follows the same approach as DSBT, but using three images (one double-labeled g and two single-labeled acceptor images c and d) as outlined in Table 7–1 and Figure 7–3. To obtain the ASBT values we averaged the ratio of elements in c and d for different range and then multiplied by the selected elements in g as shown in equation (5), (6), and (7) The acceptor bleedthrough signal in the FRET channel for all the pixel elements of the whole image was determined using the equation given below:

$$ra_{(j)} = \frac{\sum_{i=1}^{i=m}(c_i/d_i)}{m} \tag{5}$$

$$\text{ASBT}_{(j)} = \sum_{p=1}^{p=n}\left(g_p * ra_{(j)}\right) \tag{6}$$

$$\text{ASBT} = \sum_{j=1}^{j=k}\text{ASBT}_{(j)} \tag{7}$$

where j is the jth range of intensity, $ra_{(j)}$ is the acceptor bleedthrough ratio for the jth intensity range, m is the number of pixels in d for the jth range, $d_{(i)}$ is the intensity of pixel i, $\text{ASBT}_{(j)}$ is the acceptor bleedthrough factor for the range j, n is the number of pixels in g for the jth range, $g_{(p)}$ is the intensity of pixel p, k is the number of range, and ASBT is the total acceptor bleedthrough.

The results of the ASBT and DSBT corrections are applied to f (uFRET) as per equation (1) to obtain the precision (or processed) FRET (PFRET) image pixel to pixel (see Fig. 7–4). In this image, both donor and acceptor SBT were removed and the expression level variation was also corrected (Elangovan et al., 2003; Wallrabe et al., 2003a). This image was used for further data analysis, such as

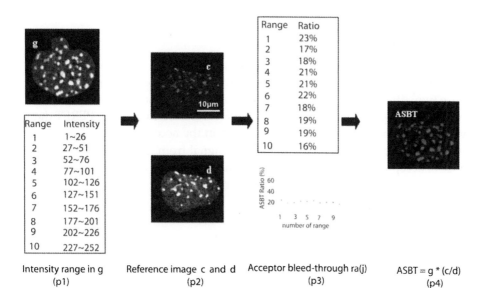

FIGURE 7–3 Demonstration of acceptor spectral bleedthrough signal (ASBT) correction.

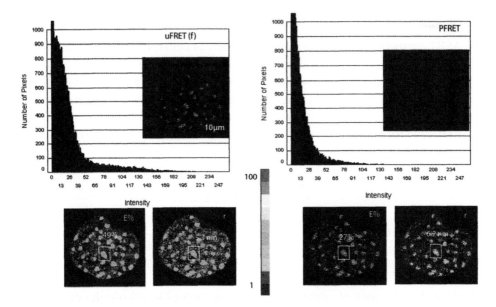

FIGURE 7–4 Precise FRET (PFRET) vs. uncorrected FRET (uFRET). The top panel shows the uFRET and PFRET images and their respective histograms. The bottom panel shows images of the two-dimensional distribution of efficiency (E) and distance (r). The rate of energy transfer efficiency (E) is 49% before correction and 27% after donor and acceptor spectral bleedthrough correction.

estimation of distance between donor and acceptor molecule as described below and the energy transfer efficiency E%. Moreover, the bandwidth of the emission filter is not an issue if one uses the PFRET algorithm for the FRET data analysis (Wallrabe et al., 2003b).

3.3 Calculation of Energy Transfer Efficiency and Distance Between Donor and Acceptor Molecule

The energy transfer efficiency (E) is calculated as the ratio of the donor image in the presence (I_{DA}) and absence (I_D) of acceptor, as shown in equation (8). Here we must use two different cells for the efficiency calculation. Therefore, we indirectly obtained the I_D image by using the PFRET image. As described in the PFRET data analysis section above, the sensitized emission in the acceptor channel is due to the quenching of donor or energy transferred signal from donor molecule in the presence of acceptor. If we add the PFRET to the intensity of the donor in the presence of acceptor, we obtain the I_D. This I_D is from the same cell used to obtain the I_{DA}. Hence, equation (8) is modified from

$$E = 1 - (I_{DA}/I_D) \qquad (8)$$

to

$$E_n = 1 - [I_{DA}/I_{DA} + \text{PFRET}] \qquad (9)$$

where

$$I_D = I_{DA} + \text{PFRET} \tag{10}$$

From equation (1),

$$\text{PFRET} = \text{uFRET} - \text{DSBT} - \text{ASBT}$$

It is important that PFRET be corrected for detector spectral sensitivity (g^{-1}) for donor and acceptor channel images (see Fig. 7–8; Appendix 7–2 and 7–3):

$$E_n = 1 - I_{DA}/[I_{DA} + g^{-1}(\text{PFRET})] \tag{11}$$

In equation (13), for calculating the distance (r) between donor and acceptor, r has changed to r_n:

$$r = R_0\{(1/E) -1\}^{1/6} \tag{12}$$

$$r_n = R_0\{(1/E_n) -1\}^{1/6} \tag{13}$$

The energy transfer efficiency was calculated and compared with a conventional method (two different cells, I_{DA}/I_D) and the new method (same cell, $I_{DA}/(I_{DA} + g^{-1}\text{PFRET})$) (Elangovan et al., 2003). The comparison of E and r for various methodologies of correction clearly demonstrate that the PFRET data analysis algorithm described here is a viable technique (see Fig. 7–4, Table 7–2, and Appendix 7–2) to study the spectroscopic nature of protein interactions for various protein molecules (Berney and Danuser, 2003). The results were significant when we compared the data for the distance calculation using the molecular size and other physical parameters.

4. ACQUISITION OF CONTROL IMAGES

Since the SBT calculation in our algorithm uses different ranges of intensity of the cellular image, it is important that the intensity distribution of the control images match the intensity distribution in the double-labeled experimental cell images, represented by e and g (see Table 7–1). As mentioned above, we use single donor–labeled and single acceptor–labeled cellular images as a reference to find the bleedthrough ratio at different levels of intensity distribution.

The case of DSBT removal is explained as follows. As shown in Figure 7–5(i), cells with different expression levels have different bleedthrough ratios. The plots in Figure 7–5(i) show the donor bleedthrough ratios for the dim (Figure 7–5(ii)) and the bright cells (Fig. 7–5(iii)). It is important to note that the dim cell (Fig. 7–5(i)(2)) has a lower bleedthrough ratio than that of the bright cell (Fig. 7–5(i)(1)). Figure 7–5 (iv) is a histogram of a double-labeled cell (image e), which must be compared with image a, according to our algorithm. For the pixels in the higher-intensity level, we

TABLE 7–2 Comparison of E and r Values for Wide-Field, Confocal, and Two-Photon-FRET Microscopy

System	E% / r (1)	E% / r (2)	E% / r (3)	E% / r (4)
Two photon	14.76/70.68	14.76/70.68	16.06/69.51	16.06/69.51
Confocal	20/66.48	20/66.48	20.76/65.97	20.76/65.97
Wide field	19.29/66.99	14.04/71.37	20.05/66.45	14.63/70.8

Back-bleed through correction did not provide much difference in E% and r (Å) values and is statistically insignificant. (1) DSBT and ASBT correction (7 images required); (2) DSBT, ASBT, and spectral sensitivity correction; (3) DSBT, ASBT, and back-bleedthrough correction (11 images required); (4) DSBT, ASBT, back-bleedthrough, and spectral sensitivity correction (see Fig. 7–8).

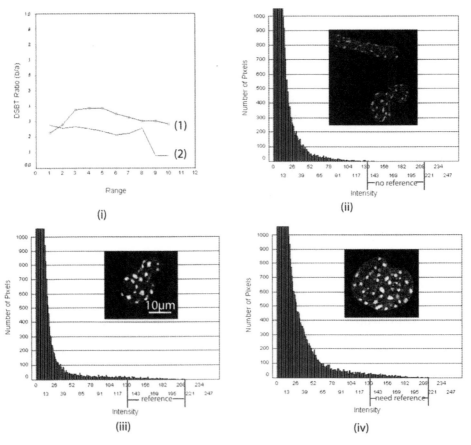

FIGURE 7–5 Demonstration of the importance of selection of control image a (Table 7–1) for spectral bleedthrough (SBT) correction. (*ii*): Histogram of single-labeled donor cell (dim cell). (*iii*): Histogram of single-labeled donor cell (bright cell). (*iv*): Histogram of double-labeled cell (image e, Table 7–1). (*i*): Donor bleedthrough ratio curve for cells (ii) and (iii). Control cell (iii) intensity distribution matches with image e (iv) as shown by the line in the horizontal axis of the histogram.

can find reference pixels in the bright cell (Fig. 7–5 (iii)) but not in the dim cell (Fig. 7–5 (ii); see the corresponding vertical lines in the histogram). Cell (iii) has an intensity distribution similar to that of double-labeled cellular image *e*. There was about 6% difference in efficiency when we used both cells (ii) and (iii) to correct for DSBT. This clearly demonstrates that to obtain a better correction, one should acquire the same control image intensity distribution as that of the double-labeled images. In the case of DSBT, the histogram of image *a* should be similar to that of *e* and for ASBT the histogram of image *d* should be similar to that of *g* to obtain a better estimation of *E* and *r*.

5. RATIONALE FOR USE OF THE ALGORITHM

5.1 Bleedthrough Ratio—Linear or Nonlinear

Recently, Berney and Danuser (2003) compared the reliability and uncertainty of several methods, including ours (Elangovan et al., 2003) and those of Tron et al. (1984) and Gordon et al. (1998). No significant difference was observed among these methods to calculate *E*% using the sensitized acceptor FRET signal and a FRET assay based on randomly organized, two-dimensional distribution of donor and acceptor fluorophores (Berney and Danuser, 2003). Moreover, the other methods listed in Berney and Danuser paper used mean intensity values of control images for the calculation of *E* and *r*. Nevertheless, our algorithm-based method should represent a significant improvement when using cell-based FRET assays because it does not assume constant crosstalk ratios or taking the mean values of the ratio. Instead, the crosstalk ratios are determined at different fluorescence intensities by generating an intensity-dependent look-up table, which is then used in the final calculation (Elangovan et al., 2003; Wallrabe et al., 2003a,b). The selection of different ranges of intensities normalizes the intensity distribution across the image and helps to correct the variation in expression level within the cellular image. This consideration is important because the SBT from donor to acceptor channel is not linear. This was consistent in the C-FRET and two photon-FRET microscopy system (see Fig. 7–6). The nonlinearity of *b/a* ratio was observed using the Nikon PCM2000, Biorad Radiance 2100, and Zeiss 510 laser scanning C-FRET microscopy. In the wide-field microscope, by contrast, linearity of the bleedthrough ratio was observed (see Fig. 7–6; see also Chapter 4).

5.2 Back-Bleedthrough

It has been reported in the literature (Gordon et al., 1998; Raheenen et al., 2004) that there could be back-bleedthrough FRET signal in image *e* or FRET signal due to acceptor excitation wavelength of the double-labeled cell. In our experience, however, we did not observe any FRET signal in *e* or in *g*. This could be attributed to the fact that the FRET occurs at peak absorption and peak emission (Herman et al., 2001). In the same literature there has also been the suggestion of some spectral back-bleedthrough, under the conditions listed below:

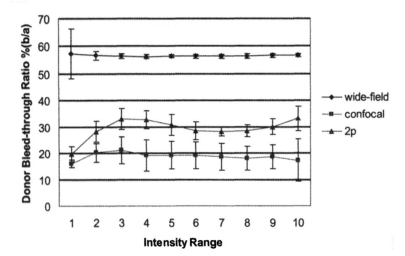

FIGURE 7–6 Comparison of *b/a* ratio for wide-field, confocal, and two-photon (2p) FRET microscopy. The ratio values for wide-field microscopy are linear in all ranges of gray-level intensity.

1. When the double-labeled cell is excited with donor excitation wavelength, the acceptor signal could back-bleedthrough in the donor channel (in image *e*).
2. When the double-labeled cell is excited with acceptor excitation wavelength, there is a possibility of donor signal bleedthrough into the acceptor channel (in image g).

We have investigated these two issues in our system and found that the excitation
intensity chosen to acquire the images did not provide any bleedthrough signal
but rather the signal was background noise of about 4–10 gray-level intensity. We
used our back-bleedthrough algorithm (Appendix 7–1) to estimate the efficiency
and found that the obtained difference of *E* values (1%–2%) was insignificant (see
Table 7–2).

6. SOFTWARE BASED ON THE ALGORITHM

User-friendly PFRET (precise or processed FRET) software was developed on the
basis of the algorithm described above. The data presented in this chapter and in
Chapters 4, 5, and 6 were processed using the PFRET software. A screen capture
of the two main functions in the software is shown in Figure 7–7. The SBT removal
window is shown on the left side of the figure; *a, b, c, d, e, f, g* represent the required
images for bleedthrough corrections listed in Table 7–1. In this window one has
the option of removing SBT by choosing the "PROCESS" button. The window on
the right side of the figure is for calculating *E*% and *r* values. This also allows calculation of the *E*% and *r* values for the whole image or for one or more regions of
interest (ROI), as shown in Figure 7–4. For more details about the software go to
www.circusoft.com.

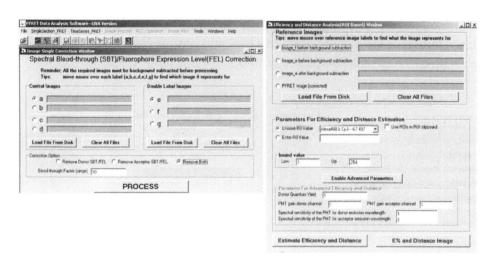

FIGURE 7–7 Screen shot of PFRET software for SBT correction and estimation of *E* and *r*. Details about the PFRET software can be obtained from www.circusoft.com.

7. CONCLUSION

In this chapter we have described the development of an algorithm to process FRET images acquired in different microscopy systems, including wide-field total internal reflection, confocal, and two-photon excitation FRET imaging microscopy systems. We analyzed the problems associated with FRET microscopy techniques in monitoring and quantifying FRET signals. Our approach for SBT correction is very effective and provides significant statistical data to substantiate the evidence for protein dimerization in live cells. The capability of this algorithm for PFRET data analysis is clearly demonstrated by comparing the data with various correction methodologies. The algorithm also corrects for the expression level differences found between cells. This is possible because we use various intensity ranges rather than a mean intensity value of the control images. This PFRET analysis provides a sensitive basis to estimate the distance between donor and acceptor molecules from the same cell used for FRET imaging. The spectral sensitivity correction provides significant data in estimating the distance between donor and acceptor molecules. We have also observed that the back-bleedthrough correction does not provide any significant difference in *E* and *r* values and it is not necessary that one include the back-bleedthrough correction in FRET data analysis. The PFRET data analysis allowed us to characterize various FRET microscopy systems for intranuclear dimer formation of CCAAT/enhancer binding protein alpha (C/EBPα) proteins in the live GHFT1-5 cell nucleus.

APPENDIX 7–1 BACK-BLEEDTHROUGH REMOVAL

The table below shows that in addition to the seven images required for PFRET data analysis, four more images (*h, i, j,* and *k*) are required to implement the back-bleedthrough correction (total, 11 images). The entire notation used in this algorithm is outlined below in the table. This back-bleedthrough equation is added to our existing PFRET data analysis algorithm.

Excitation/Sample	Donor Emission	Acceptor Emission
Donor excitation/donor	a	b
Acceptor excitation/donor	h	i
Donor excitation/acceptor	j	c
Acceptor excitation/acceptor		d
Donor excitation/double	e	f
Acceptor excitation/double	k	g

DSBT	Donor bleedthrough in f
ASBT	Acceptor bleedthrough in f
PFRET	Processed FRET signal in f
e_ASBT	Donor excitation wavelength exciting double-labeled cell– acceptor signal back-bleedthrough to donor channel
g_DSBT	Acceptor excitation wavelength exciting double-labeled cell– donor signal bleedthrough to acceptor channel
g_A	Acceptor signal in g due to acceptor excitation exciting double-labeled cell
E	Efficiency
r	Distance
e_quench	Quenched donor

Equation (1a) represents the signal from double-labeled cell excitation by donor wavelength, and emission from the donor channel (image e) contains quenched donor signal and acceptor back-bleedthrough signal:

$$e = e\_quench + e\_ASBT \tag{1a}$$

Signal from double-labeled cell excitation by acceptor wavelength and emission from the acceptor channel (image g) contains acceptor signal and donor bleedthrough because of the acceptor excitation wavelength exciting the donor, as shown in equation (2a):

$$g = g\_DSBT + g\_A \tag{2a}$$

To get g_A, which is used later to calculate acceptor bleedthrough ASBT, we need to get g_DSBT, which is calculated using equation (3a) below:

$$g\_DSBT = k * i/h \tag{3a}$$

Equation (4a) is obtained from equations (2a) and (3a). For equation (4a), we require four images (g, h, i, and k) to obtain g_A.

$$g\_A = g - k * i / h \tag{4a}$$

Now we are ready to calculate acceptor bleedthrough in image f. Comparing equations (5), (6), and (7) used to get ASBT, the only change is in equation (6), replacing g with g_A, shown in equation (5a); ASBT is calculated using equation (7) from the main text:

$$\text{ASBT}_{(j)} = \sum_{p=1}^{p=n}\left(g\_A_p * \text{ra}_{(j)}\right) \tag{5a}$$

Equation (6a) shows how to calculate back bleedthrough e_ASBT, shown in equation (1a). It is derived from equations (5a) and (5). Three images are required: g_A, j, and d.

$$e\_\text{ASBT} = \text{ASBT} * j/c = g\_A * j/d \tag{6a}$$

Equation (7a) is derived from equations (1a) and (6a) to obtain quenched donor e_quench. Four images are required: e, g_A, j, and d.

$$e\_\text{quench} = e - g\_A * j/d \tag{7a}$$

Now we are ready to calculate donor bleedthrough in image f. For DSBT correction we used equations (2), (3), and (4). Equation (8a) is obtained by replacing e with e_quench in equation (3):

$$\text{DSBT}_{(j)} = \sum_{p=1}^{p=n}\left(e\_\text{quench}_p * \text{rd}_{(j)}\right) \tag{8a}$$

Equations (8a) and (5a) were used to obtain the new ASBT and DSBT. This includes the back-bleedthrough in e and donor bleedthrough in g. Equation (1) from the main text was used to estimate the PFRET signals as shown below:

$$\text{PFRET} = f - \text{ASBT} - \text{DSBT}$$

To calculate $E\%$, we still use equation (11). But I_DA becomes e_quench instead of e. Distance r_n is calculated using equation (13).

APPENDIX 7–2 EFFICIENCY AND DISTANCE ESTIMATION

In FRET microscopy, one of the critical components is the imaging device, as it determines whether the FRET signal is be detected, the relevant structures resolved, and the dynamics of a process visualized (Aikens et al., 1988; Periasamy and Herman, 1994; Spring, 2001). A few important parameters should be considered in the selection of detectors for FRET microscopy, including the spectral sensitivity, quantum efficiency, spatial resolution, uniformity, signal/noise ratio, dynamic range, and response speed (Spring and Smith, 1987; Spring and Lowy, 1988; Tsay et al., 1990; see Chapters 4–6). Here we consider spectral sensitivity since the detector responses for donor and acceptor channels are different (Fig. 7–8). *Spectral sensitivity* is the detector signal as a function of wavelength of incident light. It is often expressed in terms of *quantum efficiency* (QE), which is the percentage of incident photons detected (Fig. 7–8a). As shown in Figure 7–8, the QE for charge-coupled devices (CCDs) is different from that for photomultiplier tubes (PMTs). For PMTs the QE depends on the chemical composition of the photocathode. The PMT photocathodes are not uniformly sensitive, and typically the photons are spread over the entire entrance window rather than in one region. Because PMTs do not store charge, unlike a CCD camera, and respond to changes in input light fluxes within a few nanoseconds, they can be used for the detection and recordings of extremely fast events (Spring, 2001).

It is important that PFRET be corrected for detector spectral sensitivity (g^{-1}) for donor and acceptor channel images, from equation (11) (see Appendix 7–3):

$$E_n = 1 - I_{DA} / [I_{DA} + g^{-1} (\text{PFRET})] \tag{1b}$$

$$E_n = 1 - I_{DA} / [I_{DA} + \text{PFRET}^*((\psi_{dd} / \psi_{aa})^* (Q_d/Q_a))] \tag{2b}$$

where ψ_{dd} and ψ_{aa} are collection efficiency in the donor and acceptor channels, Q_d is the quantum yield of the donor, and Q_a is the quantum yield of the acceptor. Equation (2b) is the final energy transfer efficiency equation.

For example, in the case of enhanced cyan fluorescent protein (ECFP) and enhanced yellow fluorescent protein (EYFP), the normalized spectral sensitivity of the donor channel is 12 and that of the acceptor channel is 8 (Biorad Radiance 2100, Multialkali PMTs; see Fig. 7–8b). Hence, the donor channel detects 3/2 times more photons than the acceptor channel. By contrast, the ratio of the quantum yield of ECFP and EYFP is about 2/3. Similarly, we need to correct for the PMT gain settings of the donor and FRET (acceptor) channels.

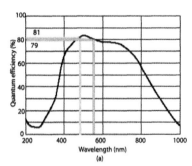
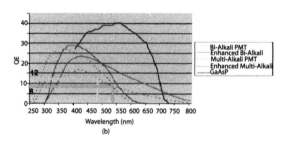

FIGURE 7–8 Spectral response characteristic of charge-coupled device (CCD) camera and photomultiplier tubes (PMT). Difference in quantum efficiency for enhanced cyan fluorescent protein and enhanced yellow fluorescent protein emission signal is larger in PMTs (Multi-alkali PMT) than CCD cameras (Orca2). The spectral sensitivity difference affects estimation of the distance between donor and acceptor molecules (see Appendix 7–2 and Table 7–2 for details). [Adapted from the data sheets at http://usa.hamamatsu.com; www.zeiss.de/cellscience]

Therefore, for particular optical configuration, the collection efficiency ratio between donor and acceptor channel is

$$(\psi_{dd} / \psi_{aa}) = \{(\text{PMT gain of donor channel / PMT gain of acceptor}$$
$$\text{channel})^*(\text{spectral sensitivity of donor channel /}$$
$$\text{spectral sensitivity of acceptor channel})\} \qquad (3b)$$

The distance between donor and acceptor molecule is estimated using equation (13):

$$r_n = R_0 \{(1/E_n) - 1\}^{1/6} \qquad (4b)$$

In the case of ECFP and EYFP, the PMT gain and spectral sensitivity ratio were 1/1 (equal gain) and 12/8, respectively (Fig. 7–8b). The quantum yield of ECFP is 0.37 and the EYFP is 0.6 (Rizzo et al., 2004). R_0 is 52.767 Å. Using one of our experimental data sets, I_{DA} is 42.72 and PFRET is 10.68. Then, from equations (2b and 4b):

$$E_n = 1 - 42.72/[42.72 + 10.68^*(((1/1)^*(12/8))^*(0.4/0.6))] = 20$$

$$r_n = 52.767 \{1/20) - 1\}^{1/6} = 66.48 \text{ Å}$$

The calculated and corrected efficiency for a PMT-based system appears to be the same, but when a CCD camera is used there is about 26% difference in the efficiency values. This 26% difference in E value for a CCD camera system would influence the distance (r) estimation between donor and acceptor molecules by about 6% (see Table 7–2).

Appendix 7–3 Correction for Detector Spectral Sensitivity (g^{-1})

According to equation (1C), the energy transfer efficiency (E) was calculated using the intensity of the donor in the presence (I_{DA}) and the absence (I_D) of the acceptor. Unfortunately, I_D and I_{DA} are two different cells with two different fluorophore contents and the value E will vary depending on the intensity of the cellular images and the detector spectral sensitivity.

$$E = 1 - (I_{DA} / I_D) \qquad (1C)$$

Here we explain the implementation of correction for detector spectral sensitivity.

The intensity I_{DA} is dependent on excitation intensity $I_{ex}D$, the number of molecules in the focal volume N_d, the collection efficiency of the detector and other optics of the donor emission in donor channel, quantum yield of donor when acceptor is present Q_{da}, quantum yield of donor alone Q_d, the absorption cross section of the donor at the donor excitation wavelength σ_{dd}, and collection efficiency in donor channel ψ_{dd} (Elangovan et al., 2003). Hence

$$I_{DA} = I_{ex}D\sigma_{dd}N_d \, \psi_{dd}Q_{da} \tag{2C}$$

$$I_D = I_{ex}D\sigma_{dd}N_d \, \psi_{dd}Q_d \tag{3C}$$

Therefore

$$E = 1 - (I_{ex}D\sigma_{dd}N_d \, \psi_{dd}Q_{da} \, / \, I_{ex}D\sigma_{dd}N_d\psi_{dd}Q_d) = 1 - Q_{da} \, / \, Q_d \tag{4C}$$

Quantum yield can be expressed in terms of the rate constants involved in the decay process of a fluorophore. If k_d is the rate constant of radiative decay when only the donor is present, k_{di} is the rate constant for nonradiative decay, and k_t is the rate constant for transfer of energy nonradiatively to the acceptor, then

$$Q_d = k_d \, / \, (k_d + k_{di}) \tag{5C}$$

$$Q_{da} = k_d \, / \, (k_d + k_{di} + k_t) \tag{6C}$$

We define

$$R_t = k_t \, / \, (k_d + k_{di} + k_t) \tag{7C}$$

where R_t is the quantum yield corresponding to nonradiative transfer. Equation (8C) was obtained after algebraic simplifications using equations (5C), (6C), and (7C):

$$Q_{da} \, / \, Q_d = Q_{da} \, /(Q_{da} + R_tQ_d) \tag{8C}$$

From equations (4C) and (8C),

$$E = 1 - Q_{da} \, / \, Q_d \tag{9C}$$

$$E = 1 - Q_{da} \, / \, (Q_{da} + R_tQ_d) \tag{10C}$$

From spectral bleedthrough correction,

$$PFRET = uFRET - DSBT - ASBT \tag{11C}$$

By using the above equations,

$$uFRET = I_{ex}D\sigma_{dd}N_d\psi_{aa}R_tQ_a + I_{ex}D\sigma_{dd}N_d\psi_{da}Q_{da} + I_{ex}D\sigma_{da} \, N_d\psi_{aa}Q_a \tag{12C}$$

where Q_a is quantum yield of acceptor alone.

$$DSBT = I_{ex}D\sigma_{dd}N_d\psi_{da}Q_{da} \tag{13C}$$

$$ASBT = I_{ex}D\sigma_{da}N_d\psi_{aa}Q_a \tag{14C}$$

Hence, by using equation (11C), (12C–14C),

$$\text{PFRET} = I_{ex}D\sigma_{dd}\,N_d\psi_{aa}R_tQ_a \tag{15C}$$

From equation (2C),

$$I_{DA} = \gamma\psi_{dd}Q_{da} \tag{16C}$$

where $\gamma = I_{ex}D\sigma_{dd}\,N_d$. Then,

$$\text{PFRET} = \gamma\psi_{aa}R_tQ_a \tag{17C}$$

We obtain equation (19C) by multiplying numerator and denominator by $\gamma\psi_{dd}$ in equation (10C). Equation (10C) can be written as

$$E_n = 1 - \gamma\psi_{dd}Q_{da}\,/\,(\gamma\psi_{dd}Q_{da} + \gamma\psi_{aa}R_t * (\psi_{dd}\,/\,\psi_{aa}) * Q_{d)} \tag{18C}$$
$$= 1 - \gamma\psi_{dd}Q_{da}\,/\,(\gamma\psi_{dd}Q_{da} + \gamma\psi_{aa}R_tQ_a *(\psi_{dd}\,/\,\psi_{aa}) * (Q_d\,/\,Q_a))$$

Then, using equations (15C–17C) in (18C),

$$E_n = 1 - I_{DA}\,/\,(I_{DA} + \text{PFRET} * ((\psi_{dd}\,/\,\psi_{aa}) * (Q_d\,/\,Q_a))) \tag{19C}$$

Equation (19C) is the final energy transfer efficiency equation. So, the correction for detector spectral sensitivity

$$g^{-1} = ((\psi_{dd}\,/\,\psi_{aa}) * (Q_d\,/\,Q_a)) \tag{20C}$$

We wish to acknowledge Dr. Richard Day for providing cells and his expert advice in preparing this chapter. We thank Dr. Margarida Barroso for her valuable help and her participation in the development of the algorithm. We also thank Ms. Cindy Booker and Mr. Horst Wallrabe for their assistance.

REFERENCES

Aikens, R., D. Agard, and J. Sedat. Solid-state imagers for optical microscopy. *Methods Cell. Biol.* 29:291–313, 1988.

Barney, C. and G. Danuser. FRET or no FRET: a quantitative comparison. *Biophys. J.* 84:3992–4010, 2003.

Bastiaens, P.I. and T.M. Jovin. Microspectroscopic imaging tracks the intracellular processing of a signal transduction protein: fluorescent-labeled protein kinase C beta I. *Proc. Natl. Acad. Sci. U.S.A.* 93:8407–8412, 1996.

Chamberlain, C., V.S. Kraynov, and K.M. Hahn. Imaging spatiotemporal dynamics of Rac activation in vivo with FLAIR. *Methods Enzymol.* 325:389–400, 2000.

Chen, Y., J.D. Mills, and A. Periasamy. Protein interactions in cells and tissues using FLIM and FRET. *Differentiation* 71:528–541, 2003.

Cubitt, A.B., R. Heim, S.R. Adams, A.E. Boyd, L.A. Gross, and Tsien, R.Y. Understanding, improving and using green fluorescent proteins. *Trends Biochem. Sci.* 20:448–455, 1995.

Day, R.N. Visualization of Pit-1 transcription factor interactions in the living cell nucleus by fluorescence resonance energy transfer microscopy. *Mol. Endosc.* 12:1410–1419, 1998.

Day, R.N., A. Periasamy, and F. Schaufele. Fluorescence resonance energy transfer microscopy of localized protein interactions in the living cell nucleus. *Methods* 25:4–18, 2001.

Day, R.N., T.C. Voss, J.F. Enwright, III, C.F. Booker, A. Periasamy, and F. Schaufels. Imaging the localized protein interactions between Pit-1 and the CCAAT/enhancer binding protein alpha (C/EBPα) in the living pituitary cell nucleus. *Mol. Endosc.* 17:333–345, 2003.

Elangovan, M., W. Horst, Y. Chen, R.N. Day, M. Barroso, and A. Periasamy. Characterization of one- and two-photon excitation energy transfer microscopy. *Methods* 29:58–73, 2003.

Gordon, G.W., G. Berry, X.H. Liang, B. Levine, and B. Herman. Quantitative fluorescence resonance energy transfer measurements using fluorescence microscopy. *Biophys. J.* 74:2702–2713, 1998.

Herman, B., G. Gordon, N. Mahajan, and V. Centonze. Measurement of fluorescence resonance energy transfer in the optical microscope. In: *Methods in Cellular Imaging*, edited by A. Periasamy. New York: Oxford University Press, 2001, pp. 257–272.

Jovin, T.M., G. Marriott, R.M. Clegg, and D.J. Arndt-Jovin. Photophysical processes exploited in digital imaging microscopy: fluorescence resonance energy transfer and delayed luminescence. *Ber. Bunsenges. Phys. Chem.* 93:387–391, 1989.

Mills, J.D., J.R. Stone, J.D. Rubin, D.E. Melon, D. O. Okonkwo, A. Periasamy, and G.A. Helm. Illuminating protein interactions in tissue using confocal and two-photon excitation fluorescent resonance energy transfer (FRET) microscopy. *J. Biomed. Opt.* 8:347–356, 2003.

Miyawaki, A., J. Llopis, R. Heim, J.M. McCaffery, J.A. Adams, M. Ikura, and R.Y. Tsien. Fluorescent indicators for Ca^{2+} based on green fluorescent proteins and calmodulin. *Nature* 388:882–887, 1997.

Periasamy, A. and B. Herman. Computerized fluorescence microscopic vision in the biomedical sciences. *J. Comput. Assist. Microsc.* 6:1–26, 1994.

Raheenen, J-v., M. Langeslag, and K. Jalink. Correcting confocal acquisition to optimize imaging of fluorescence resonance energy transfer by sensitized emission. *Biophys. J.* 86:2517–2529, 2004.

Rizzo, M.A., G.H. Springer, B. Granada, and D.W. Piston. An improved cyan fluorescent protein variant useful for FRET. *Nat. Biotech.* 22:445–449, 2004.

Roessel, P.V. and A.H. Brand. Imaging into the future: visualizing gene expression and protein interactions with fluorescent proteins. *Nat. Cell Biol.* 4:E15–E20, 2002.

Sekar, R.B. and A. Periasamy. Fluorescence resonance energy transfer microscopy imaging of live cell protein localizations. *J. Cell Biol.* 160:629–633, 2003.

Spring, K.R. Detectors for fluorescence microscopy. In: *Methods in Cellular Imaging*, edited by A. Periasamy. New York: Oxford University Press, 2001, pp. 40–52.

Spring, K.R. and R.J. Lowy. Characteristics of low light level television cameras. *Methods Cell Biol.* 29:269–289, 1988.

Spring, K.R. and P.D. Smith. Illumination and detection systems for quantitative fluorescence microscopy. *J. Microsc.* 147:265–278, 1987.

Tron, L., J. Szollosi, S. Damjanovich, S. Helliwell, D. Arndt-Jovin, and T. Jovin. Flow cytometric measurement of fluorescence resonance energy transfer on cell surfaces. Quantitative evaluation of the transfer efficiency on a cell-by-cell basis. *Biophys. J.* 45:939–946, 1984.

Tsay, T.-T., R. Inman, B. Wray, B. Herman, and K. Jacobson. Characterization of low-light-level cameras for digitized video microscopy. *J. Microsc.* 160:141–159, 1990.

Wallrabe, H., M. Elangovan, A. Burchard, A. Periasamy, and M. Barroso. Confocal FRET microscopy to measure clustering of receptor–ligand complexes in endocytic membranes. *Biophys. J.* 85:559–571, 2003a.

Wallrabe, H., M. Stanley, A. Periasamy, and M. Barroso. One- and two-photon fluorescence resonance energy transfer microscopy to establish a clustered distribution of receptor–ligand complexes in endocytic membranes. *J. Biomed. Opt.* 8:339–346, 2003b.

Wouters, F.S., P.I. Bastiaens, K.W. Wirtz, and T.M. Jovin. FRET Microscopy demonstrates molecular association of non-specific lipid transfer protein (nsL-TP) with fatty acid oxidation enzymes in peroxisomes. *EMBO J.* 17:7179–7189, 1998.

Xia, Z. and Y. Liu. Reliable and global measurement of fluorescence resonance energy transfer using fluorescence microscopes. *Biophys. J.* 81:2395–2402, 2001.

Zal, T. and N.R.J. Gascoigne. Photobleaching-corrected FRET efficiency imaging of live cells. *Biophys. J.* 86:3923–3939, 2004.

8

Photobleaching FRET Microscopy

Anne K. Kenworthy

1. Introduction

Photobleaching is a process in which, following repeated excitation, a fluorophore is irreversibly destroyed. Because photobleaching causes a loss of fluorescence signal, most quantitative fluorescence microscopy measurements are made under conditions where photobleaching is minimized. Jovin, Arndt-Jovin, and colleagues recognized that photobleaching could be used as a tool to measure Förster resonance energy transfer (FRET). They introduced two experimental methods that exploit photobleaching as a method to follow changes in the donor resulting from FRET. The first of these methods, donor photobleaching FRET (donor pbFRET), measures the kinetics of the photobleaching of donor fluorescence in the presence and absence of the acceptor (Jovin and Arndt-Jovin, 1989a, 1989b; Table 8–1). While elegant in principle, donor pbFRET has proven to be somewhat difficult in practice and has been supplanted by a second method, acceptor photobleaching FRET (acceptor pbFRET) (Table 8–2). Acceptor pbFRET allows the direct measurement of donor dequenching on a single sample by eliminating the acceptor (Bastiaens et al., 1995, 1996; Bastiaens and Jovin, 1996). Importantly, these methods can be used to directly calculate energy transfer efficiencies, without the need for extensive corrections required to quantify sensitized emission measurements of FRET. They also do not require specialized instrumentation, making them available to investigators with access to research-grade digital fluorescence microscopes. Acceptor pbFRET particularly has become widely used as a method of choice for quantitative FRET measurements, especially with confocal microscopy. For the sake of completeness, we will discuss both donor and acceptor pbFRET measurements below.

We begin this chapter with a description of the basic principles underlying donor and acceptor pbFRET measurements. We then discuss how to configure microscopy systems for pbFRET measurements. We consider the advantages and disadvantages of pbFRET, describe FRET pairs commonly used in pbFRET experiments, and discuss the application of pbFRET to the study of membrane microdomains.

1.1 Principle of Donor Photobleaching FRET

In the excited state, a fluorophore can return to the ground state by the nonradiative loss of energy, emission of a photon, or resonance energy transfer. Alternatively, it

TABLE 8–1 Applications of Donor Photobleaching FRET Microscopy

Applications	Donor/Acceptor	Reference
Protein–protein interactions		
CD8/MHC-I and LFA/ICAM-I	FITC/TRITC	Bacso et al., 1996
FcεRI/MAFA	FITC/TRITC	Jurgens et al., 1996
Cholera toxin A and B subunits	Cy3/Cy5	Bastiaens et al., 1996
Peroxisomal proteins	Cy3/Cy5	Wouters et al., 1998
Receptor oligomerization		
EGFR	FITC/ TRITC	Gadella and Jovin, 1995
Somatostatin receptors	FITC/rhodamine	Rocheville et al., 2000
Epitope mapping	FITC/phycoerythrin	Szaba et al., 1992
Analysis of protein clustering		
FcεRI	FITC/DNP	Kubitscheck et al., 1993
GPI-anchored proteins	FITC/Texas red	Hannan et al., 1993
MHC class I	FITC/TRITC	Damjanovich et al., 1995
Endocytic receptors	Alexa 488/Cy3	Wallrabe et al., 2003
Con A receptors	FITC/TRITC	Young et al., 1994
Glycolipid binding toxins or antibodies	FITC/TRITC	Vyas et al., 2001
	FITC/Cy3	Vyas et al., 2001

EGFR, epidermal growth factor receptor; GPI, glycosylphosphatidylinositol; ICAM-I, intracellular adhesion molecule I; LFA, leukocyte function antigen; MHC-I, major histocompatibity complex class I.

may undergo a chemical reaction while in the excited state, a process observed experimentally as loss of fluorescence over time—i.e., photobleaching. Photobleaching occurs in a single step for an individual fluorophore. However, for a population of fluorophores, photobleaching can be approximated as an exponential decay in fluorescence intensity I at time t,

$$I(t) = I_0 * \exp(-t/\tau_{bl}). \tag{1}$$

where I_0 is the initial intensity and τ_{bl} is the photobleaching time constant. Energy transfer shortens the time molecules spend in the excited state and therefore lowers the probability of excited-state chemical reactions. As a result, the photobleaching time constant of the donor is increased in the presence of the acceptor (τ_{bl}') according to

$$E = 1 - (\tau_{bl}/\tau_{bl}') \tag{2}$$

where E is the energy transfer efficiency (Jovin and Arndt-Jovin, 1989a, 1989b).

1.2 Principle of Acceptor Photobleaching FRET

When FRET occurs, the donor fluorescence will be quenched. In principle, this can be measured by comparing donor fluorescence in the presence and absence of the

TABLE 8–2 Applications of Acceptor Photobleaching FRET Microscopy

Application	Donor/Acceptor	Reference
Protein–protein interactions		
Cholera toxin A and B subunits	Cy3/Cy5	Bastiaens et al., 1996
Nonspecific lipid transfer protein and peroxisomal proteins	Cy3/Cy5	Wouters et al., 1998
Transcription factors	EBFP/EGFP	Day et al., 2001
	EBFP/EYFP	Day et al., 2001
Calmodulin/ Ca^{2+} channels	ECFP/EYFP	Erickson et al., 2001
LDLR-related protein and APP	FITC/Cy3	Kinoshita et al., 2001
	EGFP/Cy3	Kinoshita et al., 2001
	EGFP/DsRed	Kinoshita et al., 2001
Coactivators/nuclear hormone receptors	ECFP/EYFP	Llopis et al., 2000
nNOS/glucokinase	ECFP/EYFP	Rizzo and Piston, 2003
Translocon components	Cy3/Cy5	Snapp et al., 2004
Cell signaling		
PKC β1 processing	Cy3/Cy5	Bastiaens and Jovin, 1996
EGFR tyrosine kinase activity	EGFP/Cy3	Wouters and Bastiaens, 1999
Fas oligomerization	ECFP/EYFP	Siegel et al., 2000
Analysis of protein clustering		
GPI-anchored proteins	Cy3/Cy5	Kenworthy and Edidin, 1998
	Cy3/Cy5	Kenworthy et al., 2000
	Cy3/Cy5	Nichols, 2003
	mCFP/mCitrine	Glebov and Nichols, 2004
	mCFP/mYFP	Sharma et al., 2004
	EGFP/A1ex 568	Sharma et al., 2004
	A1ex 568/Cy5	Sharma et al., 2004
Glycolipid-binding toxins	Cy3/Cy5	Kenworthy et al., 2000
	Cy3/Cy5	Nichols, 2003
	Cy3/Cy5	Kovbasnjuk et al., 2001
Acylated proteins	mCFP/mYFP	Zacharias et al., 2002
Endocytic receptors	Alexa 488/Cy3	Wallrabe et al., 2003
MHC class I	ECFP/EYFP	Spiliotis et al., 2002
	ECFP/EYFP	Pentcheva and Edidin, 2001
Tapasin	ECFP/EYFP	Pentcheva et al., 2002

APP, amyloid precursor protein; EGFR, epidermal growth factor receptor; GPI, glycosylphosphatidy-linositol; LDLR, low-density lipoprotein receptor; MHC, major histocompatibility, nNOS, neuronal nitric-oxide synthase; PKC, protein Kinase C.

acceptor. However, in FRET microscopy experiments, this is typically not practical because of variations in protein expression levels between cells and at different intracellular compartments. Acceptor pbFRET offers a direct method to assess the level of donor quenching. In this approach, photobleaching is used as a tool to ablate the acceptor molecule (Bastiaens et al., 1995, 1996, Bastiaens and Jovin, 1996). This releases the quenching of donor fluorescence, an effect that is dramatically demonstrated in single-molecule FRET measurements (Ha et al., 1996). By direct comparison of donor fluorescence in the presence and absence of the acceptor, *E* can then be calculated as

$$E = (D_{post} - D_{pre}) / D_{post} \qquad (3)$$

where D_{pre} and D_{post} are the donor fluorescence intensity before and after photobleaching the acceptor, respectively.

2. How to Configure the Microscopy System for Photobleaching FRET

2.1 Choice of Microscope Systems

2.1.1. Wide-field
Wide-field microscopes are readily used in both donor and acceptor pbFRET experiments. As for other FRET microscopy applications, one advantage of wide-field microscopy is that excitation wavelengths are not limited to available laser lines as in laser scanning confocal microscopy, allowing for a wide range of donor and acceptor fluorophores choices. A disadvantage is that this approach does not provide much flexibility of choice of bleach regions, requiring that an entire field of view be bleached simultaneously by continuously illuminating the specimen. Furthermore, bleaching cannot be accomplished as rapidly using arc lamp excitation as a laser, making acceptor photobleaching measurements more time consuming on wide-field systems.

2.1.2 Confocal
Laser scanning confocal microscopy generates improved image contrast, compared to that of wide-field microscopy, through the use of a pinhole that rejects out-of-focus signal from the specimen. In addition to eliminating out-of-focus fluorescence, confocal microscopy offers several advantages over wide-field systems for pbFRET. Of particular value is the possibility for rapid and targeted photobleaching by repetitively scanning a desired region of interest (ROI). For example, the acceptor can be bleached from a portion of a cell, allowing for the comparison of donor fluorescence and/or donor photobleaching kinetics in the presence (unbleached) and absence (bleached) regions (Bastiaens and Jovin, 1996; Bastiaens et al., 1996; Karpova et al., 2003). This provides an important internal control, as no increase in donor fluorescence should be observed unless the acceptor has been photobleached (see Sections 2.4 and 3.2).

2.2 General Procedure for Collection of Photobleaching FRET Data

2.2.1 Sample requirements
The general requirements for sample preparation and labeling conditions are discussed elsewhere in this volume. Donor pbFRET measurements require a

donor-only–labeled sample and a double-labeled (donor plus acceptor–labeled) sample; alternatively, sequential acceptor and donor pbFRET can be performed using a double-labeled sample. For acceptor pbFRET, a donor-only, acceptor-only, and double-labeled sample are needed.

2.2.2 Define conditions for collecting images

Filter sets and excitation sources for collecting images of the donor and acceptor signal will vary depending on the microscope setup and choice of fluorophores (see individual references in Tables 8–1 and 8–2 for examples). The single-labeled samples should be used to define conditions for imaging and to ensure minimum bleedthrough in the "wrong" channels (for example, acceptor signal in a donor-only labeled sample) while maintaining as large a dynamic range as possible. The dual-labeled sample should then be imaged under the same conditions. In defining conditions for acceptor pbFRET experiments, it is important to allow for the increase in donor fluorescence that will occur following acceptor photobleaching.

2.2.3 Define conditions for bleaching the donor or acceptor

The conditions for photobleaching are different than those for image acquisition. The time required for bleaching will vary with the fluorophore, objective, and excitation source (Table 8–3). For confocal microscopy, bleaching can be accomplished by increasing the laser power, whereas for wide-field microscopes excitation intensities can be modulated between imaging and photobleaching conditions using neutral-density filters. The goal of acceptor pbFRET is to obtain complete bleaching (>95%) of the acceptor without bleaching the donor. Typically this is accomplished using high-intensity excitation to bleach the fluorescence as rapidly as possible. In contrast, in donor pbFRET measurements where the kinetics of donor bleaching are monitored, it can be advantageous to use lower excitation intensities to slow the photobleaching process so it is more easily quantified. It is also important to perform donor photobleaching under nonsaturating conditions so that the rate of photobleaching is proportional to the number of molecules in the excited state (Bastiaens and Jovin, 1998).

TABLE 8–3 Examples of Fluorophore Photobleaching Time Requirements[a]

Fluorophore	Excitation Source	Objective Lens	Bleach Time	Reference
Alexa 546	543 nm HeNe laser	100× 1.4 NA	1000 scans	Berney and Danuser, 2003
Cy3	Hg arc lamp	40× 1.3 NA	30 mm	Wouters et al., 1998
Cy3	543 nm HeNe laser	60× 1.2 NA	10 mm	Wallrabe et al., 2003
Cy5	Xenon arc lamp	63× 1.4 NA	7 mm	Kenworthy and Edidin, 1998
YFP	Hg/xenon arc lamp	60× 1.2 NA	5 min[b]	Day et al., 2001
YFP	Xenon arc lamp	40× 1.3 NA	30 mm	Erickson et al., 2001
YFP	514 nm argon laser	100× 1.3 NA	20 scans	Karpova et al., 2003

NA, numerical aperture.

[a]All are from acceptor pbFRET experiments except for Wouters et al. 1998.

[b]Reduced signal by less than fivefold.

For both donor and acceptor pbFRET experiments it is often convenient to bleach half the field of cells or, in the case of single cells, a defined ROI. This provides an internal control for overall photobleaching or focus shift during the measurement. This also provides an elegant way to measure donor and acceptor pbFRET with a single sample (Bastiaens et al., 1995, 1996; Bastiaens and Jovin, 1996, Wouters et al., 1998). Here, acceptor photobleaching is used to generate two regions, one in which both donor and acceptor are present and a second containing donor only. This allows for direct comparison of both donor photobleaching kinetics and donor de-quenching in adjacent areas of the same specimen.

2.2.4 Perform experimental measurements using the imaging and bleaching conditions defined above

Using imaging and bleaching conditions as described above, FRET experiments can then be performed as follows. An example of acceptor pbFRET data collected with a wide-field microscope is shown in Figure 8–1.

Acceptor pbFRET
1. Collect an image of the donor (Donor$_{pre}$).
2. Collect an image of the acceptor (Acceptor$_{pre}$).
3. Bleach the acceptor.
4. Collect an image of the acceptor to verify complete bleaching (Acceptor$_{post}$).
5. Collect an image of the donor, which will now be dequenched (Donor$_{post}$).

Donor pbFRET
1. Using the donor-only sample, collect an image of the donor.
2. Bleach the donor, periodically acquiring images to quantitate fluorescence over time.
3. Using the double-labeled sample, collect an image of the donor and an image of the acceptor.
4. Bleach the donor, periodically acquiring images to quantitate fluorescence over time.

Sequential acceptor and donor pbFRET
1. Collect an image of the donor for a complete field of cells.
2. Bleach the acceptor from an ROI (for example, half the field of cells).
3. Collect an image of the donor for the complete field of cells.
4. Bleach the donor, periodically acquiring images to quantitate fluorescence over time.

2.3 Data Analysis

2.3.1 Background correction

For both donor and acceptor pbFRET measurements, an unlabeled sample is useful as a control for calculating the contribution of background to the signal, especially when autofluorescence is present. For wide-field microscopes, flat-field corrections should also be performed.

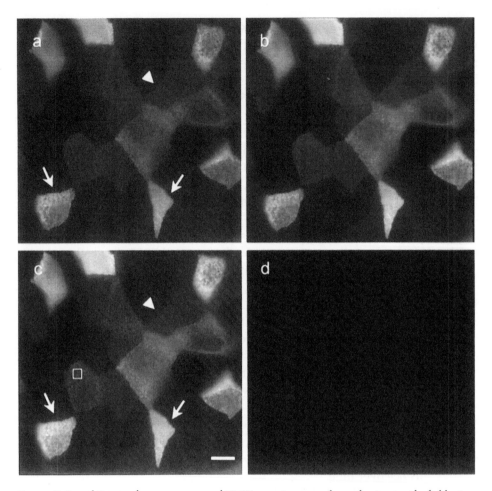

Figure 8–1 *a–d*: Images from an acceptor pbFRET experiment performed using a wide-field micro-scope illustrate an increase in donor fluorescence observed after acceptor photobleaching charac-teristic of FRET. A larger increase in donor fluorescence intensity was observed for cells expressing high levels of the labeled protein (arrows) than for cells expressing low levels (arrowheads). The box (*c*) indicates a typical region of interest sampled for subsequent data analysis. See Kenworthy and Edidin (1998) for detailed experimental protocols ([*d*] bleached acceptor image). In brief, MDCK cells were labeled with a mixture of Cy3- and Cy5-labeled IgG directed against the cell surface pro-tein 5' nucleotidase prior to fixation and visualization. Cells were imaged on a Zeiss Axiovert 135TV using a 63× 1.4 numerical aperture (NA) objective and images were collected using a charge-coupled device (CCD) camera. Fluorescence was excited using a xenon arc lamp and detected with filter sets for Cy3 and Cy5. Bleaching of >97% of Cy5 was accomplished in approximately 7 min using this setup. Bar, 10 μm. [Figure is revised from Figure 8–1 in *The Journal of Cell Biology*, 1998, volume 142, pp. 69–84, by copyright permission of The Rockefeller University Press.]

2.3.2 Image registration

Shifts in image registration may occur as a result of changing filter sets in wide-field microscopy experiments. Such shifts can lead to errors in calculation of en-ergy transfer efficiencies, especially if calculation of energy transfer efficiency images is desired. For example, if the Donor$_{pre}$ and Donor$_{post}$ images obtained in acceptor pbFRET measurements are not in register, regions of either very high or no energy transfer efficiency are measured at the edges of cells (for an example of this effect, see Kenworthy and Edidin, 1999). Registration of images can be performed with many commercially available image-processing programs.

2.3.3 Calculation of energy transfer efficiencies for acceptor photobleaching FRET

For acceptor pbFRET measurements, energy transfer efficiency is calculated according to equation (3) using background-subtracted and registered images. This calculation can be performed from donor intensities measured from individual ROI (see Fig. 8–1) using a spreadsheet or graphics program. Alternatively, it can be performed on a pixel-by-pixel basis for the entire image using an image-processing program capable of image math to generate a FRET image (Fig. 8–2). This requires the inclusion of a scaling factor to expand E over the full range of the image (Kenworthy and Edidin, 1999). Several microscope companies have included macros with their software that can be used to perform these calculations.

2.3.4 Calculation of energy transfer efficiencies for donor photobleaching FRET

Energy transfer efficiency is calculated for donor pbFRET using equation (2). This can be achieved by simply fitting exponential decays (equation (1)) to plots of intensity vs. time from selected regions of the sample (Wallrabe et al., 2003). In fitting the photobleaching data, it may be necessary to include a parameter for background fluorescence. In principle, in donor pbFRET measurements it is not necessary to follow the entire photobleaching process; assuming monoexponential decay, the time constant can be fitted from a portion of the curve. However, in practice, bleaching may follow double exponential kinetics and should be fitted accordingly (Szaba et al., 1992). Donor pbFRET data can also be calculated on a pixel-by-pixel basis for either monoexponetial or double exponetial decays (Gadella and Jovin, 1995, 1997). Such analysis yields a photobleaching time constant, amplitude, and offset for each pixel in the image.

2.4 Controls

Proteins known to interact with each other or which are in close proximity (<100Å) to one another are appropriate positive controls for FRET experiments. For example, in experiments using antibody probes, donor- and acceptor-labeled secondary antibodies can be bound to the same primary antibody, or a combination of acceptor-labeled primary antibody and donor-labeled secondary antibody can be used to label samples (Kenworthy and Edidin, 1998; Snapp et al., 2004). Constructs consisting of enhanced cyan fluorescent protein (ECFP) and enhanced yellow fluorescent protein (EYFP) separated by a linker region can be used for FRET experiments using ECFP/EYFP fusion proteins (Pentcheva and Edidin, 2001; Rizzo et al., 2004). Alternatively, interaction of the donor- and acceptor-labeled probes can be induced, for example, by cross-linking ECFP/EYFP–tagged proteins with anti-green fluorescent protein (GFP) antibodies (Glebov and Nichols, 2004), or proteins known to dimerize or interact can be used to generate ECFP and EYFP fusion proteins that

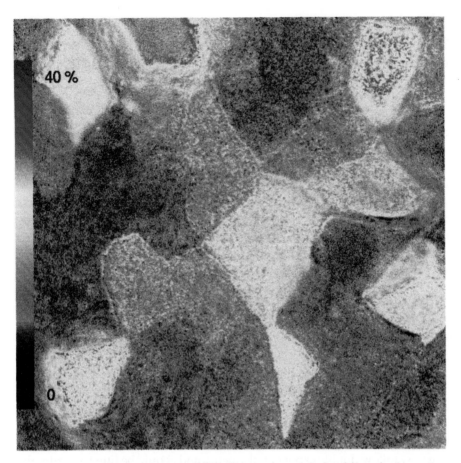

FIGURE 8–2 Example of an energy transfer efficiency image calculated from the data in Figure 8–1. The image was generated in Isee (Isee Imaging Systems) using equation (3), including a scaling factor of 10,000 to expand E over the range of a 12-bit image (0–4 096 gray levels). Energy transfer efficiencies are represented with a pseudocolor look-up table ranging from zero (purple) to 40% (red) (see scale bar, left). Note that the E varies between individual cells. See Kenworthy and Edidin (1998) for further details. [Figure is revised from Figure 2a in *The Journal of Cell Biology*, 1998, volume 142, pp. 69–84, by copyright permission of The Rockefeller University Press.]

can then be introduced into cells (Karpova et al., 2003). Ideally, the subcellular localization and expression levels of the positive control proteins should be similar to the test proteins to control for local microenvironment of the fluorophores and help evaluate the expected level of signal-to-noise ratio in the FRET measurement. Similarly, a sample containing co-localized but noninteracting/irrelevant donors and acceptors present at roughly the same expression level as the experimental pair of proteins is a useful negative control. This demonstrates that the FRET signal is specific for the proteins of interest and is not due to either dimerization of EGFP-tagged proteins or the concentration-induced proximity of proteins.

Acceptor pbFRET experiments require several specific negative controls:

1. *Donor-only–labeled sample.* This should yield E of zero and allows for determination of the noise inherent in the measurement. If E is negative,

this could indicate that bleaching of the donor occurs during the acceptor bleach or as a result of data acquisition.

2. *Acceptor-only–labeled sample.* If increase in the donor channel occurs when no donor is present, this indicates that photoconversion of the acceptor has occurred. This effect has been observed for Cy5 but is typically small compared to the magnitude of donor dequenching (Nichols, 2003; Snapp et al., 2004).

3. *Measurement of donor fluorescence on an area outside the region where the acceptor has been bleached for the experimental sample.* This provides an important internal control, as no change in donor fluorescence should occur unless the acceptor has been photobleached.

3. Advantages and Disadvantages of Photobleaching FRET

Donor and acceptor pbFRET have been widely used to address a number of biological problems, including measurements of protein–protein interactions, receptor oligomerization, and the detection of cell signaling (Tables 8–1 and 8–2). The major advantage of both acceptor pbFRET and donor pbFRET is that the energy transfer efficiency can be calculated directly and quantitatively with these techniques by comparing a property of the donor in the presence and absence of the acceptor. Their major disadvantage is that photobleaching is by nature a destructive process and potentially time consuming. The events that contribute to photobleaching are not completely understood but in general are thought to involve interactions of the fluorophore in the triplet state with oxygen species (Song et al., 1995, 1996). Depending on how the microscope is configured and the choice of donor fluorophores, it may take seconds to tens of minutes for photobleaching to occur. The time required to bleach ultimately determines the time resolution of this approach. For experiments in living cells, this should be kept in mind, as molecules may move out of the ROI or interactions may change over time. The development of photochromic acceptors that can be reversibly turned on and off (Jares-Erijman and Jovin, 2003) may ultimately make these approaches useful for repeated measurements. One example is YFP, which can be partially regenerated by ultraviolet (UV) excitation following a photobleach (Miyawaki and Tsien, 2000).

3.1 Donor Photobleaching FRET

A principal advantage of the donor pbFRET method is that photobleaching kinetics are independent of donor intensity and merely require measurement of the rate of bleaching in the presence and absence of acceptor (equation (2)). The calculation of energy transfer efficiency from donor photobleaching kinetics assumes that the only difference between the two samples is the occurrence of FRET in the double-labeled sample. However, photobleaching is a complex and poorly understood process. The rate of bleaching depends on the local environment of the fluorophores, which can differ within a given sample as well as across samples (Benson et al., 1985; Tron, 1994). It is therefore important to perform adequate control experiments on a donor-only–labeled sample to verify that this is not a significant source of heterogeneity in bleaching rates. Performing sequential acceptor and donor pbFRET measurements can help alleviate this issue by allowing for comparison of adjacent regions of the same sample (Bastiaens et al., 1995, 1996; Bastiaens and Jovin, 1996).

In principle, donor pbFRET can be performed with a single set of filters selective for donor excitation and emission. It does not even necessarily require that images of the specimen be collected, as the kinetics of donor photobleaching can be monitored with PMTs (Hannan et al., 1993). In practice, however, it is useful to visualize the acceptor fluorophores to monitor sample quality and compare regions where the donors and acceptors are co-localized and those where they undergo FRET.

3.2 Acceptor Photobleaching FRET

Acceptor pbFRET measurements can conveniently be made on a single sample and can also be readily coupled to other types of FRET measurements, including donor pbFRET (Table 8–4). Although acceptor photobleaching is essentially an end-point experiment, it can be used to retrospectively calibrate FRET obtained by another method at earlier time points with the assumption that the donor has not bleached or redistributed (Miyawaki and Tsien, 2000). Bleaching of the donor during image acquisition will result in artifactually decreased values of energy transfer efficiency. A method to correct for this effect has recently been described (Zal and Gascoigne, 2004). It is important to bleach the acceptor as fully as possible to fully quantitate energy transfer efficiency. Incomplete acceptor bleaching can give rise to a substantial underestimation of energy transfer efficiency, as recently illustrated for the case of randomly distributed molecules in membranes (Berney and Danuser, 2003).

In addition to problems with photobleaching of the donor or acceptor, other common sources of error in acceptor pbFRET measurements include cell movement or focal plane drift during the measurement and registration shifts between images. Such problems manifest as regions of either very high or negative FRET in energy transfer efficiency images. Measurements of donor fluorescence in an area

TABLE 8–4 Comparisons of Photobleaching FRET Results with Other FRET Approaches

pbFRET Method	Other FRET Method	Reference
Donor pb/acceptor pb	FLIM	Bastiaens and Jovin, 1996
Donor pb/acceptor pb	—	Bastiaens et al., 1996)
Donor pb/acceptor pb	—	Wouters et al., 1998
Acceptor pb	FLIM	Wouters and Bastiaens, 1999
Acceptor pb	Sensitized emission	Nagy et al., 1998
Acceptor pb	Sensitized emission	Erickson et al., 2001
Acceptor pb	Sensitized emission	Berney and Danuser, 2003
Donor pb/acceptor pb	Sensitized emission	Wallrabe et al., 2003
Acceptor pb	HomoFRET (anisotropy)	Sharma et al., 2004

FLIM, fluorescence lifetime imaging microscopy; pb, photobleaching.

outside the regions where the acceptor has been bleached (which ideally should remain constant) can also be used to monitor for these effects.

It is important to make many independent FRET measurements from a population of cells and to use statistical approaches to verify differences between samples or treatments. Careful analysis of control samples is also important to rule out other potential artifacts. Through such an approach, Karpova and co-workers (2003) identified pseudo-FRET in two different types of negative controls. First, in cells expressing donor (ECFP) alone, they found that E in the bleached ROIs was lower than that for control regions. This suggests that ECFP was slightly bleached by the 514 nm laser line. Second, in cells expressing both donor (ECFP) and acceptor (EYFP), they observed small positive FRET values in a region where the acceptor had not been bleached. The magnitude of this psuedo-FRET varied among experiments and its source has not yet been identified. The authors suggest that this value should be used as the baseline for determining whether FRET occurs to eliminate the risk of false positives. Other groups have reported large errors in donor-only–labeled samples using acceptor photobleaching (Sharma et al., 2004) or high scatter in donor- and acceptor-labeled specimens (Zacharias et al., 2000). Whether "pseudo-FRET" accounts for these effects as well remains to be determined.

An underlying assumption of acceptor photobleaching is that the acceptor no longer absorbs or fluoresces after photobleaching. The generation of a new fluorescent species with properties similar to the donor is a potential source of false FRET. This possibility should be checked for whenever initially setting up a new acceptor pbFRET protocol. As mentioned above, for both Cy5 and YFP, two commonly used acceptors, evidence for such photoconversion has been reported (Nichols, 2003; Glebov and Nichols, 2004; Snapp et al., 2004). The magnitude of this effect appears to be modest compared to that of a true FRET signal. Nevertheless, these observations underscore the need for thorough characterization of negative controls to ensure the validity of acceptor pbFRET measurements.

3.3 Comparisons of Acceptor and Donor Photobleaching FRET with Other Methods

In many instances, donor and acceptor pbFRET measurements have been shown to perform comparably to each other as well as to other FRET techniques (Bastiaens and Jovin, 1996; Bastiaens et al., 1996; Wouters et al., 1998; Erickson et al., 2001; Berney and Danuser, 2003; Wallrabe et al., 2003; Table 8–4). However, some notable differences in the signal-to-noise ratio across methods have been documented. For example, donor pbFRET yielded systematically higher E than measurements of sensitized emission for the same sample, an effect resulting from differences in weighing of the data (Nagy et al., 1998). Berney and Danuser (2003) estimated that even with 100% bleaching, acceptor pbFRET still provides a 10% error, which they attribute to uncorrected cross talk and the difference in the observation strategy compared to the other methods they tested. Erickson et al. (2001) also noted that the scatter in their FRET data differed depending on the method used. The signal-to-noise ratio was lowest for low acceptor fluorescence using a "three-cube" sensitized emission method, and for low donor fluorescence values via the acceptor pbFRET method.

3.4 Unusual Applications of Photobleaching in FRET Microscopy

Photobleaching can be used to provide information about the underlying organization of donors and acceptors. This is particularly valuable in situations in which FRET is expected to depend on the ratio of donor- and acceptor-labeled molecules or overall concentration of labeled molecules. One example of this approach comes from a homoFRET study of GFP-tagged major histocompatibility (MHC) class I proteins in which whole-cell photobleaching was used to test for surface density–dependent homotransfer (Rocheleau et al., 2003). Similarly, photobleaching and chemical quenching were used in a homoFRET study of glycosylphosphatidy-linositol (GPI)-anchored protein clustering to systematically decrease the overall surface density of FITC- and GFP-labeled proteins, respectively (Sharma et al., 2004). Combining this with mathematical modeling, the authors could then compare the resulting FRET as a function of fluorescence intensity with theoretical predictions for different cluster geometries, and area fraction of the membrane covered by clusters. A study of the clustering of endocytic receptors used a partial photobleaching approach to provide evidence for a clustered organization (Wallrabe et al., 2003). Here donor photobleaching was used to systematically deplete a fraction of the molecules. Since the rate of photobleaching is slower for donors engaged in FRET, partial photobleaching of a population of donors will deplete the non-FRET donors first, enriching those donors present in clusters with acceptors. This effect was evidenced by the increase in FRET signal with increased bleaching of the donors.

4. SELECTION OF FLUOREPHORES FOR PHOTOBLEACHING FRET

By definition, bleaching of either the donor or acceptor is required to perform pbFRET measurements. As such, the use of relatively photolabile fluorophores, rather than stable fluorophores, is often an advantage for these types of experiments. Acceptor pbFRET requires a photostable donor and an acceptor that can be readily bleached at a wavelength that will not lead to donor bleaching. In contrast, donor pbFRET requires an unstable donor that can be bleached without affecting the acceptor. A number of donor and acceptor pairs that fulfill these criteria have been identified and successfully used (Tables 8–1 and 8–2). As in all FRET measurements, the choice of which pair is most appropriate will depend on the nature of the experiment to be performed, the available instrumentation, and the desired Förster distance, R_0.

4.1 Conventional Fluorophores

Cy3 and Cy5 are an excellent FRET pair for acceptor pbFRET measurements, as Cy3 is relatively photostable whereas Cy5 is more easily photobleached (Bastiaens et al., 1995, 1996; Bastiaens and Jovin, 1996; Kenworthy, 2001). Suitable filters can readily separate the spectra for Cy3 and Cy5 and as a result there is little spectral bleedthrough. Further, these fluorophores can also be used for triple labeling experiments using GFP or FITC-like fluorophores. Moreover, minimal autofluorescence is typically observed over these wavelength ranges. The Cy dyes are available conjugated to secondary antibodies or as free dyes that can be used to covalently label proteins or other macromolecules of interest. Two disadvantages are that Cy5 emits in the far red and is not visible to the naked eye, and some microscopes are not equipped to image Cy5. FITC and TRITC are a good FRET pair for donor pbFRET measurements because FITC can be readily photobleached (Table 8–1).

4.2 Fluorescent Proteins

Despite the extensive spectral overlap, ECFP and EYFP, spectral variants of GFP, have been the FRET pair of choice for live-cell FRET experiments (Miyawaki and Tsien, 2000), especially for acceptor pbFRET measurements (Table 8–2 and Chapter 4). New mutants of these proteins such as cerulean and citrine may further increase the signal-to-noise ratio in FRET measurements (Hoppe et al., 2002; Glebov and Nichols, 2004; Rizzo et al., 2004). For nuclear proteins, blue fluorescent protein (BFP) can be used as a donor in combination with GFP or YFP, as the nucleus exhibits little autofluorescence (Day et al., 2001). Less frequently used but also potentially valuable for some applications is EGFP-Cy3. This combination can be used, for example, to measure FRET between a GFP-tagged protein and either a microinjected Cy3-labeled protein or a Cy3-labeled antibody (Wouters and Bastiaens, 1999). DsRed is also a potential FRET partner with ECFP or EGFP, but currently suffers from slow fluorophore maturation (Erickson et al., 2003). The presence of the immature fluorophores decreases the sensitivity of acceptor pbFRET to measurements of CFP-DsRed FRET (Erickson et al., 2003).

It is important to note that the propensity of GFP to dimerize can potentially give rise to a nonspecific FRET signal in experiments using ECFP- and EYFP-fusion proteins (Zacharias et al., 2002). GFP-induced clustering of membrane-bound proteins can be overcome by using monomeric forms of the proteins, created by a single point mutation (mCFP and mYFP) (Zacharias et al., 2002). A small but significant FRET signal in cells co-transfected with soluble ECFP and EYFP has also been observed (Erickson et al., 2003; Karpova et al., 2003). This spurious FRET could also be due to dimerization but may also arise when the molecules are present at sufficiently high concentrations that they are brought into FRET proximity by chance.

5. APPLICATION OF PHOTOBLEACHING FRET TO THE STUDY OF MEMBRANE MICRODOMAINS

In combination with fluorescence microscopy, the sensitivity of FRET to small distances (<100 Å) makes it a powerful tool for spatially and temporally resolved measurements of protein–protein interactions and cell signaling events in living cells (Tables 8–1 and 8–2). FRET microscopy has additionally proven of value in the study of membrane microdomains, especially a class of cell surface domains known as *lipid rafts*. Enriched in glycosphingolipids and cholesterol, lipid rafts are proposed to functionally concentrate and segregate raft from non-raft proteins within the plane of the membrane (Simons and Ikonen, 1997). Lipid rafts are hypothesized to play a role in a wide range of cellular and pathological functions, including regulation of membrane trafficking, cell signaling, and the entry and exit of pathogens from cells (Brown and London, 1998; Simons and Toomre, 2000; Simons and Ehehalt, 2002; Manes et al., 2003; Helms and Zurzolo, 2004). However, the structure of lipid rafts in cells is a matter of great speculation and controversy (Anderson and Jacobson, 2002; Maxfield, 2002; Edidin, 2003; Munro, 2003; Mayor and Rao, 2004). FRET offers a direct method to examine the structure of lipid rafts in cells (Jacobson and Dietrich, 1999; Kenworthy, 2002). If rafts exist as discrete structures of a finite size and lifetime, then components of rafts should be present in a clustered distribution. This contrasts with a scenario in which raft domains are not present in cell membranes. FRET can be used to distinguish between these two limiting cases, since the average distance between raft components is much closer under conditions in which they are clustered than if they were merely randomly distributed.

We used acceptor pbFRET to test these two possibilities by examining the submicron distribution of a putative raft protein, the GPI-anchored protein 5' nucleotidase (5' NT) (Kenworthy and Edidin, 1998). Our experiments showed that we could detect significant energy transfer between donor- and acceptor-labeled 5' NT molecules (Fig. 8–1). We found in particular that the magnitude of E strongly correlated with 5' NT cell surface expression (Fig. 8–2). Combining experimental observations with theoretical predictions, we showed that these data were consistent with the protein being present in a predominantly random distribution. These results suggested that 5' NT, although enriched in lipid raft domains by biochemical criteria, is not found in discrete clusters at the cell surface but is rather dispersed across the cell surface. In a subsequent acceptor pbFRET study, we showed that several additional GPI-anchored proteins were also not constitutively clustered, nor was another presumed raft marker, the glycosphingolipid-binding toxin cholera toxin B-subunit, in several different cell types (Kenworthy et al., 2000). Recent FRET measurements have likewise shown that GPI-anchored proteins are randomly distributed in T cells (Glebov and Nichols, 2004).

However, other FRET studies have supported the concept of clustering of proteins in rafts (Varma and Mayor, 1998; Kovbasnjuk et al., 2001; Zacharias et al., 2002; Nichols, 2003; Sharma et al., 2004). These discrepancies hint that the structure of lipid raft domains may be close to the limit of detection by FRET and/or vary between cell types. For example, a recent study using a combination of fluorescence polarization measurements with mathematical modeling indicates that 20%–30% of GPI-anchored proteins are clustered in cholesterol-sensitive microdomains consisting of three to four proteins (Sharma et al., 2004). However, no significant energy transfer could be detected for the same proteins using acceptor pbFRET (Sharma et al., 2004). This suggests that the sensitivity of the fluorescence polarization measurements is higher for detecting clustered proteins, perhaps because of the low probability of labeling a donor and acceptor in the same cluster. Further studies will be required to determine whether this is the case for all raft components or is a special characteristic of GPI-anchored proteins.

6. Background References

We refer the reader to the articles by Jovin and Arndt-Jovin (Jovin and Arndt-Jovin, 1989a, 1989b) for the original description of donor pbFRET method. Tron (1994) and Clegg (1996) also provide detailed discussions of the theoretical basis for these measurements. For a comprehensive review of FRET microscopy measurements and the theory behind these approaches, also see Jares-Erijman and Jovin (2003).

Several methodological studies are available that describe how to perform donor pbFRET and acceptor pbFRET measurements (Bastiaens and Jovin, 1998; Kenworthy and Edidin, 1999; Miyawaki and Tsien, 2000; Day et al., 2001; Kenworthy, 2001). A detailed description of conditions required for acceptor pbFRET measurements of ECFP/EYFP using a confocal microscope and a single laser is given in Karpova et al. (2003). Through careful analysis of control samples, this study documents several artifacts associated with acceptor pbFRET measurements. Also of technical interest is a study comparing acceptor photobleaching with a number of FRET methods using surface-bound donor- and acceptor-labeled streptavidin (Berney and Danuser, 2003). A useful discussion of the choice of acceptor vs. donor labeling of individual components of a protein complex, in addition to a detailed comparison of sensitized emission vs. acceptor pbFRET, can be found in Erickson et al. (2001).

References

Anderson, R.G.W. and K. Jacobson. A role for lipid shells in targeting proteins to caveolae, rafts and other lipid domains. *Science* 296:1821–1825, 2002.

Bacso, Z., L. Bene, A. Bodnar, J. Matko, and S. Damjanovich. A photobleaching energy transfer analysis of CD8/MHC-I and LFA-1/ICAM-1 interactions in CTL-target cell conjugates. *Immunol. Lett.* 54:151–156, 1996.

Bastiaens, P.I. and T.M. Jovin. Microspectroscopic imaging tracks the intracellular processing of a signal transduction protein: fluorescent-labeled protein kinase C beta I. *Proc. Natl. Acad. Sci. U.S.A.* 93:8407–8412, 1996.

Bastiaens, P.I.H. and T.M. Jovin. Fluorescence resonance energy transfer microscopy. In: *Cell Biology: A Laboratory Handbook,* edited by J.E. Celis, New York: Academic Press, 1998, pp. 136–146.

Bastiaens, P.I., I.V. Majoul, P.J. Verveer, H.D. Soling and T.M. Jovin. Imaging the intracellular trafficking and state of the AB5 quaternary structure of cholera toxin. *EMBO J.* 15:4246–4253, 1996.

Bastiaens, P.I.H., F.S. Wouters, and T.M. Jovin. Imaging the molecular state of proteins in cells by fluorescence resonance energy transfer (FRET). Sequential photobleaching of Forster donor-acceptor pairs. Presented at the 2nd Hamamatsu International Symposium on Biomolecular Mechanisms and Photonics: Cell–Cell Communications, 1995.

Benson, D.M., J. Bryan, A.L. Plant, A.M. Gotto, Jr., and L.C. Smith. Digital imaging fluorescence microscopy: spatial heterogeneity of photobleaching rate constants in individual cells. *J. Cell. Biol.* 100:1309–1323, 1985.

Berney, C. and G. Danuser. FRET or no FRET: a quantitative comparison. *Biophys. J.* 84:3992–4010, 2003.

Brown, D.A. and E. London. Functions of lipid rafts in biological membranes. *Annu. Rev. Cell Dev. Biol.* 14:111–136, 1998.

Clegg, R.M. Fluorescence resonance energy transfer (FRET). In: *Fluorescence Imaging Spectroscopy and Microscopy*, edited by B. Herman. New York: John Wiley, 1996, pp. 179–252.

Damjanovich, S., G. Vereb, A. Schaper, A. Jenei, J. Matko, J. P. Starink, G.Q. Fox, D.J. Arndt-Jovin, and T.M. Jovin. Structural hierarchy in the clustering of HLA class I molecules in the plasma membrane of human lymphoblastoid cells. *Proc. Natl. Acad. Sci. USA* 92:1122–1126, 1995.

Day, R.N., A. Periasamy, and F. Schaufele. Fluorescence resonance energy transfer microscopy of localized protein interactions in the living cell nucleus. *Methods* 25:4–18, 2001.

Edidin, M. The state of lipid rafts: from model membranes to cells. *Annu. Rev. Biophys. Biomol. Struct.* 32:257–283, 2003.

Erickson, M.G., B.A. Alseikhan, B.Z. Peterson, and D.T. Yue. Preassociation of calmodulin with voltage-gated Ca(2+) channels revealed by FRET in single living cells. *Neuron* 31:973–985, 2001.

Erickson, M.G., D.L. Moon, and D.T. Yue. DsRed as a potential FRET partner with CFP and GFP. *Biophys. J.* 85:599–611, 2003.

Gadella, T.W., Jr. and T.M. Jovin. Oligomerization of epidermal growth factor receptors on A431 cells studied by time-resolved fluorescence imaging microscopy. A stereochemical model for tyrosine kinase receptor activation. *J. Cell Biol.* 129:1543–1558, 1995.

Gadella, T.W.J.J. and T.M. Jovin. Fast algorithms for the analysis of single and double exponetial decay curved with a background term. Application to time-resolved imaging microscopy. *Bioimaging* 5:5–39, 1997.

Glebov, O.O. and B.J. Nichols. Lipid raft proteins have a random distribution during localized activation of the T-cell receptor. *Nat. Cell Biol.* 6:238–243, 2004.

Ha, T., T. Enderle, D.F. Ogletree, D.S. Chemla, P.R. Selvin, and S. Weiss. Probing the interaction between two single molecules: fluorescence resonance energy transfer between a single donor and a single acceptor. *Proc. Natl. Acad. Sci. U.S.A.* 93:6264–6268, 1996.

Hannan, L.A., M.P. Lisanti, E. Rodriguez-Boulan, and M. Edidin. Correctly sorted molecules of a GPI-anchored protein are clustered and immobile when they arrive at the apical surface of MDCK cells. *J. Cell Biol.* 120:353–358, 1993.

Helms, J.B. and C. Zurzolo. Lipids as targeting signals: lipid rafts and intracellular trafficking. *Traffic* 5:247–254, 2004.

Hoppe, A., K. Christensen, and J.A. Swanson. Fluorescence resonance energy transfer-based stoichiometry in living cells. *Biophys. J.* 83:3652–3664, 2002.

Jacobson, K. and C. Dietrich. Looking at lipid rafts? *Trends Cell Biol.* 9:87–91, 1999.

Jares-Erijman, E.A. and T.M. Jovin. FRET imaging. *Nat. Biotechnol.* 21:1387–1395, 2003.

Jovin, T.M. and D.J. Arndt-Jovin. FRET microscopy: digital imaging of fluorescence resonance energy transfer. Application in cell biology. In: *Cell Structure and Function by Microspectrofluorimetry*, edited by E. Kohen, J.S. Ploem, and J.G. Hirschberg, Orlando: Academic Press, 1989a, pp. 99–117.

Jovin, T.M. and D.J. Arndt-Jovin. Luminescence digital imaging microscopy. *Annu. Rev. Biophys. Biophys. Chem.* 18:271–308, 1989b.

Jurgens, L., D. Arndt-Jovin, I. Pecht, and T.M. Jovin. Proximity relationships between the type I receptor for Fcε (FcεRI) and the mast cell function-associated antigen (MAFA) studied by donor photobleaching fluorescence resonance energy transfer microscopy. *Eur. J. Immunol.* 26:84–91, 1996.

Karpova, T.S., C.T. Baumann, L. He, X. Wu, A. Grammer, P. Lipsky, G.L. Hager, and J.G. McNally. Fluorescence resonance energy transfer from cyan to yellow fluorescent protein detected by acceptor photobleaching using confocal microscopy and a single laser. *J. Microsc.* 209:56–70, 2003.

Kenworthy, A.K. Imaging protein–protein interactions using fluorescence resonance energy transfer microscopy. *Methods* 24:289–296, 2001.

Kenworthy, A.K. Peering inside lipid rafts and caveolae. *Trends Biochem. Sci.* 27:435–438, 2002.

Kenworthy, A.K. and M. Edidin. Distribution of a glycosylphosphatidylinositol-anchored protein at the apical surface of MDCK cells examined at a resolution of <100 Å using imaging fluorescence resonance energy transfer. *J. Cell Biol.* 142:69–84, 1998.

Kenworthy, A.K. and M. Edidin. Imaging fluorescence resonance energy transfer as a probe of membrane organization and molecular associations of GPI-anchored proteins. In: *Protein Lipidation Protocols,* edited by M.H. Gelb, Totowa, NJ: Humana Press, 1999, pp. 37–49.

Kenworthy, A.K., N. Petranova, and M. Edidin. High-resolution FRET microscopy of cholera toxin B-subunit and GPI-anchored proteins in cell plasma membranes. *Mol. Biol. Cell* 11:1645–1655, 2000.

Kinoshita, A., C.M. Whelan, C.J. Smith, I. Mikhailenko, G.W. Rebeck, D.K. Strickland, and B.T. Hyman. Demonstration by fluorescence resonance energy transfer of two sites of interaction between the low-density lipoprotein receptor–related protein and the amyloid precursor protein: role of the intracellular adapter protein Fe65. *J. Neurosci.* 21:8354–8361, 2001.

Kovbasnjuk, O., M. Edidin, and M. Donowitz. Role of lipid rafts in *Shiga* toxin 1 interaction with the apical surface of Caco-2 cells. *J. Cell Sci.* 114:4025–4031, 2001.

Kubitscheck, U., R. Schweitzer-Stenner, D.J. Arndt-Jovin, T.M. Jovin, and I. Pecht. Distribution of type I Fcε-receptors on the surface of mast cells probed by fluorescence resonance energy transfer. *Biophys. J.* 64:110–120, 1993.

Llopis, J., S. Westin, M. Ricote, J. Wang, C.Y. Cho, R. Kurokawa, T.M. Mullen, D.W. Rose, M.G. Rosenfeld, R.Y. Tsien, and C.K. Glass. Ligand-dependent interactions of coactivators steroid receptor coactivator-1 and peroxisome proliferator-activated receptor binding protein with nuclear hormone receptors can be imaged in live cells and are required for transcription. *Proc. Natl. Acad. Sci. U.S.A.* 97:4363–4368, 2000.

Manes, S., G. del Real, and A.C. Martinez. Pathogens: raft hijackers. *Nat. Rev. Immunol.* 3:557–368, 2003.

Maxfield, F.R. Plasma membrane microdomains. *Curr. Opin. Cell Biol.* 14:483–487, 2002.

Mayor, S. and M. Rao. Rafts: scale-dependent, active lipid organization at the cell surface. *Traffic* 5:231–240, 2004.

Miyawaki, A. and R.Y. Tsien. Monitoring protein conformations and interactions by fluorescence resonance energy transfer between mutants of green fluorescent protein. *Methods Enzymol.* 327:472–500, 2000.

Munro, S. Lipid rafts: elusive or illusive? *Cell* 115:377–388, 2003.

Nagy, P., G. Vamosi, A. Bodnar, S.J. Lockett, and J. Szollosi. Intensity-based energy transfer measurements in digital imaging microscopy. *Eur. Biophys. J.* 27:377–389, 1998.

Nichols, B.J. GM1-containing lipid rafts are depleted within clathrin-coated pits. *Curr. Biol.* 13:686–690, 2003.

Pentcheva, T. and M. Edidin. Clustering of peptide-loaded MHC class I molecules for endoplasmic reticulum export imaged by fluorescence resonance energy transfer. *J. Immunol.* 166:6625–6632, 2001.

Pentcheva, T., E.T. Spiliotis, and M. Edidin. Cutting edge: Tapasin is retained in the endoplasmic reticulum by dynamic clustering and exclusion from endoplasmic reticulum exit sites. *J. Immunol.* 168:1538–1541, 2002.

Rizzo, M.A. and D.W. Piston. Regulation of beta cell glucokinase by *S*-nitrosylation and association with nitric oxide synthase. *J. Cell Biol.* 161:243–248, 2003.

Rizzo, M.A., G.H. Springer, B. Granada, and D.W. Piston. An improved cyan fluorescent protein variant useful for FRET. *Nat. Biotechnol.* 22:445–449, 2004.

Rocheleau, J.V., M. Edidin, and D.W. Piston. Intrasequence GFP in class I MHC molecules, a rigid probe for fluorescence anisotropy measurements of the membrane environment. *Biophys. J.* 84:4078–4086, 2003.

Rocheville, M., D.C. Lange, U. Kumar, R. Sasi, R.C. Patel, and Y.C. Patel. Subtypes of the somatostatin receptor assemble as functional homo- and heterodimers. *J. Biol. Chem.* 275:7862–7869, 2000.

Sharma, P., R. Varma, R.C. Sarasij, Ira, K. Gousset, G. Krishnamoorthy, M. Rao, and S. Mayor. Nanoscale organization of multiple GPI-anchored proteins in living cell membranes. *Cell* 116:577–589, 2004.

Siegel, R.M., J.K. Frederiksen, D.A. Zacharias, F.K. Chan, M. Johnson, D. Lynch, R.Y. Tsien, and M.J. Lenardo. Fas preassociation required for apoptosis signaling and dominant inhibition by pathogenic mutations. *Science* 288:2354–2357, 2000.

Simons, K. and R. Ehehalt. Cholesterol, lipid rafts, and disease. *J. Clin. Invest.* 110:597–603, 2002.

Simons, K. and E. Ikonen. Functional rafts in cell membranes. *Nature* 387:569–572, 1997.

Simons, K. and D. Toomre. Lipid rafts and signal transduction. *Nat. Rev. Mol. Cell Biol.* 1:31–41, 2000.

Snapp, E.L., G.A. Reinhart, B.A. Bogert, J. Lippincott–Schwartz, and R.S. Hegde. The organization of engaged and quiescent translocons in the endoplasmic reticulum of mammalian cells. *J. Cell Biol.* 164:997–1007, 2004.

Song, L., E.J. Hennink, I.T. Young, and H.J. Tanke. Photobleaching kinetics of fluorescein in quantitative fluorescence microscopy. *Biophys. J.* 68:2588–2600, 1995.

Song, L., C.A. Varma, J.W. Verhoeven, and H.J. Tanke. Influence of the triplet excited state on the photobleaching kinetics of fluorescein in microscopy. *Biophys. J.* 70:2959–2568, 1996.

Spiliotis, E.T., T. Pentcheva, and M. Edidin. Probing for membrane domains in the endoplasmic reticulum: retention and degradation of unassembled MHC class I molecules. *Mol. Biol. Cell* 13:1566–1581, 2002.

Szaba, G., Jr., P.S. Pine, J.L. Weaver, M. Kasari, and A. Aszalos. Epitope mapping by photobleaching fluorescence resonance energy transfer measurements using a laser scanning microscope system. *Biophys. J.* 61:661–670, 1992.

Tron, L. Experimental methods to measure fluorescence resonance energy transfer processes. In: *Mobility and Proximity in Biological Membranes*, edited by L. Tron. Boca Raton: CRC Press, 1994, pp. 1–47.

Varma, R. and S. Mayor. GPI-anchored proteins are organized in submicron domains at the cell surface. *Nature* 394:798–801, 1998.

Vyas, K.A., H.V. Patel, A.A. Vyas, and R.L. Schnaar. Segregation of gangliosides GM1 and GD3 on cell membranes, isolated membrane rafts, and defined supported lipid monolayers. *Biol. Chem.* 382:241–250, 2001.

Wallrabe, H., M. Elangovan, A. Burchard, A. Periasamy, and M. Barroso. Confocal FRET microscopy to measure clustering of ligand–receptor complexes in endocytic membranes. *Biophys. J.* 85:559–571, 2003.

Wouters, F.S. and P.I. Bastiaens. Fluorescence lifetime imaging of receptor tyrosine kinase activity in cells. *Curr. Biol.* 9:1127–1130, 1999.

Wouters, F.S., P.I.H. Bastiaens, K.W.A. Wirtz, and T.M. Jovin. FRET microscopy demonstrates molecular assocation of non-specific lipid transfer protein (nsL-TP) with fatty acid oxidation enzymes in peroxisomes. *EMBO J.* 17:7179–7189, 1998.

Young, R.M., J.K. Arnette, D.A. Roess, and B.G. Barias. Quantitation of fluorescence energy transfer between cell surface proteins via fluorescence donor photobleachig kinetics. *Biophys. J.* 67:881–888, 1994.

Zacharias, D.A., G.S. Baird, and R.Y. Tsien. Recent advances in technology for measuring and manipulating cell signals. *Curr. Opin. Neurol.* 10:416–421, 2000.

Zacharias, D.A., J.D. Violin, A.C. Newton, and R.Y. Tsien. Partitioning of lipid-modified monomeric GFPs into membrane microdomains of live cells. *Science* 296:913–916, 2002.

Zal, T. and N.R. Gascoigne. Photobleaching-corrected FRET efficiency imaging of live cells. *Biophys. J.* 86:3923–3939, 2004.

9

Single-Molecule FRET

Sungchul Hohng and Taekjip Ha

1. Introduction

Technical advances in optical spectroscopy and microscopy during the past dozen years have made it possible to detect even a single dye molecule under biologically relevant conditions. Observing single molecules instead of ensemble populations has many advantages. It allows us to explore heterogeneity hidden in ensemble measurements. More importantly, it provides a unique opportunity to observe real-time movements of individual biomolecules while they are functioning and enables us to determine the detailed kinetics and mechanisms of complex biochemical reactions without the need for synchronization. Many excellent reviews have described the development of single-molecule fluorescence detection techniques and their applications to physics, chemistry, and biology (Moerner and Orrit, 1999; Weiss, 1999, 2000).

Among the different single-molecule fluorescence techniques, such as localization (Yildiz et al., 2003), fluorescence polarization (Ha et al., 1999a), lifetime (Jia et al., 1997), and quenching (Lu et al., 1998), single-molecule Förster resonance energy transfer (FRET) is the most popular and arguably the most general approach. In a typical FRET experiment a biological molecule is labeled with a donor fluorophore and an acceptor fluorophore at two different positions. The internal motion of the molecule brings the two fluorophores closer to or farther from each other, resulting in distance changes. The FRET efficiency E is defined as the fraction of donor excitation events that lead to the excitation of the acceptor. This quantity is a strong function of the distance R between the two fluorophores and is calculated according to

$$E = 1/(1 + (R/R_0)^6)$$

Here, R_0 is the distance at which E is 0.5 and is typically between 3 and 7 nm, an ideal length for many macromolecules. Hence, single-molecule FRET measurements can provide information on the conformational changes of biological molecules in real time that would be difficult to observe using conventional ensemble studies. The power of single-molecule FRET in biophysical applications has been well established. Starting from the proof-of-concept experiments in which FRET between a single pair of donor and acceptor molecules was reported (Ha et al., 1996), the field went through a validation process, which showed that the technique indeed works (Schütz et al., 1998; Deniz et al., 1999; Ha et al., 1999b, 1999c). We are now in the discovery period when new and interesting information continues to be uncovered via this technique. The ever-expanding list of biological systems that have been studied by single-molecule FRET include RNA (Zhuang et al., 2000, 2002; Bokinsky et al., 2003; Tan et al., 2003; Hohng et al., 2004), DNA (McKinney et al., 2003; Ying et al., 2003), proteins that interact with DNA (Ha et al., 2002), protein folding (Deniz et al., 2000; Talaga et al., 2000; Schuler et al., 2002; Rhoades et al., 2003), protein conformational changes (Börsch et al., 2002; Margittai et al., 2003), and protein–protein interactions (Weninger et al., 2003). Some of these single-molecule FRET applications have been reviewed elsewhere (Weiss, 1999; Ha, 2001b, 2004). In this chapter, we focus on the practical aspects of single-molecule FRET experiments, based mostly on practices in our laboratory (Ha, 2001a).

2. FLUORESCENT PROBE SELECTION, LABELING, AND OXYGEN SCAVENGER SYSTEM

2.1 Dye Selection

Ideal dyes for single-molecule fluorescence studies possess as many of the following characteristics as possible: they should be (1) photostable; (2) bright (high extinction coefficient and quantum yield of emission); (3) stable, showing little intensity fluctuation, at least in the timescale of events under study; (4) excitable and emit in the visible spectrum; (5) relatively small so that they introduce minimum perturbations to the host molecule; and (6) commercially available in a form that can be conjugated to biomolecules. In addition, there must be a large spectral overlap between the donor emission and acceptor absorption, which is necessary for efficient FRET (see Chapter 2). Further, ideal single-molecule FRET dyes have large Stokes shifts to minimize the amount of donor emission that bleeds through into the acceptor channel and to reduce the amount of direct excitation of acceptor by the laser. Finally, comparable emission quantum yields for the donor and acceptor are desirable. This is useful because they guarantee anticorrelated intensity changes of the donor and acceptor when FRET changes occur.

Fluorescent probes for single-molecule spectroscopy and their bioconjugation chemistries have been reviewed (Kapanidis and Weiss, 2002). A list of fluorophores often used for single-molecule studies, along with their characteristic properties, is provided in Table 9.1. Cy3 and Cy5 have been arguably the most popular donor and acceptor pair for single-molecule FRET because their spectral separation is large (~100 nm) and their quantum yields are comparable. Additionally, they are photostable in oxygen-free environments and commercially available in both amino- and thiol-reactive forms, making them easily attachable to many biological molecules. When Cy5 is used as the acceptor at a 1:1 stoichiometry with a donor, there is a fraction of donor–acceptor complexes that show only donor even when other assays indicate that the donor is very close to Cy5 ($R \ll R_0$). This is presumably due to a fraction of inactive Cy5, likely because of prebleaching, and accounts

TABLE 9–1 Characteristic Properties of Selected Fluorophores Used
for Single-Molecule Spectroscopy

Dye	Absorption (nm)	Extinction Coefficient (cm⁻¹ M⁻¹)	Emmission (nm)	Quantum Yield	Laser (nm)	Photostability
GFP	395	30,000	509	0.8	457	Good
	470	7000				
Fluo	490	67,000	520	0.5–0.9	488	Poor
Alex488	495	80,000	520	0.5–0.9	488	Moderate
TMR	554	85,000	585	0.2–0.5	514, 532	Good
Cy3	550	150,000	570	>0.15	514, 532	Good
Cy3.5	581	150,000	596	0.15	532	Good
TR	596	85,000	620	0.5	514, 532	Good
Cy5	650	250,000	670	>0.15	633	Moderate
Cy5.5	675	250,000	694	>0.28	633	Moderate

Values were taken from Kapanidis (2002) and Amersham Bioscience. Fluo, Fluorescein; TMR, tetramethyl-rhodamine;
TR, Texas red.

for approximately 15%–55% of total population depending on the batch of Cy5 used. Fortunately, for most experiments we could easily identify such complexes because a truly zero FRET was not seen when the two dyes were on the same host molecule.

2.2 Labeling

The labeling of DNA and RNA is straightforward, as many dyes are available in phosphoramidite form that can be readily incorporated during nucleic acid synthesis. For site-specific labeling of proteins, "Cys-light" proteins can be prepared that contain reactive cysteines only at a desired position(s), allowing the attachment of thiol-reactive fluorophores (Rasnik et al., 2004). Because introduction of a second modification (chemically orthogonal to cysteine as far as dye labeling is concerned) to the protein is very difficult, a more practical approach for attaching both a donor and an acceptor to the same protein is to have two reactive cysteine residues, which can be labeled with an equal mixture of donor and acceptor molecules. Through single-molecule measurements, one should be able to easily distinguish a donor–acceptor–labeled complex from complexes labeled with two donors or two acceptors using their spectroscopic signatures. Another promising approach that has been used for single-molecule protein folding studies is peptide ligation. In this method, one of the two peptides to be ligated is labeled prior to the ligation, followed by labeling of the other residue (Deniz et al., 2000).

2.3 Oxygen Scavenger System

Photobleaching limits the time during which single-molecule FRET measurements can be made. Photobleaching is the result of an excited-state (probably from triplet state) reaction of dye with highly reactive species in solution. Singlet oxygen is the prime culprit and, indeed, removal of oxygen molecules from solution with an enzymatic oxygen scavenging system (for example, 0.4% [w/v] glucose, 0.1 mg/ml glucose oxidase [Sigma], and 0.02 mg/ml catalase [Roche]) can lengthen the photobleaching lifetime of several dyes. We have observed at least a 30-fold increase in Cy5 photostability and a smaller but significant effect for Cy3 and tetramethylrhodamine. Therefore, more than 10^5 photons can be detected from single molecules of Cy3 or Cy5 before bleaching. Since 100 photons are enough to give a signal-to-noise ratio of 10:1, which is adequate for most applications, 10^3 data points can be obtained before photobleaching. This is often enough for observing multiple conformational changes. But the effect of oxygen on photostability is not universal, and for other dyes removing oxygen has only a small effect on the photostability (Texas red and Alexa 488) or an adverse effect (fluorescence intensity is significantly reduced for fluorescein). We have also found that sealing the sample chamber from the ambient air is important, and that the addition of mercaptoethanol (1% v/v), which reduces reactive species, can increase the amount of time before bleaching of Cy5.

3. EXPERIMENT

3.1 Experimental Setup

There are two classes of fluorescence microscopy tools for single-molecule fluorescence studies. The first involves point detection using a single-element detector that is most commonly in the confocal microscope configuration (Nie et al., 1994; Macklin et al., 1996). The point detection method is good for observing fast dynamics up to 1 ms. The time resolution of this method is usually limited by detector shot noise and background luminescence, not by instrument response. Its long data acquisition time, however, makes it impractical for studies of very rare events or irreversible processes.

The second class of tools for single-molecule studies consists of wide-field microscopy with a two-dimensional detector such as a charge-coupled device (CCD) camera. Hundreds of single molecules can be detected simultaneously, performing hundreds of single-molecule experiments in parallel, which is especially useful for irreversible reactions or for very rare biological events (Zhuang et al., 2000; Ha et al., 2002). Another advantage of the wide-field technique is that two- or three-dimensional diffusion of molecules can be directly visualized (Schmidt et al., 1995; Dickson et al., 1996). However, because an arrayed detector has to be used, the time resolution and sensitivity are not as good as in the point detection method. Clearly, confocal and wide-field microscopy techniques are mutually complementary.

In confocal microscopy (Fig. 9–1), the laser excitation is focused on a diffraction-limited spot using a high numerical aperture (NA) objective. Usually, a few microwatts of excitation power are enough for single-molecule fluorescence detection. Fluorescence emission is collected using the same objective. A pinhole is used to block the out-of-focus background light and achieve single-molecule sensitivity. Silicon avalanche photodiodes (APD) have become the de facto standard detector for a number of reasons: a high detection quantum yield (>70%) for visible light, low dark

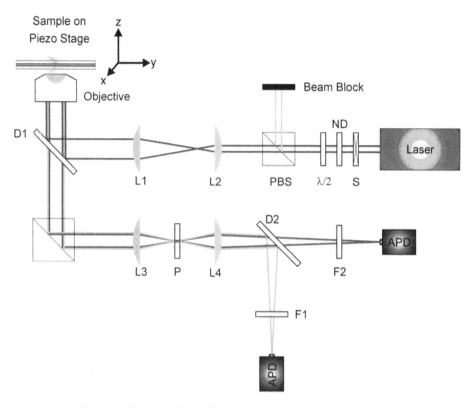

FIGURE 9–1 Schematic diagram of a confocal setup for single-molecule observation of FRET between Cy3 and Cy5. The setup is based on an inverted microscope (Olympus, IX50). A solid-state laser (532 nm, Crystalaser) is used for excitation through an oil-immersion objective (PlanApo 100×/ 1.4 Oil, Olympus). Collected luminescence is split by a dichroic mirror (D2:635dclp, Chroma) and filtered out by bandpass filters (F1: HQ580/60m, F2:670DF40, Chroma). For imaging, the sample is scanned using a piezo stage (P-527–3CL, PI). The combination of a polarizing beam splitter (PBS) and a half-wave plate is used for beam intensity control (λ/2). Identification of other components are D1, dichroic (545drlp, Chroma); L1 and L2, lens set for beam expansion; P, pinhole (300 μm), L3, tube lens; L4, achromat ($D = 30$ mm, $f = 140$ mm, Melles Griot); APD, Avalanche Photo Diode (SPCM-AQR-14, Perkin Elmer); ND, neutral density filter; S, shutter (LS6–T2, Uniblitz). The APD signal is collected using a Counter/Timer board (PCI-6602, National Instruments).

count rate (<50 per second), photon counting ability, and a signal-to-noise ratio limited by the statistical photon number noise (and can exceed 20:1). The APD has a small active area (~200 μm) that acts as an additional pinhole to reject the out-of-focus background. The fluorescence emission is typically split by a dichroic mirror, allowing two APDs to record the donor and acceptor signals simultaneously. The fluorescence signal can be further divided on the basis of polarization, providing information on the polarization anisotropy as well as FRET (Rothwell et al., 2003). The use of additional bandpass filters greatly reduces background luminescence from cover glass and is critical for detecting single-molecule fluorescence in a confocal setup. Pulsed excitations can be combined with time-correlated photon counting to yield fluorescence lifetime information (Jia et al., 1997; Rothwell et al., 2003).

Wide-field microscopy can be done via either epi-illumination (Sase et al., 1995) or evanescent field excitation (Funatsu et al., 1995). An excitation laser whose power is larger than 20 mW is usually desirable for single-molecule detection with wide-field illumination. Evanescent field excitation does not permit excitation light to propagate toward the detector and hence can reduce the autofluorescence to an undetectable level. This evanescent field excitation is generated by total internal reflection (TIR) of the excitation light at the glass–water interface. A large incident angle is required to achieve TIR. For instance, an incident angle (relative to the surface normal) larger than 61° is needed for glass–water interface. This angle can be achieved through the edge of a high NA objective (NA ≥ 1.4) (Tokunaga et al., 1997; (Fig. 9–2) or by using a prism (Fig. 9–3). To obtain donor and acceptor images or images of different polarizations simultaneously using one CCD camera, a dual-view scheme can be used in which two images are projected onto two halves of the camera (Suzuki et al., 1995; Zhuang et al., 2000). A sample image obtained using the dual-view scheme is shown in Figure 9–4. The left image is the donor channel and the right corresponds to the acceptor channel. To obtain intensity time

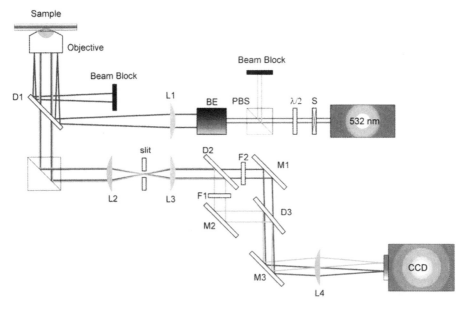

FIGURE 9–2 Objective-type total internal reflection (TIR) microscope. For wide-area excitation the laser beam is expanded by means of a beam expander (BE05, Thorlabs) and focused at the edge of the back focal plane of the oil-immersion objective (PlanApo, 60×/1.45 Oil TIRFM ∞/ 0.17). The slit adjusts the imaging area on a charge-coupled device (CCD) camera (I-PentaMax, Roper Scientific or DV-887–BI, Andor). The CCD pixel size should be considered when the magnification of L3 and L4 are selected. For example, for L3; $f = 100$ mm, $D = 50$ mm; and for L4, $f = 260$ mm, $D = 50$ mm are used for both I-PentaMAX (Roper Scientific) and DV-887-BI (Andor). Selection of dichroics and bandpass filters depends on dyes used for the FRET pair. For the Cy3–Cy5 pair, we use D1, dichroic (545drlp, chroma); D2, dichroic (645dclp or 645dcxr, Chroma); F1, bandpass filter for Cy3 (HQ580/60m, Chroma); and F2, bandpass filter for Cy5 (HQ670/40m, Chroma). M, dielectric mirror (BB1–E02, Thorlabs); PBS, polarizing beam splitter; L1, focusing lens; L2, tube lens of the microscope.

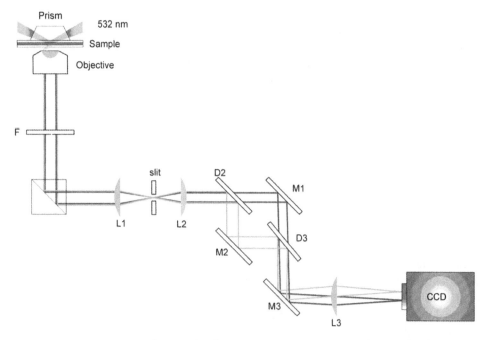

FIGURE 9–3 Prism-type total internal reflection (TIR) microscope. Total internal reflection occurs on the upper surface of the channel. The detection optics are the same as in Figure 9–2 except for the use of a water-immersion objective (UplanApo, 60×/1.20W, ∞/0.13–0.21) to obtain a long working distance. Prism, Pellin broca prism (PLBC-5.0–79.5–SS, CVI Laser); F, long pass filter (E550LP, Chroma).

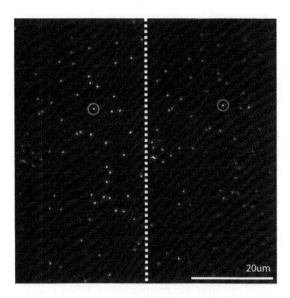

FIGURE 9–4 Example CCD image of beads in wide-field microscopy. The left channel is the donor image and the right corresponds to the acceptor image. To get intensity traces from recorded frames, a mapping process between donor and acceptor spots is required.

traces mapping between corresponding donor (right circle) and acceptor (left circle) spots is required. Even a four-view scheme has been used successfully in which the fluorescence emission is divided twice on the basis of color and then polarization (Cognet et al., 2000). For CCD cameras, the readout noise becomes significant for high frame rates (>1 frames per second). This problem can be overcome by amplifying the signal by using either an intensifier coupled to the CCD or a newer CCD camera that employs an on-chip gain mechanism. Through use of these methods it is possible to achieve frame rates exceeding 30 Hz. However, the intrinsic noise coming from the amplification process can become dominant, limiting the signal-to-noise ratio to 5:1. It is convenient to use highly fluorescent beads to align the optics and, for wide-field microscopy, the mapping between the two channels.

Optimum FRET resolution requires minimal cross talk between the two detection channels—i.e., donor emission leakage into the acceptor detector and acceptor image leakage into the donor detector. The former is usually more significant because the fluorescence emission spectrum is asymmetric, tailing off into the longer wavelengths. In a typical experiment using Cy3 as the donor and Cy5 as the acceptor, approximately 10%–15% of the donor signal is detected at the acceptor channel when a dichroic mirror with a cutoff at 635–645 nm is used. Although the overall signal may be decreased, it is often advantageous to use bandpass filters to reduce the cross talk.

3.2 Surface Immobilization

To study the dynamic changes of individual molecules tagged with a FRET pair over extended time periods (>1 ms), molecules need to be localized in space. This is often achieved by surface immobilization. An ideal surface would allow specific immobilization of nucleic acids or proteins while rejecting nonspecific adsorption. For nucleic acid studies, a glass coated with biotinylated bovine serum albumin (BSA) and streptavidin has been widely used for its simplicity (Fig. 9–5; Ha et al., 1999c). Biotinylated DNA and RNA molecules can be immobilized with high specificity (> 500:1) and their bulk solution activities faithfully reproduced. An procedure for surface immobilization in RNA experiments is explained in the caption of Figure 9-5.

The BSA surface can be too adhesive for most studies involving proteins, and PEG (polyethylene glycol)-coated surfaces can reduce the protein adsorption to an undetectable level (Ha et al., 2002). A small fraction of PEG polymers end-modified by biotin is added to facilitate the immobilization of biotinylated macromolecules. PEG surfaces have been essential for helicase studies and other systems, including ribosomes.

Biomolecules can also be immobilized nonspecifically. For instance, DNA can be attached to a charged surface such as aminopropylsilane-coated surface via electrostatic interactions (Ha et al., 1996). Although this method avoids DNA aggregation and works in water, it is unlikely that the properties of DNA and its interaction with other molecules can be studied reliably under these conditions. Dyes on DNA immobilized in this way often display polarized emission, meaning that they transiently stick to the surface. This complicates the interpretation of FRET

FIGURE 9–5 Surface immobilization of biomolecules using streptavidin (yellow) and biotin (black rectangle) interaction. As an example, an RNA four-way junction (gray double helices) is attached to the surface of a narrow channel made between a cleaned microscope slide and a coverslip by successive addition of 40 μl of 1 mg/ml biotinylated bovine serum albumin (BSA; Sigma), 40 μl of 0.2 mg/ml streptavidin (Molecular Probes), and 40 μl of 10–50 pM biotinylated RNA in 10 mM Tris·HCl (pH 8.0), 50 mM NaCl (TN buffer). Each addition was incubated for 10 min and followed by washing with TN buffer. The concentration of the RNA solution needed to be adjusted to give a good surface density for the single-molecule experiments. After checking that fluorescent spots were well separated from one another, the imaging buffer containing the oxygen scavenger (6% (w/v) glucose, 0.1 mg/ml glucose oxidase, and 0.02 mg/ml catalase), 1% (v/v) 2–mercaptoethanol, and other chemicals is injected for single-molecule experiments.

signal changes. Another nonspecific immobilization method used successfully for single-molecule fluorescence studies is the trapping of molecules inside the pores formed in polyacrylamide gels (Dickson et al., 1996) or agarose gel (Lu et al., 1998). While gel immobilization has the merit of not requiring any special modification of the biomolecule, it has some disadvantages. Concentrations of other small molecules, such as enzyme substrates and ions, are difficult to change in a short time. In addition, it is not easy to study macromolecular interactions in gel because of limited molecular diffusion. More recently, a new method was developed in which a vesicle encapsulating a molecule of interest is tethered to a surface (Rhoades et al., 2003). This allows the effective localization of the molecule in a volume smaller than the resolution of the optical microscope, while minimizing the surface interactions of the molecules.

Immobilization can be bypassed altogether by measuring FRET from molecules diffusing through the focused laser light (Deniz et al., 1999). This approach is completely free from surface artifacts, but dynamics that take longer than molecular diffusion cannot be measured. In addition, only a limited number of photons, typically less than 50, can be detected from each molecule. Thus the statistical photon number noise dominates and FRET resolution becomes worse than that measured in immobilization studies.

3.3 Controllable Parameters

Experiments are usually performed in aqueous solution while molecules are immobilized. Therefore, ions or other chemicals can be easily added or exchanged during sample preparation or the experiment. The use of a flow system is necessary to see the effect of chemicals on molecular dynamics in real time. The temperature of a sample can be controlled from 4°C to ~50°C by cooling down or heating up the microscope objective and sample chamber by means of either a thermoelectric cooling or water-cooling system. Single-molecule experiments are also possible under high pressure. Using a pressure cell consisting of a fused silica microcapillary tube, florescence correlation spectroscopy under ~3 Kbar was achieved (Müller and Gratton, 2003). The dynamics of RNA four–way junctions have been observed at 1 Kbar using single-molecule FRET (Hohng, unpublished observation).

4. DATA ACQUISITION AND ANALYSIS

Single-molecule FRET experiments yield two time traces, a donor and an acceptor trace, for each molecule. Although the glass surface is cleaned before sample preparation to reduce extraneous fluorescence spots, not all fluorescent spots are from the molecules under investigation. Some arise from impurities whereas others are the result of mislabeled molecules. In most cases, however, traces representing the desired molecule can be selected with relative ease by noting anticorrelated behavior between the donor and acceptor signal due to acceptor bleaching or conformational changes. Such selected traces usually show fluctuations between different FRET states (Fig. 9–6a)). Mechanisms of biochemical processes can be deduced after identifying conformational states for each FRET level. This is often assisted by available information obtained in conventional biochemical or structural studies. Kinetic information can be obtained by performing a dwell-time analysis as shown in Figures 9–6b and 9–6c). When the dynamics are too fast for dwell-time analysis to be done reliably, auto- or cross-correlation analysis can be used, which gives the sum of the forward and backward rates for two-state fluctuations.

5. SINGLE-MOLECULE FRET: THE ROAD AHEAD

5.1 Combined Force and Fluorescence Measurements

As has been discussed thus far, single-molecule fluorescence measurements can be used to piece together the motion of biomolecules step by step. Yet biomolecules in a cell perform their functions under a variety of environmental conditions, such as electrochemical potential for ion channels and force and load for molecular motors. Hence, to understand their behavior we must emulate these environmental variables. Techniques such as electrical patch clamping or optical trapping apply these forces and are now well established. Yet, by themselves, they do not have the selectivity to isolate those parts of the biomolecules that are undergoing the shape changes respon-

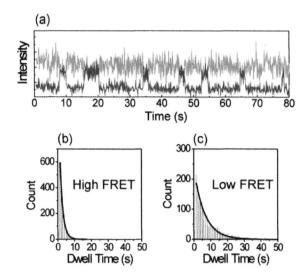

FIGURE 9–6 Example of single-molecule FRET data and analysis. *a:* Intensity time traces of typical single-molecule FRET experiments. Anticorrelation between donor signal (green) and acceptor signal (red) is an important criterion for discriminating real molecules from junk. High and low FRET values correspond to different conformations. The transition rates of each state are obtained from exponential fitting of dwell-time histograms of a high-FRET state (*b*) or a low-FRET state (*c*).

sible for the molecule's behavior. The combining of fluorescence with mechanical manipulation makes this connection. Such a combination, while ambitious, is feasible: Yanagida and colleagues performed heroic experiments in which force and fluorescence measurements were combined in single molecules (Ishijima et al., 1998; Harada et al., 1999). However, the fluorescence measurements were limited to the binding and dissociation of fluorescent ATP analogues (Ishijima et al., 1998) onto fluorescently labeled enzymes (Harada et al., 1999). More recently, a combination of FRET measurements and single-channel recording was demonstrated for gramicidin (Borisenko et al., 2003; Harms et al., 2003). In these works, FRET was used to report on the assembly of gramicidin monomers that form the active dimeric species. In another significant step, the Block laboratory combined optical tweezers with single-molecule fluorescence measurements (Lang et al., 2003). Using this approach, they measured the mechanical unzipping of duplex DNA via force and fluorescence dequenching, marking the first example of force-induced "structural change" of single molecules measured via fluorescence. Clearly the same approach should work for single-molecule FRET, and it will be interesting to test how the folding properties of many other biomolecules are influenced by applied force. In contrast to mechanical measurements, which do not allow detection of conformational changes at low forces because of the flexibility of the long DNA tethers used, FRET measurements can be performed at arbitrarily low forces.

5.2 Better Probes: Quantum Dots?

Although several popular FRET pairs, such as Cy3 and Cy5, are acceptable for single-molecule experiments, further improvements in dye properties are desirable to extend the FRET technique to resolve faster and smaller movements for longer times. In addition, the difficulty of standard FRET pairs for imaging single molecules within cells is an important limitation of the FRET technique. The main difficulty in intracellular single-molecule imaging lies in the rapid photobleaching of conventional probes (organic dyes and fluorescent proteins) and high cellular autofluorescence. Colloidal semiconductor quantum dots are promising solutions to these problems because they are brighter (10–20 times) and far more photostable than organic dyes. However, severe intermittence in emission (also known as *blinking*) has been universally observed from single dots (Nirmal et al., 1996) and has been considered to be an intrinsic limitation. For instance, in a recent application of single–quantum dot imaging, the tracking of membrane receptors was frequently interrupted because of the stroboscopic nature of recording (Dahan et al., 2003). A surprising discovery was recently reported (Hohng and Ha, 2004) in which quantum dot blinking was nearly completely suppressed. This may allow continuous observation of molecular movements as long as necessary, both in vitro and in vivo. It remains to be seen whether quantum dots can be made small enough for observing internal motion of single molecules via FRET.

5.3 Multicolor FRET

Thus far, single-molecule FRET has been measured only between two fluorophores. For complex molecular dynamics or multicomponent binding interactions, it is beneficial to have more than one FRET pair to observe correlated changes. Despite the obvious advantages of three-color FRET over the regular two-color FRET technique, its realization has been hindered because of contradictory requirements: a clear spectral separation of three fluorophores' signals and an appreciable amount of FRET between them. For clear signal separation, the spectral overlap should be small, but this leads to weaker FRET. Realization of multicolor single-molecule FRET is expected to bring deeper understanding of complex biochemical processes.

We thank Chirlmin Joo and Michelle Nahas for their comments on the manuscript.

REFERENCES

Bokinsky, G., D. Rueda, V.K. Misra, M.M. Rhodes, A. Gordus, H.P. Babcock, N.G. Walter, and X. Zhuang. Single-molecule transition-state analysis of RNA folding. *Proc. Natl. Acad. Sci. U.S.A.* 100:9302–9307, 2003.

Borisenko, V., T. Lougheed, J. Hesse, E. Fureder-Kitzmuller, N. Fertig, J.C. Behrends, G.A. Woolley, and G.J. Schutz. Simultaneous optical and electrical recording of single gramicidin channels. *Biophys. J.* 84:612–622, 2003.

Börsch, M., M. Diez, B. Zimmermann, R. Reuter, and P. Gräber. Stepwise rotation of the γ-subunit of EF_0F_1–ATP synthase observed by intramolecular single-molecule fluorescence resonance energy transfer. *FEBS Lett.* 527:147–152, 2002.

Cognet, L., G.S. Harms, G.A. Blab, P.H.M. Lommerse, and T. Schmidt. Simultaneous dual-color and dual-polarization imaging of single molecules. *Appl. Phys. Lett.* 77:4052–4054, 2000.

Dahan, M., S. Levi, C. Luccardini, P. Rostaing, B. Riveau, and A. Triller. Diffusion dynamics of glycine receptors revealed by single-quantum dot tracking. *Science* 302:442–445, 2003.

Deniz, A.A., M. Dahan, J.R. Grunwell, T. Ha, A.E. Faulhaber, D.S. Chemla, S. Weiss, and P.G. Schultz. Single-pair fluorescence resonance energy transfer on freely diffusing molecules: observation of Förster distance dependence and subpopulations. *Proc. Natl. Acad. Sci. U.S.A.* 96:3670–3675, 1999.

Deniz, A.A., T.A. Laurence, G.S. Beligere, M. Dahan, A.B. Martin, D.S. Chemla, P.E. Dawson, P.G. Schultz, and S. Weiss. Single-molecule protein folding: diffusion fluorescence resonance energy transfer studies of the denaturation of chymotrypsin inhibitor 2. *Proc. Natl. Acad. Sci. U.S.A.* 97:5179–5184, 2000.

Dickson, R.M., D.J. Norris, Y.L. Tzeng, and W.E. Moerner. Three-dimensional imaging of single molecules solvated in pores of poly(acrylamide) gels. *Science* 274:966–969, 1996.

Funatsu, T., Y. Harada, M. Tokunaga, K. Saito, and T. Yanagida. Imaging of single fluorescent molecules and individual ATP turnovers by single myosin molecules in aqueous solution. *Nature* 374:555–559, 1995.

Ha, T. Single-molecule fluorescence resonance energy transfer.*Methods* 25:78–86, 2001a.

Ha, T. Single-molecule fluorescence methods for the study of nucleic acids. *Curr. Opin. Struct. Biol.* 11:287–292, 2001b.

Ha, T. Structural dynamics and processing of nuleic acids revealed by single-molecule spectroscopy. *Biochemestry* 43:4055–4063, 2004.

Ha, T., T. Enderle, D.F. Ogletree, D.S. Chemla, P.R. Selvin, and S. Weiss. Probing the interaction between two single molecules: fluorescence resonance energy transfer between a single donor and a single acceptor. *Proc. Natl. Acad. Sci. U.S.A.* 93:6264–6268, 1996.

Ha, T., T.A. Laurence, D.S. Chemla, and S. Weiss. Polarization spectroscopy of single fluorescence molecules. *J. Phys. Chem. B* 103:6839–6850, 1999a.

Ha, T., I. Rasnik, W. Cheng, H. Babcock, G.H. Gauss, T.M. Lohman, and S. Chu. Initiation and re-initiation of DNA unwinding by the *Escherichia coli* Rep helicase. *Nature* 419:638–641, 2002.

Ha, T., A.Y. Ting, J. Liang, W.B. Caldwell, A.A. Deniz, D.S. Chemla, P.G. Schultz, and S. Weiss. Single-molecule fluorescence spectroscopy of enzyme conformational dynamics and cleavage mechanism. *Proc. Natl. Acad. Sci. U.S.A.* 96:893–898, 1999b.

Ha, T., X. Zhuang, H.D. Kim, J.W. Orr, J.R. Williamson, and S. Chu. Ligand-induced conformational changes observed in single RNA molecules. *Proc. Natl. Acad. Sci. U.S.A.* 96:9077–9082, 1999c.

Harada, Y., T. Funatsu, K. Murakami, Y. Nonoyama, A. Ishihama, and T. Yanagida. Single-molecule imaging of RNA polymerase–DNA interactions in real time. *Biophys. J.* 76:709–715, 1999.

Harms, G.S., G. Orr, M. Montal, B.D. Thrall, S.D. Colson, and H.P. Lu. Probing conformational changes of gramicidin ion channels by single-molecule patch-clamp fluorescence microscopy. *Biophys. J.* 85:1826–1838, 2003.

Hohng, S. and T. Ha. Near-complete suppression of quantum dot blinking in ambient conditions. *J. Am. Chem. Soc.* 126, 1324–1325, 2004.

Hohng, S., T.J. Wilson, E. Tan, R.M. Clegg, D.M.J. Lilley, and T. Ha. Conformational flexibility of four-way junctions in RNA. *J. Mol. Biol.* 336:69–79, 2004.

Ishijima, A., H. Kojima, T. Funatsu, M. Tokunaga, H. Higuchi, H. Tanaka, and T. Yanagida. Simultaneous observation of individual ATPase and mechanical events by a single myosin molecule during interaction with actin. *Cell* 92:161–171, 1998.

Jia, Y., A. Sytnik, L. Li, S. Vladimirov, B.S. Cooperman, and R.M. Hochstrasser. Nonexponential kinetics of a single tRNA[Phe] molecule under physiological conditions. *Proc. Natl. Acad. Sci. U.S.A.* 94:7932–7936, 1997.

Kapanidis, A.N. and S. Weiss. Fluorescent probes and bioconjugation chemistries for single-molecule fluorescence analysis of biomolecules. *J. Chem. Phys.* 117:10953–10964, 2002.

Lang, M.J., P.M. Fordyce, and S.M. Block. Combined optical trapping and single-molecule fluorescence. *J. Biol.* 2:6, 2003.

Lu, H.P., L. Xun, and X.S. Xie. Single-molecule enzymatic dynamics. *Science* 282:1877–1882, 1998.

Macklin, J.J., J.K. Trautman, T.D. Harris, and L.E. Brus. Imaging and time-resolved spectroscopy of single molecules at an interface. *Science* 272:255–258, 1996.

Margittai, M., J. Widengren, E. Schweinberger, G.F. Schröder, S. Felekyan, E. Haustein, M. König, D. Fasshauer, H. Grubmüller, R. Jahn, and C.A.M. Seidel. Single-molecule fluorescence resonance energy transfer reveals a dynamic equilibrium between closed and open conformations of syntaxin 1. *Proc. Natl. Acad. Sci. U.S.A.* 100:15516–15521, 2003.

McKinney, S.A., A.-C. Declais, D.M.J. Lilley, and T. Ha. Structural dynamics of individual Holliday junctions. *Nat. Struct. Biol.* 10:93–97, 2003.

Moerner, W.E.and M. Orrit. Illuminating single molecules in condensed matter. *Science* 283:1670–1676, 1999.

Müller, J.D. and E. Gratton. High-pressure fluorescence correlation spectroscopy. *Biophys. J.* 85:2711–2719, 2003.

Nie, S.M., D.T. Chiu, and R.N. Zare. Probing individual molecules with confocal fluorescence microscopy. *Science* 266:1018–1021, 1994.

Nirmal, M., B.O. Dabbousi, M.G. Bawendi, J.J. Macklin, J.K. Trautman, T.D. Harris, and L.E. Brus. Fluorescence intermittency in single cadmium selenide nanocrystals. *Nature* 383:802–804, 1996.

Rasnik, I., S. Myong, W. Cheng, T.M. Lohman, and T. Ha. DNA-binding orientation and domain conformation of the *E. coli* Rep helicas monomer bound to a partial duplex juntion: Single-molecule studies of fluorescently labeled enzymes. *J. Mol. Biol.* 336:395–408, 2004.

Rhoades, E., E. Gussakovsky, and G. Haran. Watching protein fold one molecule at a time. *Proc. Natl. Acad. Sci. U.S.A.* 100:3197–3202, 2003.

Rothwell, P.J., S. Berger, O. Kensch, S. Felekyan, M. Antonik, B.M. Wöhrl, T. Restle, and R.S. Goody. Multiparameter single-molecule fluorescence spectroscopy reveals heterogeneity of HIV-1 reverse transcriptase: primer/template complexes. *Proc. Nat. Acad. Sci. U.S.A.* 100:1655–1660, 2003.

Sase, I., H. Miyata, J.E. Corrie, J.S. Craik, and K. Kinosita. Real time imaging of single fluorophores on moving actin with an epifluorescence microsocpe. *Biophys. J.* 69:323–328, 1995.

Schmidt, T., G.J. Schutz, W. Baumgartner, H.J. Gruber, H. Schindler. Characterization of photophysics and mobility of single molecules in a fluid lipid membrane. *J. Phys. Chem.* 99:17662–17668, 1995.

Schuler, B., E.A. Lipman, and W.A. Eaton. Probing the free-energy surface for protein folding with single-molecule fluorescence spectroscopy. *Nature* 419:743–747, 2002.

Schütz, G.J., W. Trabesinger, and T. Schmidt. Direct Observation of ligand colocalization on individual receptor molecules. *Biophys. J.* 74:2223–2226, 1998.

Suzuki, K., Y. Tanaka, Y. Nakajima, K. Hirano, H. Itoh, H. Miyata, T. Hayakawa, and K. Kinosita. Spatiotemporal relationships among early events of fertilization in sea urchin eggs revealed by multiview microscopy. *Biophys. J.* 68:739–748, 1995.

Talaga, D.S., W.L. Lau, H. Roder, J. Tang, Y. Jia, W.F. DeGrado, and R.M. Hochstrasser. Dynamics and folding of single two-stranded coiled-coil peptides studied by fluorescent energy transfer confocal microscopy. *Proc. Natl. Acad. Sci. U.S.A.* 97:13021–13026, 2000.

Tan, E., T.J. Wilson, M.K. Nahas, R.M. Clegg, D.M.J. Lilley, and T. Ha. A four-way junction accelerates hairpin ribozyme folding via a discrete intermediate. *Proc. Natl. Acad. Sci. U.S.A.* 100:9308–9313, 2003.

Tokunaga, M., K. Kitamura, K. Saito, A.H. Iwane, and T. Yanagida. Single molecule imaging of fluorophores and enzymatic reactions achieved by objective-type total internal reflection fluorescence microscopy. *Biochem. Biophs. Res. Co.* 235:47–53, 1997.

Weiss, S. Fluorescence spectroscopy of single biomolecules. *Science* 283:1676–1683, 1999.

Weiss, S. Measuring conformational dynamics of biomolecules by single molecule fluorescence spectroscopy. *Nat. Struct. Biol.* 7:724–728, 2000.

Weninger, K., M.E. Bowen, S. Chu, and A.T. Brunger. Single-molecule studies of SNARE complex assembly reveal parallel and antiparallel configurations. *Proc. Natl. Acad. Sci. U.S.A.* 100:14800–14805, 2003.

Yildiz, A., J.N. Forkey, S.A. McKinney, T. Ha, Y.E. Goldman, and P.R. Selvin. Myosin V walks hand-over-hand: single fluorophore imaging with 1.5-nm localization. *Science* 300:2061–2065, 2003.

Ying, L., J.J. Green, H. Li, D. Klenerman, and S. Balasubramanian. Studies on the structure and dynamics of the human telomeric G quadruplex by single-molecule fluorescence resonance energy transfer. *Proc. Natl. Acad. Sci. U.S.A.* 100:14629–14634, 2003.

Zhuang, X., L.E. Bartley, H.P. Babcock, R. Russell, T. Ha, D. Herschlag, and S. Chu. A single-molecule study of RNA catalysis and folding. *Science* 288:2048–2051, 2000.

Zhuang, X., H. Kim, M.J.B. Pereira, H.P. Babcock, N.G. Walter, and S. Chu. Correlating structural dynamics and function in single ribozyme molecules. *Science* 296:1473-1476, 2002.

10

FRET Measurements Using Multispectral Imaging

Raad Nashmi, Scott E. Fraser, Henry A. Lester, and Mary E. Dickinson

1. Introduction

Over the last several years there has been increased interest in multicolor fluorescence experiments. The development of additional dyes, more fluorescent protein variants and novel fluorescent labels such as quantum dots gives us far more opportunities than ever before for fluorescence imaging in live cells and animals. This has heightened the need for more advanced methods to discriminate between different dyes and fluorescent markers simultaneously. In response to this need, multispectral imaging has become an important tool for many biologists.

A common problem in multicolor fluorescence imaging of cells is signal cross talk created by the inability to fully separate overlapping emission spectra. In many cases, cross talk can be avoided through fluorochrome choice, but in Förster resonance energy transfer (FRET) experiments spectral overlap between the donor and acceptor fluorochrome is often inevitable, as has been discussed in other chapters. Quantifying the separate signal contribution from the donor and acceptor represents one of the most significant difficulties for determining FRET efficiencies. Other chapters in this volume have discussed this issue and most methods rely on the ratio of signal contribution to different imaging channels that use specific bandpass filters. Using these methods, one can often calculate the likelihood of cross talk and compensate for signal cross talk in FRET calculations (see Chapters 4–9). Multispectral imaging goes a step further by using the entire spectral response from cells. By acquiring data from multiple wavelengths, the relative contribution of individual molecular components can be determined through mathematical analysis.

2. Basics of Spectral Imaging

Multispectral imaging, sometimes referred to as hyperspectral imaging, has more recently become an important tool for confocal microscopy. This approach has long been used in other fields, however, such as remote sensing (Chang, 2003), in which detailed images of landscape and structural variations of the earth's surface are produced by analyzing different spectral signals produced by scattered and reflected sunlight. Variations in spectral information are determined by measuring intensity differences at different wavelengths from the same field of view, much like using a spectrometer to measure fluorescence from a cuvette at different wavelengths. Image data sets are analyzed for individual spectral frequencies and pixels with matching spectral information, often showing similar structures, are classified to produce an image. Because different terrains interact with light differently, spectral variations can be used to resolve details in the field of view that cannot be seen in single images taken when all the light from the visible spectrum is used or with single filters.

In fluorescence imaging, similar approaches have been used to analyze multiple dyes within a sample. Multispectral imaging was first used for wide-field fluorescence microscopy (Eng et al., 1989; Wiegmann et al., 1993; Morris et al., 1994; Schrock et al., 1996; Zangaro et al., 1996; Mooney et al., 1997; Richmond et al., 1997; Wachman et al., 1997; Garini et al., 1999; Ornberg et al., 1999; Zeng et al., 1999; Levenson and Hoyt, 2000; Tsurui et al., 2000; Farkas, 2001; Ford et al., 2001; Yang et al., 2003), but has recently been implemented as a tool for confocal and multiphoton laser scanning microscopy (Lansford et al., 2000; Dickinson et al., 2002, 2003; Haraguchi et al., 2002; Hiraoka et al., 2002; Bertera et al., 2003). Similar to remote sensing, a three-dimensional stack is collected (x,y,λ), and information from the third spectral dimension is used to classify the identity of pixels in the x,y dimension. Biological samples often have the advantage of the user knowing precisely which dyes have been added to the preparation so the individual spectral components are known or can be easily characterized through control samples. With known reference spectra from the dyes used in the experiment, it is then possible to determine the source of the signal in each pixel, even with a high degree of spectral and spatial overlap. The mixed spectral information in each pixel can be unmixed using linear unmixing algorithms, and images representing the contributions of single fluorochromes can be created (Fig. 10–1).

3. SYSTEM REQUIREMENTS FOR MULTISPECTRAL IMAGING

As mentioned above, images are collected at a series of wavelength bands (x,y,λ) in multispectral imaging. Images acquired in a wavelength series, similar to a time series or a z-series, are called a *lambda stack*, *image cube*, or *spectral cube*. The spectrum of any field of view can be revealed by graphing how the average intensity changes with respect to wavelength. Special hardware is required to acquire lambda stacks. Multiple approaches have emerged for this purpose (Dickinson, 2005) but we have primarily used the LSM 510 META detector to analyze emission signal. The META detector consists of diffraction grating paired with a 32-element multichannel photomultiplier tube (PMT) to separate emission signal into 10.7 nm bands, each collected in a separate PMT channel. By means of the META detector, multiple bands of data can be acquired in parallel instead of in series, greatly reducing the amount of time needed to collect a lambda stack. It should also be noted that variations in excitation spectra can also be used for this purpose with tunable

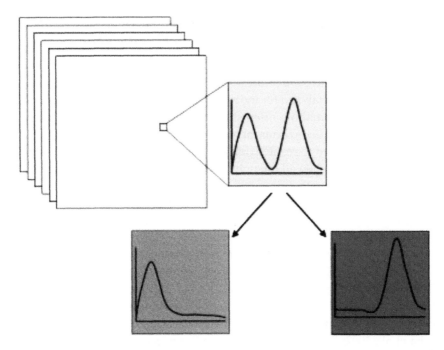

Figure 10–1 Schematic representation showing the results of linear unmixing. A pixel with mixed spectral information is resolved into two images representing the contribution of each component.

excitation sources, such as those used for multiphoton excitation (Dickinson et al., 2003), or even with separate individual laser lines (Zimmermann et al., 2003).

By collecting lambda stacks, it is possible to measure the spectral response in cells. The spectrum of any field of view can be revealed by graphing how the average intensity changes with respect to wavelength. Many pixels in the image will contain signal from a mixed contribution of different fluorochromes. To resolve how different fluorochromes contribute to the total signal in each image pixel, linear unmixing algorithms are used. The fluorescence intensity in a pixel in which fluorochromes are co-localized and have overlapping emission spectra will be the sum of the intensity of each label (Lakowicz, 1999). Thus, the pixel is said to be linearly mixed, where the sum spectrum, $S(\lambda)$, equals the proportion or weight, A, of each individual spectrum $R(\lambda)$:

$$S(\lambda) = A_1R_1(\lambda) + A_2R_2(\lambda) + A_3R_3(\lambda) \ldots \ldots \text{etc.}$$

Individual component spectra are recorded by generating lambda stacks from control cells that have a single fluorochrome signal. Sum spectra can be predicted for possible combinations of individual component spectra. Therefore, to determine the spectral content of each pixel in an image, the shape of the sum spectrum from each pixel is matched to a known sum spectrum in a library by means of a least-squares fit, much like matching a fingerprint in a database. This method is often referred to as *emission* or *excitation fingerprinting*, depending on whether emission or excitation spectra are collected. Matrix algorithms are used to very quickly identify the best fit and thus determine the signal contribution for each dye in a pixel. Once the weight of each spectral component is determined, the lambda stack can be separated into individual images for each fluorescent label. The intensity of each pixel in the unmixed images is the total collected intensity of the pixel multiplied by the proportion of spectral signals in the pixel. Thus linear unmixing is used to convert the lambda stack into individual images representing the signal for each fluorochrome. These images are then used for further quantitative analysis, such as calculations of FRET efficiencies.

4. Multispectral FRET Imaging

Multispectral imaging has been used in several FRET experiments to determine donor and acceptor concentrations (Haraguchi et al., 2002; Zimmermann et al., 2002; Amiri et al., 2003; Nashmi et al., 2003; Ecker et al., 2004). While the ability to unambiguously detect donor and acceptor emission signals without cross talk is an advantage, multispectral imaging does not distinguish between acceptor emission signal generated through resonance transfer and emission signal originating from direct excitation of the acceptor. Therefore, multispectral imaging is generally used in conjunction with the acceptor photobleaching method or is used to study dynamic changes in FRET efficiencies, such as those seen with FRET-based calcium indicators (Hiraoka et al., 2002).

In our experiments, we have used the acceptor photobleaching (donor dequenching) method to determine FRET efficiencies between cyan fluorescent protein (CFP) and yellow fluorescent protein (YFP) fusion proteins (see Chapters 4, 5, and 8). The approach we have used to measure FRET between subunits of $\alpha 4\beta 2$ nicotinic acetylcholine receptor (nAChR) is described below (Nashmi et al., 2003).

4.1 Image Collection Parameters

For all images, 12-bit, 512 × 512 pixel images were taken with a Zeiss 63× IR Achroplan 0.9 numerical aperture (NA) water immersion objective lens.

Photobleaching was performed with the argon laser line of 514 nm set at 50% transmission, using a pixel dwell time of 12.8 μs and a total scan time of 7.86 s (scan speed 5 on the Zeiss LSM 510 META). The amount of power needed for photobleaching will vary depending on the system. To prevent damage to live cells, 2 mM ascorbic acid was added to cultures as a free-radical scavenger. Without this additive, damage to cells and processes was sometimes observed.

We took great care to locate the cell of interest quickly, without excessive scans, to minimize photobleaching before the experiment was performed. A single fast (maximum speed) scan was used to center the cell in the field of view, at low 514 nm laser power. Next, a quick lambda scan was taken to adjust the image settings so that the pixel intensities would not be saturated.

FRET was recorded by examining the dequenching of CFP during incremental photobleaching of YFP by the 514 nm argon laser line. The full field of view was bleached with the 514 nm line at the same power level for each scan. A single lambda scan at 458 nm excitation was acquired between each bleaching event to record the spectral response. Lambda stacks consisting of data from 13 channels of the META detector were collected over a spectrum of wavelengths between 462.9 and 602 nm with bandwidths of 10.7 nm. Spectral emission was collected by means of a 3× zoom with a pinhole setting of 1.32 airy units. The detector gain and amplifier gain adjustments were varied with the signal strength. The offset was adjusted so that the average lowest black level was just above zero. Images were scanned with a pixel dwell time of 12.8 µs and 458 nm excitation. Laser excitation was kept to the minimum amount necessary to provide a good signal-to-noise level and limit cell damage. Table 10–1 is an example of typical values for YFP signal that were obtained using this approach.

Note that this paradigm serves as a guideline; it is not critical to repeat exactly this protocol for lambda stack scans or the number of photobleaching scans. This approach is used to minimize the error from individual measurements by fitting the changes in intensity to a standard curve. Values for the maximum increase in CFP intensity can be extrapolated from this curve, even though YFP signals are never bleached fully. Thus, damage to cells and inadvertent bleaching of CFP can be avoided.

4.2 Multispectral FRET Data Analysis

There are a series of steps to calculate FRET efficiencies: (1) linear unmixing of CFP and YFP components, (2) measurement of mean CFP and YFP intensities, and (3) calculation of CFP and YFP changes to determine FRET efficiencies.

Following the acquisition of lambda stack of images, linear unmixing is used to generate CFP and YFP images from the lambda stack. Figure 10–2 shows an example from a typical experiment. Lambda stacks were analyzed using the linear unmixing algorithm in the Zeiss LSM 510 META software. The success of linear unmixing depends on acquiring images with a good signal-to-noise ratio and using reference spectra that faithfully represent the spectra in the sample to ensure the best fit. Controls must be imaged under the same conditions as those for the mixed sample, pixel saturation should be avoided, and lambda stack acquisition must be

TABLE 10–1. Yellow Fluorescence Protein
(Acceptor) Photobleaching—A Typical Example

Photobleaching Scans Signal	Cumulative PB Time (min)	% YFP
0	0	100
1	0.13	78
1	0.26	62
2	0.52	49
4	1.05	42
4	1.57	35
4	2.09	30
36	6.81	25

PB, photobleaching.

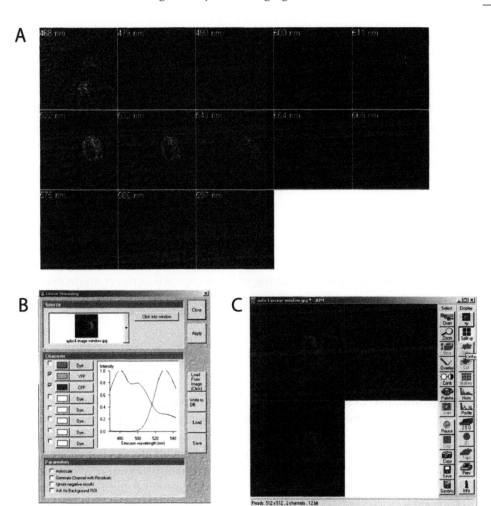

FIGURE 10–2 Results from a typical FRET experiment using multispectral imaging. *A:* A lambda stack shows images generated from different channels of the META detector. *B:* Curves of the reference spectra for cyan fluorescent protein (CFP) and yellow fluorescent protein (YFP) are in the linear unmixing dialogue box. *C:* Results of linear unmixing produces two separate images of the CFP and YFP contributions to the image.

reliable. The best reference spectra are those acquired in exactly the same way the lambda stack is acquired (same objective, offset, amplifier gain, wavelength range, dichroics, etc). It is often helpful to determine the image parameters for the lambda stack and then acquire reference spectra in exactly the same way. Images were collected with minimal signal saturation and with the average minimum pixel intensity set to just above zero. The range indicator look-up table provided in the LSM 510 software can be very helpful in setting the offset and gain values properly.

Calculations of mean CFP and YFP intensities from regions of interest (ROI) were performed with ImageJ software. An ImageJ macro was written to automate some steps of the analysis (Appendix 10–1). CFP and YFP images at all time points were imported in ImageJ as an image sequence. An ROI adjacent to the fluorescent cell was selected to determine the background signal. An ROI around the fluorescent cell was then selected at each time point by tracing the edges of the cell. Then a macro was run that subtracted the mean intensity of the background ROI from the whole field of view for both CFP and YFP images. Thus, a series of background-subtracted images were created for both CFP and YFP at all time points. Then mean intensities of ROIs of the fluorescent cell were calculated for both CFP and YFP for all time points. The intensities at different time points were normalized to that at the beginning, which was taken as 100%.

Measurements of FRET efficiencies were calculated on the basis of the equation:

$$E = 1 - (I_{DA}/I_D)$$

where I_{DA} and I_D represent normalized intensities of the donor before photobleaching and after 100% photobleaching of the acceptor, respectively, and E is FRET efficiency. Because 100% photobleaching of the acceptor was never attainable experimentally, the theoretical value of I_D was extrapolated from a linear fit of a plot of percent increase in CFP vs. percent decrease in YFP (Fig. 10–3).

5. POSSIBLE SOURCES OF ERROR

For any quantitative live-cell imaging experiment, there can often be many possible sources of error. One source of error that we encountered was focus drift. This can be minimized if the system is allowed to warm up for at least an hour before imaging. Environmental control chambers used to maintain physiological temperatures on the microscope stage can also help to eliminate drift. Overlay of images at the beginning and end of each experiment were confirmed to rule out movement and drift artifacts. Negative control constructs can help to determine if this is a significant issue. Both large negative or positive FRET efficiencies calculated from negative control samples can be a sign of movement or drift artifacts.

Multispectral imaging relies on the ability to determine signal fluctuations that occur at different defined wavelength bands. The number and spectral bandwidth of different bands needed for robust separation of overlapping fluorochromes has

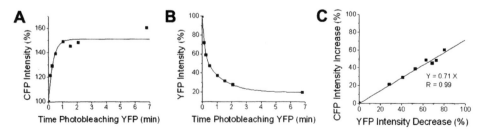

FIGURE 10–3 Graphical representation of intensity changes from a typical acceptor photobleaching experiment. *A:* Increases in cyan fluorescent protein (CFP) signal following each round of acceptor photobleaching. *B:* Decreases in yellow fluorescent protein (YFP) signal following each round of acceptor photobleaching. *C:* Linear relationship between the increase in CFP and decrease in YFP.

been empirically determined in several cases. The 10.7 nm bandwidth offered by individual META channels is more than sufficient for dyes with as little as 4–7 nm of separation (Dickinson et al., 2002) and most FRET pairs offer even less overlap. As few as two channels have been used for linear unmixing of FRET pairs (Zimmermann et al., 2002), but a greater number of spectral channels can lead to more robust results (Ecker et al., 2004).

Detector noise and background can significantly affect linear unmixing results. Background signal from laser lines, mounting media, or autofluorescence can be considered component spectra and included in the unmixing, which can often reduce the error associated with these signals.

Low signal can also be a difficulty and can adversely affect the results produced from nearly every quantitative method discussed in this book. This is also true for linear unmixing and can be a complication of dispersing the emission signal over a broad spectral range. Measurements of cells with a range of signal intensities but consistent donor and acceptor proportions can be used to determine the effect of the signal-to-noise level on FRET efficiency calculations.

6. Biological Applications

We have used the methods described here to examine α4β2 nAChR subunit assembly in different subcellular regions and to examine changes in nicotinic subunit assembly with chronic nicotine application. We constructed α4-YFP- and β2-CFP-labeled subunits. Through FRET, we have determined that surface nicotinic receptors and receptors in dendrites display a significantly greater degree of assembly than intracellular nicotinic receptors in the soma. This suggests that receptors are first assembled in the soma before trafficking to the surface or to the dendrites. Furthermore, we measured increased FRET between α4-YFP and β2-CFP nAChR subunits in neurons exposed to nicotine, compared with controls, and the results indicated increased nicotinic subunit assembly with nicotine treatment (Nashmi et al., 2003).

For these experiments, we relied on a number of controls to establish FRET constructs and ensure the most robust results. Several CFP and YFP fusion proteins were constructed and tested for proper localization in the cell, wild-type expression levels, and normal activity. In our study of α4-YFP and β2-CFP nAChRs, we found that there was little alteration in function when placing the fluorescent proteins in the M3–M4 intracellular loop. When inserting fluorescent proteins into the cytoplasmic region of the receptor, we tried to avoid regulatory sequences, including phosphorylation motifs, trafficking motifs, and regions important for ion permeation. We assessed function by means of whole-cell electrophysiology and Fura-2 calcium imaging. Fusion proteins with abnormally low nicotinic responses were rejected. Immunocytochemistry was used to study the localization and expression level of endogenous receptor proteins and transfected cells were chosen on the basis of whether constructs mimicked normal expression. Overexpressing cells often contained brightly fluorescent subcellular compartments, where presumably an abnormal excess of protein had accumulated. We avoided imaging these cells. Only cells with moderate or low levels of expression, but with fluorescence significantly above background, were used for FRET analysis. It is also important to ensure that the acceptor/donor ratios are nearly equal or that the acceptor is in excess to the donor so that each donor has a FRET acceptor.

On the basis of structural data about the protein of interest, one can strategically insert fluorescent labels at sites where robust FRET efficiency can be predicted since

$$E = R^6 / (R^6 + R_0{}^6)$$

where R_0 is the Förster's distance. One would ideally like a distance between fluorophores that would lie in the middle of the declining slope portion of this relationship so that any change in protein conformation could be detected as FRET changes. On the basis of the Unwin cryoelectromagnetic structure of the torpedo nicotinic receptor (Miyazawa et al., 1999), the cytoplasmic loops of adjacent subunits, which are the sites of insertion of YFP and CFP for α4 and β2 subunits, are separated by approximately 30–40 Å, making it optimal for measuring FRET because the YFP and CFP have a Förster's distance of 49.2 Å.

Zacharias et al. (2002) reported that fluorescent proteins can form dimers. In addition, overexpression or inclusion into subcellular domains can often force aggregation. Therefore, controls are necessary to show that FRET between fluorescent proteins is not an artifact resulting from nonspecific association. For a negative control, we used proteins of similar structure that are known not to co-assemble but do target similar compartments. The control experiment we performed was to express cells with α4-YFP nAChR and GluClα-CFP subunits (Nashmi et al., 2003). Both subunits are members of the nicotinic receptor superfamily with similar structures. However, GluClα is a glutamate-activated chloride channel subunit and does not co-assemble with nicotinic receptors. Our control experiment showed no FRET between the two subunits.

7. Conclusion

Here we have outlined the methods and controls we have used to look at relative changes in FRET efficiencies. Although there are many ways to evaluate FRET samples, as evidenced by the extensive chapters in this book, multispectral imaging combined with linear unmixing offers a robust way to analyze samples with significant spectral cross talk, such as that in FRET experiments. Analysis of FRET efficiencies from image data can be a challenging but rewarding way to determine relative changes in protein–protein interactions.

Appendix 10–1 Image J Macro "Automate FRET Analysis"

```
// written on
// April 10, 2003
```

```
// program to analyze FRET data
// 1) subtracts background from raw linear unmixed im-
ages
// 2) saves the subtracted images into directory
// 3) measures mean pixel intensity of all images using
ROIs
// aybc2_1yfp.tif an example of a yfp image name, 2=cell
number,
//      and 1=time point#; aybc2_1cfp.tif is the cfp image
name
// 2-1.roi = name given to ROI of cell 2, time point#1
// 2back.roi =name given to background ROI for back-
ground subtraction

run ("Clear Results");

cellnum=getNumber("Cell Number", 1);
cellroi=getString("Region of Interest", "roi")

// subtract background from cfp and yfp
for (x=1; x<9; x++) {
  run("Clear Results");
  // Process CFP Raw image
  run("Open ...", "path='D:\\temp\\raw images and
ROI\\aybc" + cellnum + "_" + x + "cfp.tif'");
  run("Open ...", "path='D:\\temp\\raw images and ROI\\"
+ cellnum + "back.roi'");
  run("Measure");
  run("Measure");
  mean=getResult("Mean",1);
  run("Select All");
  run("Subtract ... ", "value=" + mean + "");
  run("Tiff    ... ",   "path='D:\\temp\\subtracted
images\\aybc" + cellnum + "_" + x + "cfps.tif'");
  run("Close");

  run("Clear Results");
  // Process YFP Raw image
  run("Open ... ", "path='D:\\temp\\raw images and
ROI\\aybc" + cellnum + "_" + x +  "yfp.tif'");
```

```
    run("Open . . . ", "path='D:\\temp\\raw images and ROI\\"
+ cellnum  + "back.roi'");
    run("Measure");
    run("Measure");
    mean´getResult("Mean",1);
    run("Select All");
    run("Subtract . . . ", "value=" + mean + "");
    run("Tiff   . . .   ",   "pathx'D:\\temp\\subtracted
images\\aybc" + cellnum + "_" + x + "yfps.tif'");
    run("Close");
}

// loop to make measurements on roi of linear unmixed files
run ("Clear Results");

for (x=1; x<9; x++) {
    run("Open   . . .   ",   "path='D:\\temp\\subtracted
images\\aybc" + cellnum + "_" + x + "cfps.tif'");
    run("Open . . . ", "path='D:\\temp\\raw images and ROI\\"
+ cellroi + "-" + x + ".roi'");
    run("Measure");
    run("Close");

    run("Open   . . .   ",   "path='D:\\temp\\subtracted
images\\aybc" + cellnum + "_" + x + "yfps.tif'");
    run("Open . . . ", "path='D:\\temp\\raw images and ROI\\"
+ cellroi + "-" + x + ".roi'");
    run("Measure");
    run("Close");
}
}
```

References

Amiri, H., G. Schultz, and M. Schaefer. FRET-based analysis of TRPC subunit stoichiometry. *Cell Calcium* 33:463–470, 2003.

Bertera, S., X. Geng, Z. Tawadrous, R. Bottino, A.N. Balamurugan, W.A. Rudert, P. Drain, S.C. Watkins, and M. Trucco. Body window-enabled in vivo multicolor imaging of transplanted mouse islets expressing an insulin-timer fusion protein. *Biotechniques* 35:718–722, 2003.

Chang, C.-I. *Hyperspectral Imaging: Techniques for Spectral Detection and Classification.* New York: Kluwer Academic/Plenum Publishers, 2003.

Dickinson, M.E. Multiphoton, multispectral laser scanning microscopy. In: *Live Cell Imaging: A Laboratory Manual*, edited by D.L. Goldman, Cold Spring Harbor, NY: Cold Spring Harbor Press, 2005.

Dickinson, M.E., E. Simbuerger, B. Zimmermann, C.W. Waters, and S.E. Fraser. Multiphoton excitation spectra in biological samples. *J. Biomed. Opt.* 8:329–338, 2003.

Dickinson, M.E., C.W. Waters, R. Wolleschensky, G. Bearman, S. Tille, and S.E. Fraser. Sensitive imaging of spectrally overlapping fluorochromes using the LSM 510 META. In: *Multiphoton Microscopy in the Biomedical Sciences*, vol. 4620, edited by A. Periasamy and P.T. So, Washington, DC: SPIE Publications, 2002, pp. 123–136.

Ecker, R.C., R. de Martin, G.E. Steiner, and J.A. Schmid. Application of spectral imaging microscopy in cytomics and fluorescence resonance energy transfer (FRET) analysis. *Cytometry* 59A:172–81, 2004.

Eng, J., R.M. Lynch, and R.S. Balaban. Nicotinamide adenine dinucleotide fluorescence spectroscopy and imaging of isolated cardiac myocytes. *Biophys. J.* 55:621–630, 1989.

Farkas, D.L. Spectral microscopy for quantitative cell and tissue imaging. In: *Methods in Cellular Imaging*, edited by A. Periasamy, New York: Oxford University Press, 2001, pp. 345–361.

Ford, B.K., C.E. Volin, S.M. Murphy, R.M. Lynch, and M.R. Descour. Computed tomography-based spectral imaging for fluorescence microscopy. *Biophys. J.* 80:986–993, 2001.

Garini, Y., A. Gil, I. Bar-Am, D. Cabib, and N. Katzir. Signal-to-noise analysis of multiple color fluorescence imaging microscopy. *Cytometry* 35:214–226, 1999.

Haraguchi, T., T. Shimi, T. Koujin, N. Hashiguchi, and Y. Hiraoka. Spectral imaging fluorescence microscopy. *Genes Cells* 7:881–887, 2002.

Hiraoka, Y., T. Shimi, and T. Haraguchi. Multispectral imaging fluorescence microscopy for living cells. *Cell Struct. Funct.* 27:367–374, 2002.

Lakowicz, J.R. *Principles of Fluorescence Spectroscopy.* 2nd edition. New York: Kluwer Academic/Plenum, 1999.

Lansford, R., G. Bearman, and S.E. Fraser. Resolution of multiple green fluorescent protein color variants and dyes using two-photon microscopy and imaging spectroscopy. *J. Biomed. Opt.* 6:311–318, 2000.

Levenson, R.M. and C.C. Hoyt. Spectral imaging and microscopy. *Am. Lab.* 32:26–34, 2000.

Miyazawa, A., Y. Fujiyoshi, M. Stowell, and N. Unwin. Nicotinic acetylcholine receptor at 4.6 Å resolution: transverse tunnels in the channel wall. *J. Mol. Biol.* 288:765–786, 1999.

Mooney, J.M., V.E. Vickers, M. An, and A.K. Brodzik. High-throughput hyperspectral infrared camera. *J. Opt. Soc. Am. A Opt. Image Sci. Vis.* 14:2951–2961, 1997.

Morris, H.R., C.C. Hoyt, and P.J. Treado. Imaging spectrometers for fluorescence and Raman microscopy—acoustooptic and liquid-crystal tunable filters. *Appl. Spectrosc.* 48:857–866, 1994.

Nashmi, R., M.E. Dickinson, S. McKinney, M. Jareb, C. Labarca, S.E. Fraser, and H.A. Lester. Assembly of $\alpha 4 \beta 2$ nicotinic acetylcholine receptors assessed with functional fluorescently labeled subunits: effects of localization, trafficking, and nicotine-induced upregulation in clonal mammalian cells and in cultured midbrain neurons. *J. Neurosci.* 23:11554–11567, 2003.

Ornberg, R.L., B.M. Woerner, and D.A. Edwards. Analysis of stained objects in histological sections by spectral imaging and differential absorption. *J. Histochem. Cytochem.* 47:1307–1314, 1999.

Richmond, K.N., S. Burnite, and R.M. Lynch. Oxygen sensitivity of mitochondrial metabolic state in isolated skeletal and cardiac myocytes. *Am. J. Physiol.* 273:C1613–1622, 1997.

Schrock, E., S. du Manoir, T. Veldman, B. Schoell, J. Wienberg, M.A. Ferguson-Smith, Y. Ning, D.H. Ledbetter, I. Bar-Am, and D. Soenksen. Multicolor spectral karyotyping of human chromosomes. *Science* 273:494–497, 1996.

Tsurui, H., H. Nishimura, S. Hattori, S. Hirose, K. Okumura, and T. Shirai. Seven-color fluorescence imaging of tissue samples based on Fourier spectroscopy and singular value decomposition. *J. Histochem. Cytochem.* 48:653–662, 2000.

Wachman, E.S., W. Niu, and D.L. Farkas. AOTF microscope for imaging with increased speed and spectral versatility. *Biophys. J.* 73:1215–1222, 1997.

Wiegmann, T.B., L.W. Welling, D.M. Beatty, D.E. Howard, S.S. Vamos, and S.J. Morris. Simultaneous imaging of intracellular [Ca^{2+}] and pH in single MDCK and glomerular epithelial cells. *Am. J. Physiol.* 265:C1184–1190, 1993.

Yang, V.X., P.J. Muller, P. Herman, and B.C. Wilson. A multispectral fluorescence imaging system: design and initial clinical tests in intraoperative Photofrin-photodynamic therapy of brain tumors. *Lasers Surg. Med.* 32:224–232, 2003.

Zacharias, D.A., J.D. Violin, A.C. Newton, and R.Y. Tsien. Partitioning of lipid-modified monomeric GFPs into membrane microdomains of live cells. *Science* 296:913–916, 2002.

Zangaro, R.A., L. Silveira, R. Manoharan, G. Zonios, I. Itzkan, R.R. Dasari, J. VanDam, and M.S. Feld. Rapid multiexcitation fluorescence spectroscopy system for in vivo tissue diagnosis. *Appl. Opt.* 35:5211–5219, 1996.

Zeng, H.S., A. Weiss, C. MacAulay, and R.W. Cline. System for fast measurements of in vivo fluorescence spectra of the gastrointestinal tract at multiple excitation wavelengths. *Appl. Opt.* 38:7157–7158, 1999.

Zimmermann, T., J. Rietdorf, A. Girod, V. Georget, and R. Pepperkok. Spectral imaging and linear un-mixing enables improved FRET efficiency with a novel GFP2–YFP FRET pair. *FEBS Lett.* 531:245–249, 2002.

Zimmermann, T., J. Rietdorf, and R. Pepperkok. Spectral imaging and its applications in live cell microscopy. *FEBS Lett.* 546:87–92, 2003.

11

Real-Time Fluorescence Lifetime Imaging and FRET Using Fast-Gated Image Intensifiers

GLEN REDFORD AND ROBERT M. CLEGG

1. INTRODUCTION

This chapter discusses practical features and guidelines for Fluorescence lifetime imaging (FLI) and Förster resonance energy transfer (FRET) that maybe helpful in reading the literature and in interpreting FLI and FRET measurements of complex biological systems. Several reviews are already available (Clegg and Schneider, 1996; Clegg et al., 1996, 2003; Periasamy et al., 1996; So et al., 1996, 1998; Schneider and Clegg, 1997; Gadella et al., 2001; Cubeddu et al., 2002; Hink et al., 2002; Peter and Ameer-Beg, 2004). However, we feel that the general reader and aspiring user of FLI with applications to FRET would benefit from a concise, coherent presentation of fundamental aspects of these measurements and interpretations of the data. We will not include particular results from a biological system, nor specifics of new instrumentation. Instead, we focus on basic descriptions of the photophysical measurements and the general characteristics of the instrumentation and analyses. Regarding the instrumentation, we concentrate mostly on what is needed to achieve real-time frequency-domain measurements of lifetime imaging that are useful for FRET measurements of biological and in vivo samples. Our aim is to introduce readers unfamiliar with the particulars of FLI or FRET to some important quantitative aspects of real-time FLI measurements that are significant for FRET. We present quantitative details in sufficient depth to be useful to those interested in quantitative applications and in pursuing these techniques in more detail. We discuss some details that are critical when making FRET measurements of complex biological samples because one expects a distribution of energy transfer efficiencies (which will produce a distribution of lifetimes), as well as a mixture of doubly and singly labeled molecules and widely varying, usually unknown, concentrations. We hope that after reading this chapter the uninitiated reader will be able to go directly to the latest literature and gain an appreciation of the combination of FRET and FLI.

2. LIFETIMES, FRET, AND FLUORESCENCE LIFETIME-RESOLVED IMAGING

2.1 Fluorescence Lifetimes and FRET

The average fluorescence lifetime ($\langle\tau\rangle$) of a single molecule is the average time the molecule remains in an excited electronic state before returning to the ground state (Hercules, 1966; Parker, 1968; Stepanov and Gribkovskii, 1968; Becker, 1969; Birks, 1970; Harris and Bertolucci, 1978; Craig and Thirunamachandran, 1984; Turro, 1991; Chen and Kotlarchyk, 1997; Lakowicz, 1999; Valeur, 2002). The fluorescence emission is a convenient signal that provides a measure of this excited-state dwell time. A single molecule decays from the excited state as an exponential process

$$f_{sm}(t) = ae^{-t/\tau_{sm}}$$

where τ_{sm} is the decay time for the single molecule and a is the pre-exponential factor dependent on the instrumentation and absorption coefficient. In the case of a single molecule, the value τ_{sm} is the average dwell time in the excited state, averaged over many excitation processes. Single-molecule exponential decay times may differ for different conformations of the molecule, or for differing molecular environments.

The emission of a photon is not the only pathway for exiting the excited state. All kinetic pathways out of the excited state (fluorescence, energy transfer, internal conversion, intersystem crossing, etc.) are assumed to be first-order processes over the short time the molecule is in the excited state (Förster, 1951; Parker, 1968; Birks, 1970; Clegg and Schneider, 1996; Valeur, 2002; Clegg et al., 2003). Therefore, since all pathways out of the excited state are in direct kinetic competition with each other, the fluorescence decay process from a single molecule is still described by an exponential. The measured dwell time in the excited state, which is usually in the nanosecond time range, will decrease as additional kinetic pathways to exit the excited state become available; and this is accompanied by a decrease in the measured number of emitted photons (intensity). By measuring the fluorescence lifetime in specific experimental circumstances, we can obtain quantitative information about any particular kinetic pathway, specifically FRET.

Usually the fluorescence decay signal from an ensemble of many molecules is recorded, or from a single molecule over a longer time. The decay function for an ensemble of molecules may not be (and is usually not) a single exponential or sometimes not even a sum of exponentials. This is also true for a single molecule measured over a longer time, because the molecule, or its environment, could change during the data acquisition time. For such an ensemble (or measured decay curves of a single molecule over a longer time), the "measured average fluorescence lifetime" is determined by multiplying the normalized decay curve $f(t)\big/\int_0^\infty f(t)dt$ by time and integrating it over the complete decay curve

$$\langle\tau\rangle = \int_0^\infty tf(t)dt \Big/ \int_0^\infty f(t)dt$$

where $f(t)$ is the *measured* time-dependent decay function (from the excited state), assuming that the excitation pulse is very short compared to the molecular decay processes. Fluorescence decay curves provide a much more detailed view of the molecular processes competing on the fluorescence timescale than static steady-state measurements (Förster, 1951; Birks, 1970, 1975; Cundall and Dale, 1983; Lippert and Macomber, 1995). For this reason, dynamic fluorescence measurements play an essential, highly informative role in spectroscopy and all its applications. Detection of fluorescence lifetimes enables us to observe directly dynamic molecular processes, without losing information by time-averaging as in steady-state measurements.

New experimental techniques allow direct measurement of the fluorescence lifetime while imaging (Marriott et al., 1991; vande Ven and Gratton, 1992; Wang et al., 1992; Gadella et al., 1993; Oida et al., 1993; Lakowicz and Szmacinski, 1996; So et al., 1996; French et al., 1997; Bastiaens and Squire, 1999; Gadella et al., 2001; Gerritsen et al., 2002; Hink et al., 2002; Mitchell et al., 2002; Clegg et al., 2003; Schneckenburger et al., 2004)—that is, the lifetime can be found for every pixel in a computer-generated image. This experimental technique is called *fluorescence lifetime-resolved imaging* (FLI). Currently, lifetime-resolved images can be recorded nearly as conveniently as making a standard fluorescence intensity imaging measurement.

2.2 The Advantages of Measuring FRET with Fluorescence Lifetimes

Lifetime measurements are independent of the concentration of the fluorophores. This is especially significant in biological experiments, as intensities can vary widely from sample to sample.

The fluorescence lifetime of a molecule is sensitive to the microenvironment surrounding the molecule. Lifetimes have been used to image the presence of specific ions (Rabinovich et al., 2000), the pH (Szmacinski and Lakowicz, 1993; Sanders et al., 1995; Hanson et al., 2002), metabolites such as glucose (DiCesare and Lakowicz, 2001), conformational changes and aggregation of molecules (Bacskai et al., 2003; Berezovska et al., 2003; Calleja et al., 2003; Vermeer et al., 2004), fluorescent proteins (GFPs) (Periasamy et al., 2002), FRET (Holub et al., 2000; Jakobs et al., 2000; Harpur et al., 2001; Chen et al., 2003; Chen and Periasamy, 2004), changes in the immediate environment of the fluorophore (Clayton et al., 2002), fluorescence from DNA arrays (Valentini et al., 2000), and histological samples (Eliceiri et al., 2003). Lifetime-resolved spectra are very useful for separating fluorescing molecular components in the presence of mixed species or titrations (Koti and Periasamy, 2003). Lifetimes are attractive for biological samples because they do not require the extensive experimental controls and normalization procedures that are essential for quantifying steady-state fluorescence intensity measurements. The capability to directly measure the dynamic process is especially advantageous when determining parameters such as FRET in images in biological samples, where it is difficult to determine the control parameters necessary to interpret intensity-based FRET measurements.

The total rate (probability per unit time) for an excited molecule to exit the excited state is the sum of the rates of all the possible kinetic pathways out of the excited state (Förster, 1951; Birks, 1970; Clegg, 1996; Lakowicz, 1999; Valeur, 2002; Clegg et al., 2003). Consider a donor molecule with a characteristic lifetime τ_0 without FRET. The inverse of this lifetime, $1/\tau_0$, which is the rate (probability per unit time) that the molecule will exit from the excited state, includes all de-excitation rate processes from the excited state other than FRET. The lifetime of the fluorophore in the presence of FRET (written as a rate, $1/\tau$) will be

$$\frac{1}{\tau} = \frac{1}{\tau_0} + k_{\text{FRET}} \tag{1}$$

where k_{FRET} is the rate at which the excitation energy is transferred to the accepter molecule from the donor. We define $k \equiv 1/\tau_0 = k_F + k_{nr}$ where k_F is the spontaneous rate of photon emission (this is an inherent property of the molecule called the Einstein spontaneous rate of emission [Parker, 1968; Birks, 1970] and k_{nr} is the summation of all nonradiative rates of de-excitation from the excited state other than k_{FRET}. The energy transfer efficiency, E, is the fraction of times the energy is transferred from the excited donor molecule to the acceptor and not expended in any other rate process (including being emitted as fluorescence by the donor),

$$E = \frac{k_{\text{FRET}}}{k_{\text{FRET}} + k_F + k_{nr}} = k_{\text{FRET}}\tau = 1 - \frac{\tau}{\tau_0} \tag{2}$$

k_{FRET} is just the rate (probability per unit time) for the excited molecule to deactivate through the energy transfer pathway times the average time that a donor is in the excited state. Thus, when measuring lifetimes, only two measurements must be made to determine the energy transfer efficiency: (1) the characteristic lifetime of the donor (donor only), and (2) the lifetime of the donor in the presence of FRET (donor and acceptor). Commonly, the donor-only lifetime is known, and then only one measurement is needed to obtain the FRET efficiency.

2.3 Fundamentals of Measuring Fluorescence Lifetimes

Details of the data analysis are being addressed here, at the beginning of the chapter, to simplify the later description of instrumentation for lifetime imaging measurements. Readers already acquainted with the basic theoretical formulations of time- and frequency-domain measurements of fluorescence and with the effect of a distribution of fluorescence species on both the steady-state and time-varying fluorescence measurements can skip the following section and proceed to Section 3.

2.4 Intensity Distribution Functions for Steady-State Fluorescence Measurements

Often a system of interest is not just a collection of fluorophores with a single lifetime. A fluorescence species can have a distribution of components with different lifetimes (and therefore different fluorescence contributions) because of a spatially varying molecular environment (such as would be expected for donors and acceptors separated by different distances). For simplicity, in the following discussion we will consider only one chemical species of fluorescent molecule. This description can easily be extended to multiple species by just adding each species to the detected signal (Jameson et al., 1984; Clegg and Schneider, 1996). Even though we are describing steady-state intensity in this section, the steady-state intensity and distribution of fluorescence components are addressed in terms of their lifetimes. This is natural because the measured steady-state intensity is a result of a kinetic competition between many different pathways of leaving the excited state; only one of these pathways will lead to the emission of a photon.

Consider a fluorophore with a distribution of lifetimes; the molecular number distribution is represented by a distribution function $p(\tau)$, where $\int p(\tau)d\tau = 1$. $p(\tau)$ represents the *fraction* of molecules with lifetime τ within an incremental lifetime interval $d\tau$. For any shape or time course of excitation light, the average steady-state contribution of each fluorescence molecule to the total measured intensity depends on its lifetime, τ. To calculate the contribution of each lifetime component to the steady-state intensity we consider the following kinetic equations for a two-state system. N_1 is the number of donor fluorophores in the ground state and N_2 is the number of excited donor fluorophores. The differential equations representing this kinetic process are

$$dN_1/dt = -\sigma N_1 + kN_2 + k_{FRET}N_2 \tag{3}$$

$$dN_2/dt = +\sigma N_1 - kN_2 - k_{FRET}N_2 \tag{4}$$

σ is the rate constant of excitation of the fluorophores (including the cross section, excitation intensity, etc.), k is the rate constant of de-excitation from the excited state in the absence of the FRET de-excitation pathway (remember, $k = k_F + k_{nr} = 1/\tau_0$), and k_{FRET} is the rate constant of de-excitation through the FRET de-excitation pathway. We have specifically selected FRET as the particular pathway under discussion because this review deals with FRET; but it could be any other kinetic pathway for exiting the excited state, such as dynamic quenching. k_F is defined above as the spontaneous rate of fluorescence emission ($k_F = 1/\tau_F$) of a single excited molecule. The steady-state solution, assuming a constant intensity of excitation light ($\partial N_1/\partial t = \partial N_2/\partial t = 0$), for this system is

$$N_2 = N\sigma/(k + k_{FRET} + \sigma) \tag{5}$$

where N is the total number of fluorophores of this fluorescent species.

The total fluorescence intensity (number of photons emitted per unit time) from a fluorophore, I, is $k_F N_2$ If we consider the nonsaturation limit where $\sigma << k$ and k_{FRET}, which is usually the case for most steady-state fluorescence measurements, we find

$$I = \left(N\sigma \frac{1}{k + k_{FRET}}\right)k_F = N\sigma\frac{\tau}{\tau_F} \tag{6}$$

This equation states that the rate of exciting molecules ($N\sigma$) times the emission quantum yield ($qy = k_F/(k + k_{FRET})$) is the total intensity of emitted photons. We have also made use of the fact that measured lifetime τ is the inverse of the total rate of leaving the excited state, which is the sum of the rate in the absence of FRET $k = 1/\tau_0$ plus the rate of energy transfer; that is,

$$\frac{1}{\tau} = k + k_{FRET} = \frac{1}{\tau_0} + k_{FRET}$$

The component of the steady-state intensity of emitted light with lifetime τ is then

$$A(\tau) = N(\tau)\sigma(\tau)\tau/\tau_F \qquad (7)$$

where $N(\tau) = Np(\tau)$. We have specifically indicated that the rate of excitation $\sigma(\tau)$ may be different for each component with a different τ because the absorption coefficient can depend on the environment (Förster, 1951; Strickler and Berg, 1962; Birks, 1970; Valeur, 2002). Similarly, different fluorescence lifetimes often occur for a single species of fluorophore, even in the same sample, because of different molecular environments. $A(\tau)$ is an *intensity distribution function*. $A(\tau)$ is the contribution to the *intensity* per $d\tau$; that is, it is the contribution to the total fluorescence intensity from those fluorescent components with lifetime τ. Note that since, $\int p(\tau)d\tau = 1$ then $\int A(\tau)d\tau$ is the total fluorescence intensity. $A(\tau)$ contains the absorption cross section (probability of becoming excited by the light pulse) for the chromophores with lifetimes in this range, their concentrations (or number of molecules), and quantum yields, etc.

It is easy to see that this formalism can be easily extended to a collection of different fluorescing species. For a single lifetime component with no distribution, $A(\tau)$ = $A\delta(\tau - \tau')$, where the delta function ($\delta(\tau - \tau') = 0$ if $\tau \neq \tau'$; $\delta(\tau - \tau') = 1$ if $\tau = \tau$) is used; in this case we would get a single exponential decay as expected. If the fluorescence emission is due to a distribution of lifetimes (from a single fluorescent chemical species) we would observe a distribution of lifetimes; for instance, the distribution could be a single Gaussian centered on a particular lifetime, $\tau_{i,0}$,

$$A(\tau_i) = A_{Gi}e^{-(\tau-\tau_{i,0})^2/2w(\tau_{i,0})}$$

In this case the measured steady-state fluorescence intensity would be weighted according to this distribution.

In general, for cases in which there is more than a single lifetime, $A(\tau)$ is either a series of delta functions at specific values of lifetimes,

$$A(\tau) = \sum_j A_j\delta(\tau - \tau_j)$$

a series of distributions (e.g. Gaussians)

$$A(\tau) = \sum_i A(\tau_i) = \sum_i A_{Gi}e^{-(\tau-\tau_{i,0})^2/2w(\tau_{i,0})}$$

or a mixture of singular lifetimes and distributions,

$$A(\tau) = \sum_{j} A_{j}\delta(\tau - \tau_{j}) + \sum_{i} A_{Gi}e^{-(\tau - \tau_{i,0})^{2}/2w(\tau_{i,0})}$$

A_{Gi} is the appropriate amplitude of the *i*th Gaussian.

2.5 Time-Resolved Fluorescence

Fluorescence lifetime imaging is used to spatially distinguish areas of varying lifetime that can be resolved (see Fig. 11–1). There may also be multiple components to the lifetimes measured within a pixel of the fluorescence image; that is, often the different lifetime components cannot be separated spatially. Any lifetime analysis should take into account these heterogeneities. It is easiest to think of the time response of fluorescence emission from an ensemble of fluorophores in terms of the response of the sample to an infinitely narrow excitation pulse, or a delta function. For a single lifetime component, the time course of the fluorescence following the pulse would decay as an exponential, $F(t) \propto e^{-t/\tau}$ The time-dependent fluorescence signal, F, from a collection of fluorophores with a distribution of lifetimes that have been excited by a delta function light pulse (very short compared to all fluorescence lifetimes), can be expressed as

$$F^{\delta}(t) = \int_{0}^{\infty} a(\tau)e^{-t/\tau}d\tau \tag{8}$$

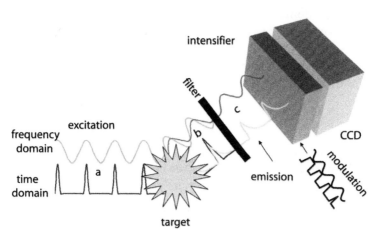

FIGURE 11–1 Schematic of frequency- and time-domain lifetime instrumentation. Modulated light (frequency domain) or pulsed light (time domain) is used (a) to illuminate the target. An optical filter separates the unwanted fluorescence and scattered emission and passes the desired spectral components of the fluorescence image. The detector depends on the method of data acquisition (see text). For full-field acquisition the light is detected first with a rapid modulated image intensifier, and the image output from the intensifier is focused onto a charge-coupled device (CCD) camera. The averaged CCD data are digitized and rapidly read out into a computer, where the data are processed to determine the lifetime information. For scanning single-point acquisition the detector is either a photomultiplier or an avalanche photodiode (APD). Frequency-domain data are collected by means of continuously modulated detectors and homo- or heterodyne detection. Time-domain data are acquired with time-gated detectors or single-photon counting methods (see text). Essentially any optical imaging apparatus can be used, such as a microscope, direct camera optics, or endoscope.

$a(\tau)$ is the pre-exponential amplitude density for the components with lifetime τ. The total signal is summed (integrated) over all the components with different τ. Often the excitation is not a very short pulse, but is some function of time that is not short compared to the fluorescence lifetimes. In this case, the fluorescence response for an arbitrary time-dependent excitation function, $E(\tau)$ would be (Cundall and Dale, 1983; Clegg and Schneider, 1996)

$$F(t) = \int_{-\infty}^{\infty} E(t')F^\delta(t-t')dt' = \int_{-\infty}^{\infty} E(t')\left(\int_0^{\infty} a(\tau)e^{-(t-t')/\tau}d\tau\right)dt' \tag{9}$$

This equation (written in terms of a convolution of the fluorescence response to a delta function excitation and the time-dependent excitation process) just says that the measured fluorescence at time t is the sum (integral) of signals resulting from all previous excitation events at times t'. We can reverse the order of the integrals ($a(\tau)$ does not depend on t' and $E(t')$ does not depend on τ to give a more useful form:

$$F(t) = \int_0^{\infty} a(\tau) \int_{-\infty}^{\infty} E(t')e^{-(t-t')/\tau}dt'd\tau \tag{10}$$

This will be the starting equation for the following theoretical analyses.

2.6 Time-Domain Measurement

The description in the time domain is straightforward. Consider a very short (essentially a delta function compared to the fluorescence lifetimes) excitation pulse, $E\delta(t') = E_0\delta(t')$ at time $t' = 0$; we arbitrarily set the zero time to be the excitation pulse. Now the integral in equation (10) is greatly simplified, resulting in a single exponential decay for every lifetime component as we had before:

$$F(t) = E_0 \int_0^{\infty} a(\tau)e^{-t/\tau}d\tau \tag{11}$$

For an excitation pulse that has a significant width, the excitation pulse must be deconvoluted from the data, as indicated above (see equation 9). Some details of fitting procedures of time-domain measurements are given below.

2.7 Frequency Domain

Since the frequency domain is not as familiar to many, and the fast FLI instrumentation we discuss below is done in the frequency domain, we describe this in more detail.

Consider a sinusoidal modulated excitation source,

$$E(t) = E_0 + E_\omega \cos(\omega t - \varphi_0) \tag{12}$$

Putting this into equation (10) and evaluating the integral gives

$$F(t) = \int_0^{\infty} a(\tau)\left(E_0\tau + E_\omega\tau \frac{1}{1+(\omega\tau)^2}\left(\cos(\omega t - \varphi_0) - \omega t \sin(\varphi_0 - \omega t)\right)\right)d\tau \tag{13}$$

We can use the trigonometric relation

$$[\cos(a-b)-c\sin(b-a)]/\sqrt{1+c^2} = \cos(b-a-\arctan(c)) \tag{14}$$

to simplify, and get

$$F(t) = \int_0^\infty a(\tau)\left[E_0\tau + E_\omega\tau\frac{1}{\sqrt{1+(\omega\tau)^2}}\cos(\omega t - \varphi_0 - \arctan(\omega\tau))\right]d\tau \tag{15}$$

or, normalizing to the amplitude of the excitation pulse, E_0,

$$\frac{F(t)}{E_0} = \int_0^\infty a(\tau)\tau\left[1 + \frac{E_\omega}{E_0}\frac{1}{\sqrt{1+(\omega\tau)^2}}\cos(\omega t - \varphi_0 - \arctan(\omega\tau))\right]d\tau \tag{16}$$

$\arctan(\omega\tau)$ is the resultant phase shift of the fluorescence signal with lifetime τ relative to the phase of the excitation ϕ_0, and

$$\frac{E_\omega}{E_0}\frac{1}{\sqrt{1+(\omega\tau)^2}} \tag{17}$$

is the modulation depth of the fluorescence component with lifetime τ (the modulation depth is the fraction of a signal that is modulated relative to the average signal). This is the modulation of the emitted fluorescence signal *at the high frequency*. Thus, for a single lifetime, the modulation depth of the fluorescence is decreased by a demodulation factor of $\frac{1}{\sqrt{1+(\omega\tau)^2}}$ relative to the depth of modulation of the excitation light $\frac{E_\omega}{E_0}$. The higher the frequency, the lower the modulation depth of the high-frequency fluorescence signal. This demodulation factor and the phase shift, arctan ($w\tau$), provide the principal information needed to estimate the fluorescence lifetime (Birks, 1970; Gratton and Limkeman, 1983; Jameson et al., 1984; Clegg and Schneider, 1996; Lakowicz, 1999). At frequencies close to $\omega \approx 1/\tau$, these two parameters vary notably in a characteristic way. For a single lifetime, τ', where $a(\tau) = a\cdot\delta(\tau - \tau')$,

$$\frac{F(t)}{E_0} = a\cdot\tau'\left[1 + \frac{E_\omega}{E_0}\frac{1}{\sqrt{1+(\omega\tau)^2}}\cos(\omega t - \varphi_0 - \arctan(\omega\tau'))\right] \tag{18}$$

The fluorescence signal shown in equation (18) is modulated in time at a very high frequency (in the frequency range of megahertz to hundreds of megahertz) to measure lifetimes in the nanosecond range (see Fig. 11–2). It is difficult to analyze data taken at such high frequencies because of the very large noise bandwidth at these high frequencies. Two foremost techniques of frequency-domain measurements are used to lower the frequency bandwidth of the signal. They are the *homodyne* and the *heterodyne* techniques (these are common measurement techniques for high-frequency phase detection, or lock-in detection). The same information is available by performing the heterodyne and homodyne operations as that from the original high-frequency signal, but the signal-to-noise level is improved a great deal,

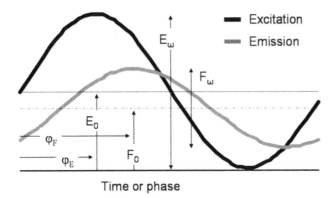

FIGURE 11–2 Frequency-domain signal (see text for definitions). The excitation light has a certain phase φ_e and modulation $M_E = E_w/E_0$ set by the instrumentation. As in equation (12), we define F_0 and F_w to represent the time-independent average and the time-varying part of the fluorescence signal. The fluorescence signal is phase shifted by $\varphi_F - \varphi_E$, and demodulated, $M_F = F_w/F_0$. The phase shift and demodulation (the decrease in the depth of modulation) increase with increasing lifetime of the fluorophore at a certain set frequency. This signal is present at every pixel of the recorded image. This figure depicts directly the high-frequency signal; the homodyne or heterodyne parameters are equivalent.

and the electronics is greatly simplified because the frequency of the data acquisition is appreciably lower.

2.7.1 Homodyne method for measuring fluorescence lifetime components
In the homodyne mode, the amplification level of the detector is modulated *at the same frequency* as the modulation of the excitation light (of course, the fluorescence is modulated at the same frequency as the excitation light). The time-averaged measured signal, S/T (the cumulative summed average of the signal, S, divided by the time of averaging, T), is time independent. This time-normalized homodyne signal is an integral over time of the multiplication of the detector's time-dependent high-frequency amplification modulation, $R(t)$ equation (20), and the fluorescence signal, $F(t)$, equation (13),

$$\frac{S}{T} = \frac{1}{T}\int_T R(t)F(t)dt \tag{19}$$

The modulation of the detector gain is

$$R^{\text{homo}}(t) = R_0 + R_w \cos(\omega t + \varphi_R) \tag{20}$$

The fluorescence signal after being modulated by the detector is averaged (integrated) over the time T, to give the following homodyne signal (just substituting into equation (19)):

$$S = \int_0^\infty a(\tau)\left(\int_0^T R^{\text{homo}}(t)\left(E_0\tau + E_w\tau\frac{1}{\sqrt{1+(\omega\tau)^2}}\cos(\omega t - \varphi_0 - \arctan(\omega\tau))\right)dt\right)d\tau \tag{21}$$

T is long enough so that all of the terms that do not scale with T drop out (that is, T is long enough so that the transient terms have subsided). The integrated signal is then divided by the time of integration, as in equation (19), to give the final measured, averaged, time-independent steady-state signal:

$$\frac{S}{T} = \int_0^\infty a(\tau)\tau \left[E_0 R_0 + E_\omega R_\omega \frac{1}{2\sqrt{1+(\omega\tau)^2}} \cos\left(\varphi_0 + \varphi_R + \arctan(\omega\tau)\right) \right] d\tau \qquad (22)$$

We normalize measured homodyne signal by the amplitude of the excitation light, E_0, and the amplification factor of the detector, R_0, to give

$$S_{tot}^{homo} \equiv \left(\frac{1}{E_0 R_0}\right)\frac{S}{T} = \int_0^\infty a(\tau)\tau \left[1 + \frac{E_\omega R_\omega}{2E_0 R_0} \frac{1}{\sqrt{1+(\omega\tau)^2}} \cos\left(\varphi_0 + \varphi_R + \arctan(\omega\tau)\right) \right] d\tau$$

$$\qquad (23)$$

$$S_{tot}^{homo} = \int_0^\infty A(\tau) \left[1 + M_0 \frac{1}{\sqrt{1+(\omega\tau)^2}} \cos\left(\varphi(\tau)\right) \right] d\tau$$

In the second line of equation (23) we have defined the modulation and phase for each τ-component:

$$M(\tau) \equiv \frac{E_\omega R_\omega}{2E_0 R_0} \frac{1}{\sqrt{1+(\omega\tau)^2}} \equiv M_0 \frac{1}{\sqrt{1+(\omega\tau)^2}} \qquad (24)$$

and

$$\varphi(\tau) \equiv \varphi_0 + \varphi_R + \arctan(\omega\tau) \qquad (25)$$

$M(\tau)$ is the *experimentally determined* relative modulation, which is the ratio of the frequency-dependent term to the static term of equations (22) and (23). The modulation is determined experimentally by fitting the acquired data to an expression similar to equation (23) (see below). The lifetime can be determined by comparing the experimentally determined modulation and phase to the theoretical expressions in equations (24) and (25).

As mentioned above, every separate measurement of the normalized homodyne signal is time independent. The measured homodyne signal depends on $\cos(\varphi(\tau))$ $= \cos(\varphi_0 + \varphi_R + \arctan(\omega\tau))$. If we vary φ_0 or φ_R over a complete period, and record measurements at these different relative phases, we can determine $\varphi(\tau)$ and, therefore, $\arctan(\omega\tau)$. We can determine $\varphi_0 + \varphi_R$ by measuring a standard (where we know the fluorescence lifetime, or light scattering, where $\tau = 0$). By measuring a fluorescence (or light scattering) standard we can also determine the modulation factor, $M_0 = E_\omega R_\omega / 2E_0 R_0$, which is τ independent.

For the case of a single lifetime, that is, assuming that $a(\tau) = a \cdot \delta(\tau - \tau')$, we can extract two separate estimates for the lifetime, which for a single exponential decay should be equal: we get τ_ϕ from the phase of the fluorescence signa $\varphi(\tau) = \varphi_0 + \varphi_R + \arctan(\omega\tau_\phi) = \varphi_0 + \varphi_R + \varphi_F$, and τ_M from the demodulation factor

$$\frac{E_\omega R_\omega}{E_0 R_0} \frac{1}{2\sqrt{1+(\omega\tau_M)^2}}$$

These two independent estimates are

$$\tau_\varphi = \frac{1}{\omega}\tan(\varphi_F) = \frac{1}{\omega}\tan(\varphi(\tau) - \varphi_0),\tag{26a}$$

and

$$\tau_M = \frac{1}{\omega}\sqrt{\left(\frac{M_0}{M(\tau)}\right)^2 - 1}\tag{26b}$$

2.7.2 Heterodyne method for measuring fluorescence lifetime components

The heterodyne technique is similar to the homodyne method, except that the amplification gain of the detector is modulated at a different frequency, ω', than the frequency of the excitation light, ω,

$$R^{\text{hetero}}(t) = R_0 + R_\omega \cos(\omega' t + \varphi_R)\tag{27}$$

So,

$$S = \int_0^\infty a(\tau)\left[\int_t^{+T} R^{\text{hetero}}(t)\left(E_0\tau + E_\omega\tau\frac{1}{\sqrt{1+(\omega\tau)^2}}\cos(\omega t - \varphi_0 - \arctan(\omega\tau))\right)dt\right]d\tau\tag{28}$$

The total solution of this integration will produce terms with multiple frequencies at ω, ω', $\omega + \omega'$, and $\omega - \omega'$ By setting ω and ω' to be nearly equal (they are usually within a few Hertz to 10 kHz), the $(\omega - \omega')$ frequency will be much lower than the others and a simple low-pass filter will remove all frequencies higher than $(\omega - \omega')$ from the recorded signal. This can be done by setting the integration time T to be much longer than ω^{-1}, ω'^{-1}, $(\omega + \omega')^{-1}$ to avoid any contribution of these high-frequency terms to the measured time-varying signal. This leaves us with only the $(\omega - \omega')$part:

$$S_{\text{tot}}^{\text{hetero}}(t) \equiv \frac{S(t)}{E_0 R_0 T} = \int_0^\infty \tau a(\tau)\left[1 + \frac{E_\omega R_\omega}{2E_0 R_0}\frac{1}{\sqrt{1+(\omega\tau)^2}}\cos((\omega'-\omega)t + \varphi_0 + \varphi_R + \arctan(\omega\tau))\right]d\tau$$

$$S_{\text{tot}}^{\text{hetero}}(t) = \int_0^\infty A(\tau)\left[1 + M_0\frac{1}{\sqrt{1+(\omega\tau)^2}}\cos((\omega'-\omega)t + \varphi(\tau))\right]d\tau\tag{29}$$

which has the same form for $M_0\dfrac{1}{\sqrt{1+(\omega\tau)^2}}$ and $\varphi(\tau)$ as the homodyne method, equations (23–25), but the signal varies in time with the slow frequency $|\omega' - \omega|$.

There are several ways of measuring this time-dependent signal and determining the phase and demodulation ratio. In general, the time-dependent signal is acquired in either an analogue or digital fashion, and the phase and modulation are ascertained relative to a reference signal to determine $\varphi_0 + \varphi_R$ and the demodulation ratio. In order to increase the signal-to-noise level, the signal is passed through a very narrow bandpass filter (analogue, or digitally by Fourier transform techniques with a computer) with the center frequency at $|\omega' - \omega|$ to quantify accurately the time-varying part of the signal (Gratton and Limkeman, 1983; Clegg and Schneider, 1996; Lakowicz, 1999). The signal is also passed through a very low pass filter (or just averaged over a much longer time than that for $|\omega' - \omega|^{-1}$) to extract an accurate measure of the time- independent part of the signal (the time average). The heterodyne method is the preferred method of measurement when measuring lifetimes in nonimaging modes (for instance, a cuvette measurement, or single- point acquisition of an image) because very narrow–band active filters can be used for the measurement of the time-varying signal with synchronous phase detection methods, either analogue (lock-in detection) or digital (digital Fourier transform methods; Brigham, 1974).

2.7.3 Interpreting experiments with multiple lifetimes

We can derive explicit closed expressions for experimentally determined *effective* τ_0 and τ_M for data taken at a single frequency. These are just the values of the modulation and phase lifetimes that one would calculate from the separate phase and demodulation data at one frequency assuming only a single exponential decay. These are only apparent lifetimes and are not equal unless there is only one lifetime present in the fluorescence signal. If this simple data analysis shows that $\tau_M > \tau_\varphi$, one can be sure that multiple lifetimes are present (Gratton et al., 1984; Jameson et al., 1984; Clegg and Schneider, 1996; Lakowicz, 1999).

If more than one exponential lifetime is present, the fluorescence signal is still a sinusoidal signal at the same frequency as the excitation modulation, but the analysis of the phase and modulation for multiple lifetimes is more complex than for a single lifetime, equation (23). We can write a general expression for the experimentally determined measured signal (using the frequency-domain homodyne method) from any fluorescence sample at a single frequency as (see equation (23)

$$S_{\text{tot}}(\omega) = S_{0,\text{tot}} + S_{\omega,\text{tot}} \cos(\varphi_{\text{tot}}) \tag{30}$$

The signal at a single frequency for a distribution of lifetimes is given under the integral in equation (23), where we would assume that $A(\tau) = A_\tau \delta(\tau - \tau')$:

$$S_{\text{tot}} = A_\tau \left[1 + M_0 \frac{1}{\sqrt{1+(\omega\tau)^2}} \cos(\varphi_0 + \varphi_R + \varphi_F) \right] \tag{31}$$

Or, in terms of the measured parameters,

$$S_{\text{tot}} = A_\tau \left[1 + M(\tau)\cos(\varphi(\tau)) \right] \tag{32}$$

The complete theoretical expressions for multiple components relating the theoretical expressions for the different components and the experimental data, equation (30), are quite complex but are well known in the literature (Jameson et al., 1984; Clegg and Schneider, 1996). Here we will focus on how to fit the phase angle and the modulation depth to the measured quantities and how to relate these fitted parameters to the distribution of lifetime components. Care must be taken in the choice of fitting procedures because we are conducting rapid data analysis for imaging purposes.

Consider a collection of fluorophores with an intensity distribution of lifetime components, $A(\tau)$. For each lifetime, there would be an associated modulation and phase (equations (24) and (25)) and an associated total response of the form of equation (30). We have expressed the frequency-dependent term of the measured signal in terms of a cosine function. Once we have determined the phase of the cosine, φ_{tot}, that fits the data, we can also define the following equations, where we simply define $\cos(\varphi_{tot})$ and $\sin(\varphi_{tot})$ in terms of a sum (integral) over $\cos(\varphi(\tau))$ and $\sin(\varphi(\tau))$ terms of our chosen model.

$$S_{\omega,tot}\cos\varphi_{tot} = \int_\tau A(\tau)M(\tau)\cos(\varphi(\tau))d\tau \qquad (33)$$

and

$$S_{\omega,tot}\sin\varphi_{tot} = \int_\tau A(\tau)M(\tau)\sin(\varphi(\tau))d\tau \qquad (34)$$

The integral or sum of sines or cosines of the same periodicity are always sine or cosine functions. The original choice to fit the data with a cosine is arbitrary; we could just as well have fit the data to a sine function, and the resulting phase would have been different by $\pi/2$. One simply has to be consistent throughout the analysis.

Equations (33) and (34) can be put in the form,

$$\tan\varphi_{tot} = \frac{\int_\tau A(\tau)M(\tau)\sin\varphi(\tau)d\tau}{\int_\tau A(\tau)M(\tau)\cos\varphi(\tau)d\tau}, \qquad (35)$$

and

$$S_{\omega,tot}^2 = \left(\int_\tau A(\tau)M(\tau)\sin\varphi(\tau)d\tau\right)^2 + \left(\int_\tau A(\tau)M(\tau)\cos\varphi(\tau)d\tau\right)^2 \qquad (36)$$

$M(\tau)$ and $\varphi(\tau)$ are related for any given τ. Assuming M_0, φ_0, and φ_R are known, we can determine two independent parameters to describe the distribution functions $M(\tau)$ and $\varphi(\tau)$ from a single frequency measurement of M_{tot} and φ_{tot} (two equations, two unknowns). More measurements must be carried out at different frequencies to solve for more parameters to describe more complex distributions.

2.7.4 Two lifetimes
Consider two lifetime components with relative fractional intensities α and β; that is, $A(\tau_1) + A(\tau_2) = \alpha + \beta = 1$. Equations (35) and (36) would reduce to

$$\tan\varphi_{tot} = \frac{\alpha M(\tau_a)\sin\varphi(\tau_a) + \beta M(\tau_b)\sin\varphi(\tau_b)}{\alpha M(\tau_a)\cos\varphi(\tau_a) + \beta M(\tau_b)\cos\varphi(\tau_b)}, \qquad (37)$$

$$S_{\omega,tot}^2 = N^2\left(\alpha^2 M(\tau_a)^2 + \beta^2 M(\tau_b)^2 + 2\alpha\beta M(\tau_a)M(\tau_b)\cos(\varphi(\tau_b) - \varphi(\tau_a))\right) \qquad (38)$$

There are only three independent variables in the above two equations. This means that if one of the variables is known (a lifetime of one component, for example) the other two can be determined from a single frequency measurement. Additional measurements at more than one frequency (Jameson et al., 1984; Clegg and Schneider, 1996; Lakowicz, 1999; Valeur, 2002) would be required to solve for (or fit numerically) all of the variables with no known parameters or to solve a more complex intensity distribution function.

2.8 Distributions of τ for Distributions in Donor and Acceptor Positions

If we have information about a continuous distribution of lifetimes, as often happens with FRET (Rolinski and Birch, 2000, 2002; Koti and Periasamy, 2003; Duncan et al., 2004), we can sometimes determine the distribution parameters from a single frequency experiment. In FRET, the donor and acceptor are often not held at particular permanent positions in space, but have some degree of freedom relative to each other. Since the efficiency of energy transfer is a strong function of the distance between each individual molecular pair of donor (D) and acceptor (A), $E = \left(1 + (r/R_0)^6\right)^{-1}$ and the efficiency of transfer is linearly related to the measured lifetime, $E = k_{FRET}\tau = 1 - \tau/\tau_0$, a distribution of distances between the D and A molecules will correspond to a distribution of lifetimes that will be skewed toward shorter lifetimes; that is, the population of D-A pairs that are closer will transfer much more efficiently, emphasizing the shortening of the shorter lifetimes (greater energy transfer) (Förster, 1951; Clegg, 1996). Since we are measuring the fluorescence lifetime, it is useful to present a general analysis of the effect of this spatial distribution of D and A on the measured lifetime distribution.

It is easy to write formally the probability distribution for distances between D and A molecules if we know the spatial distribution of the donor and acceptor individually. Assume that the position of the centers of the distributions of the donor and acceptor are located at positions $r_{0,D}$ and $r_{0,A}$, and we denote the probability of the donor and acceptor being at positions r_D and r_A as $p(r_D)$ and $p(r_A)$ (see Fig. 11–3). Note that these probabilities are functions of distances from some molecular reference positions (usually the average position of the probes)—that is, $p(r_D) = p(r_D - r_{0,D})$ and $p(r_A) = p(r_A - r_{0,A})$. The positions of the D and A molecules are also considered independent of each other (no intermolecular interaction). Then the probability distribution for finding the donor at r_D and the acceptor simultaneously at r_A is $p(r_D, r_A) = [p(r_D) \times p(r_A)]$. The probability of finding a distance Δr between the D and A molecules is (we assume that only the magnitude of the distance is important)

$$p(\Delta r) = \int [p(r_D) \times p(r_A)]\delta(|r_D - r_A| - \Delta r)dr_A dr_D / \left[\int p(r_A)dr_A \times \int p(r_D)dr_D\right]$$
$$= \int [p(r_D) \times p(r_A)]\delta(|r_D - r_A| - \Delta r)dr_A dr_D \qquad (39)$$

The last equality is because $\int p(r_D)dr_D = \int p(r_A)dr_A = 1$. The delta function $\delta(|r_D - r_A| - \Delta r)$ in the integral selects only the parts of the distributions such that $|r_D - r_A| = \Delta r$.

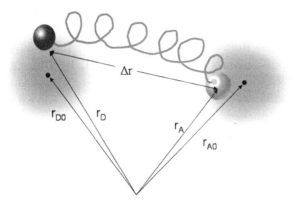

FIGURE 11–3 Picture depicting relative distributed locations of a FRET pair according to the model described in the text. It is assumed that each molecule (the donor [D] and acceptor [A]) is distributed in space according to some probability distribution of position about the average central positions (r_{A0} and r_{D0}). The distance between the two molecules Δr will then have a probability distribution as described in the text, equation (39).

We can now easily find the distribution of lifetimes if we know the lifetime as a function of the distance between the D and A molecules. This is,

$$\tau = \tau_0(1-E) = \tau_0\left(1-\left(\frac{1}{1+(\Delta r/R_0)^6}\right)\right) = \tau_0\left[\frac{(\Delta r/R_0)^6}{1+(\Delta r/R_0)^6}\right] \tag{40}$$

So, in terms of the supposed distribution of distances between D and A molecules (Δr), we can write the unnormalized average measured lifetime from this distribution of D and A molecules as

$$
\begin{aligned}
p(\tau)_{\substack{\text{not}\\\text{normalized}}} &= p(\Delta r)\tau_0\left[\frac{(\Delta r/R_0)^6}{1+(\Delta r/R_0)^6}\right] \\
&= \left\{\int[p(r_D)\times p(r_A)]\delta(|r_D - r_A|-\Delta r)dr_A dr_D\right\}\tau_0\left[\frac{(\Delta r/R_0)^6}{1+(\Delta r/R_0)^6}\right]
\end{aligned} \tag{41}
$$

By integrating this over all values of Δr, we can normalize the probability of measuring a lifetime τ:

$$p(\tau) = p(\Delta r)\tau_0\left[\frac{(\Delta r/R_0)^6}{1+(\Delta r/R_0)^6}\right] \Bigg/ \int_0^\infty p(\Delta r)\tau_0\left[\frac{(\Delta r/R_0)^6}{1+(\Delta r/R_0)^6}\right]d(\Delta r) \tag{42}$$

where now $\int p(\tau) = 1$. These equations can easily be manipulated numerically for any known distribution of D and A molecules or inverted (usually numerically) to find specific model parameters of the D and A distribution functions $p(r_D)$ and $p(r_A)$.

2.9 Gaussian Lifetime Distribution

If we know a specific function for the lifetime distribution, we can substitute that distribution for $p(\tau)$ in the equations. $P(\tau)$ contains intensity information and will most likely be some distribution of species concentrations multiplied by τ. For example, assume that the concentration of molecules with different lifetimes follows the equation

$$p(\tau) = \frac{2}{w\sqrt{\pi}} e^{-(\tau - \tau_c)^2/w^2} \tag{43}$$

where τ_c is the peak of the distribution. $A(\tau)$ would then contain an intensity factor, $I(\tau)$, which is proportional to τ. If we assume that the intensity factor is some constant, I_0, multiplied by τ, then we can write the distribution as

$$A(\tau) = I_0 \frac{2\tau}{w\sqrt{\pi}} e^{-(\tau - \tau_c)^2/w^2} \tag{44}$$

Note that $A(\tau)$ has only two parameters: w, the "width" of the distribution, and τ_c. Therefore, with a single measurement both parameters can be found. The integrals can be solved numerically to get the relationship between the distribution parameters and the measured values.

Now let us consider a more realistic distribution for FRET. One can envision a FRET pair system in which the probability distribution of the location of the fluorophores is Gaussian. We can write the distance distribution between the two molecules as (see equation (39))

$$p(\Delta r) = \int \left[\frac{1}{w\sqrt{\pi}} e^{-(r_D - r_{0,D})^2/w_D^2} \frac{1}{w_A\sqrt{\pi}} e^{-(r_A - r_{0,A})^2/w_A^2} \right] \delta\big(|r_D - r_A| - \Delta r\big) dr_A dr_D \tag{45}$$

The distribution of lifetime is then

$$p(\tau) = p(\Delta r)\tau_0 \left[\frac{(\Delta r/R_0)^6}{1 + (\Delta r/R_0)^6} \right] \bigg/ \int_0^\infty p(\Delta r)\tau_0 \left[\frac{(\Delta r/R_0)^6}{1 + (\Delta r/R_0)^6} \right] d(\Delta r) \tag{46}$$

This can be solved numerically, and because there is a total of five parameters defining the distribution it would require three different separate frequency measurements to solve. Also, because there are five parameters, not much confidence could be given to any fit to the data for the parameters. If any or most of the parameters are known, this distribution would be useful for accurately representing the data. Note that an ordinary two-Gaussian fit or even perhaps a two–discrete lifetime fit would probably fit the data equally well, but may not give an accurate measure of the physical system.

3. INSTRUMENTATION

3.1 Motivation for Measuring with Fast Frame Rates

The method we use to achieve our fastest data acquisition operates in the frequency domain and uses the homodyne principle (Gadella et al., 1993; Clegg and Schneider, 1996; Holub et al., 2000). Although the fluorescence lifetime is intrinsically independent of the fluorophore concentration, changes in the intensity of fluorescence parameters during the time required to acquire one period of data will distort the data. For example, significant photobleaching during data acquisition leads to artifacts in determining the modulation and phase, possibly leading to values of modulation and phase that are theoretically impossible to occur from a distribution of lifetimes (e.g., where $\tau_M < \tau\varphi$, which is physically impossible, barring artifacts such as photobleaching). Other factors that can affect the signal during data acquisition are rotation of fluorophores, time-dependent changes in the signal due to photoreactions, or simple movement of the target in the image. To minimize these effects, short acquisition times are required (Holub et al., 2000). A typical integration time for collecting one data point at a single phase for most fluorescent targets is on the order of milliseconds with our present instrument. Three data points (three phases in homodyne; three time points in heterodyne) are minimally required to determine a lifetime; however, sometimes even two are sufficient (Clegg et al., 2003). Of course, collecting more phase points or increasing the time for averaging at every phase setting of a homodyne measurement increases accuracy.

Typical lifetime image acquisition rates for scanning setups in both the frequency and time domains are much longer, typically a few per second and usually longer (tens of seconds) (Dowling et al., 1998a; Sytsma et al., 1998; Gerritsen et al., 2002; Becker et al., 2003; Duncan et al., 2004). For our full-field homodyne device, a typical data acquisition rate for the complete series necessary for calculating a lifetime image and creating the display is tens of frames per second. The whole process is able to run at video rates (30 Hz).

The ability to acquire, calculate, and display lifetime images in real time offers many advantages. It becomes possible to measure live biological samples and to determine lifetime images from moving objects. In addition, the rapid observation of multiple objects is advantagous when searching for appropriate biological objects (e.g., cells) in a complex biological system. Our instrumentation is user friendly, even when imaging nonstatic targets. New biomedical applications are being designed around our real-time system to add performance enhancement over traditional intensity-only fluorescence measurements.

A major application of lifetime imaging in biological systems is FRET, and lifetime imaging is a very attractive alternative for imaging FRET signals. Combined fluorescence lifetime imaging microscopy (FLIM)–FRET is becoming increasingly popular because it avoids complex normalization, subtraction, and spectral correction procedures necessary for steady-state methods of FRET measurements. This is true especially now that the methods and instrumentation have become readily available.

3.2 Instrumentation for Microscope Imaging

A typical fluorescence lifetime instrumentation system consists of three components: the excitation light source, the fluorescence optics system, and the detector. Systems based on a standard fluorescence microscope benefit from the convenience of using commercial microscopes. Confocal, two-photon, or other scanning microscopes can be used with a single channel detector, allowing rejection of out-of-focus light and three-dimensional imaging. Except for the synchronization of the scanning beam and the spectroscopic measurements, lifetime measurements are carried out similarly to cuvette type measurements (Cundall and Dale, 1983; Gratton and Limkeman, 1983). Both frequency- and time- domain measurements are carried out with single-channel scanning measurements (Becker et al., 2003). Frequency-domain scanning single-point FLIM is typically carried out using heterodyne techniques (Gratton et al., 2003). Often, to increase the speed of data acquisition in imaging situations, fewer separate phase measurements are made. Instrumentation is becoming commercially available (LaVision GmbH, Goettingen; PicoQuant, Berlin; Lambert Instruments, Leutingewolde). Full-field, time-correlated, single-photon counting instrumentation with simultaneous multiple position possibilities has also been demonstrated (McLoskey et al., 1996). Frequency-domain measurements with multiple detectors have also been shown (Rabinovich et al., 2000).

New commercial instrumentation has improved the performance characteristics of the time-domain scanning photon-counting FLIM (Becker et al., 2003). The time-domain experiment can also be carried out with a full-field (or full-frame) microscope; the advantage here is that every point in the image is acquired simultaneously, speeding up the process considerably. In this case only a few (sometimes just two or three) time delays are acquired (Agronskaia et al., 2003).

Table 11–1 is a short summary of instrumentation, along with the respective advantages and disadvantages.

3.3 Full-Field Mode of Real-Time Image Acquisition with Fluorescence Lifetime Imaging

We now discuss details of our rapid FLI acquisition, analysis, and display (see Fig. 11–1).

3.3.1 Fast intensifiers

The full-field mode of FLI requires a fast-gated high-frequency modulated image intensifier. An image intensifier functions by converting a light signal (individual photons) into an electrical signal (emitted electrons), applying a gain, and then converting this back into a light signal (Saleh and Teich, 1991; Sturz, 1999; see also http://www.proxitronic.de/prod/bv/eein.htm). A cathode emits electrons upon bombardment with light (in FLI, the primary image is focused on the cathode of

TABLE 11–1 Comparison of Lifetime Instruments

Factor	Description
Two-photon, time-domain, single-photon, scanning microscope	
Source	Infrared pulsed laser
Optics	Scanning system with fluorescence microscope
Detector	Single photon discriminating PMT and TCSPC electronics
Advantages	High accuracy and sensitivity, 3D imaging
Disadvantages	Big and expensive, slow image capture time
Full-field, time-domain microscope	
Source	Pulsed laser
Optics	Standard fluorescence microscope
Detector	Gated intensifier running in a box car mode
Advantages	Fast image capture time
Disadvantages	Requires a pulsed laser, less sensitive
Full-field, frequency-domain microscope	
Source	CW laser diode and Pockel's cell
Optics	Standard fluorescence microscope
Detector	Modulated intensifier
Advantages	Fastest image capture and display time, small package
Disadvantages	Less sensitive, requires calibration

CW, continuous wave; PMT, photomultiplier tube; TCSPC, time-correlated single-photon counting; 3-D, three-dimensional.

the intensifier). These electrons are sent through a high voltage microchannel plate (MCP). The electrons are accelerated in the microchannels by the high voltage. The electrons collide with the walls of the microchannels in the MCP, causing an avalanche of electrons as they knock multiple electrons free for every collision. In this manner a small photon signal is amplified into a large electrical signal. The free electrons are confined in the micro-channels, preserving the spatial resolution of the signal. A phosphor is located at the backside of the MCP. A very large voltage is placed between the output of the microchannels and the phosphor to accelerate the electrons emerging from the microchannels; this also retains the spatial resolution as the electrons are accelerated onto the phosphor. When the electrons collide at the phosphor, the electrical signal is converted by the photoelectric effect into visible photons, which can be detected by a charge-coupled device (CCD) or camera.

New technology in image intensifiers has produced fast-gated intensifiers that can modulate the gain on the cathode with GHz frequencies or tens of picoseconds gate width. For example, the HRI or the GOI high-speed gated intensifiers from Kentech Instruments Ltd. (South Moreton) are recommended for full-frame lifetime systems (Dowling et al., 1998b, 1999; Webb et al., 2002). This allows fluorescence lifetime measurements to be performed with a full-field setup; that is, every location of a fluorescence image falling on the cathode can be modulated simultaneously. Because every pixel in the image can be processed simultaneously, the data acquisition is very efficient. A rapidly decaying phosphor is required for rapid data acquisition (Gadella et al., 1993, 2001; Clegg et al., 1994).

3.3.2 Full-frame homodyne systems

With such fast modulating intensifiers in the frequency domain, one can use either the homodyne or the heterodyne mode of measurement to carry out full-field, time-resolved acquisition. The homodyne technique has the great advantage that the output signal is time independent, and this is more convenient for rapid full-field applications. At each phase setting, the signal can be integrated (averaged) for arbitrary lengths of time. The timing and synchronization are under software control, and the data acquisition is easy to coordinate (Holub et al., 2000). This is especially useful for computer systems with non–real-time operating systems (e.g., Windows), which do not guarantee processor time for user applications. Homodyne systems provide the opportunity to improve significantly the signal-to-noise ratio of the measurement by lengthening the integration time or increasing the number of phases measured.

A full-field heterodyne system requires that the image be read out rapidly compared to the low-frequency signal. The image data output from a CCD is difficult to pass through a narrow band filter for every pixel; this eliminates the primary signal-to-noise benefit for using a heterodyne system. Boxcar techniques can be used (Clegg et al., 1994), but this lowers considerably the data throughput rate. A heterodyne technique works well for scanning systems because the single-channel detectors can easily accommodate kilohertz signals. We will not discuss these systems here but refer the reader to the literature (Piston et al, 1989).

Our homodyne technique employs a digital phase shifter, which is controlled by software to shift the relative phase between the excitation and the detection sinusoidal modulations in the consecutive images. Digital phase shifters vary widely in performance and cost. High bandwidths and intelligent software are required to support multiple-frequency measurements. An example of a 9–bit digital phase shifter is the D-P-2-9 (Lorch Microwave, Salisbury), which can shift the phase within microseconds, preventing delays between images. This digital phase shifter is a switched series of digital delay lines that have been tuned to provide constant attenuation at the different delays.

3.4 Illumination Options

An important consideration for any lifetime instrument is the excitation light source. Lifetime measurements rely on structured time-varying excitation light with high-frequency response times at least on the order of the lifetimes measured. Following is a list of possible excitation sources with their advantages, disadvantages, and applications.

3.4.1 Pulsed laser

This is the source of choice for most time-domain measurements, which require picosecond or narrower pulse widths (see http://industrial-lasers.globalspec.com/Industrial-Directory/pulsed_laser). A high–repetition rate pulsed laser (for instance, the Tiger line of lasers from Time-Bandwidth Products, Zurich) can also be used for frequency-domain measurements. Pulsed-laser systems usually deliver excellent performance in a lifetime system because of their stability, narrow spectral bandwidth, power, and choice of wavelengths. The major drawback is the high cost and large laser systems for even a single laser line. It is even more expensive to use multiple-wavelength excitations.

3.4.2 Modulated lasers

Laser systems are available with directly modulated laser light, which is useful for frequency-domain measurements. These setups are very convenient and usually available with multiple laser wavelengths (consider the PDL-800B from PicoQuant; Steigmiller et al., 2004). The cost is moderate. However, currently available systems in the moderate price range suffer from low power. Low-power light systems are unsuitable for full-field setups because the light is spread over an area and a higher intensity is required to create enough signal over the whole field of view.

3.4.3 Light modulators

Continuous-wave (CW) laser light can be modulated to give the high-frequency structured light needed for frequency-domain measurements. There are several different types of light modulators, the primary ones being acousto-optic modulators (AOMs) and Pockel's cell modulators. AOMs are inexpensive and easy to set up (Piston et al., 1989), but they usually only operate at particular frequencies and suffer from low depth of modulation and low throughput (amount of the source light in the modulated beam). Often great care is required to keep the AOM phase stable and the temperature must be carefully controlled. A.A. (St-Rémy-Lès-Chevreuse) makes AOMs for many different applications. New Pockel's cell systems are able to meet the high-frequency requirements and operate at different frequencies. They have excellent depth of modulation and all of the light is modulated (two beams are formed, with 50% of the input CW light in each beam, the second beam could be used as a reference for systems that do not need calibration). They suffer from phase drift over a longer time period, but this can be resolved in the data analysis. Conoptics, Inc., makes several different modulation systems that have the high-frequency response needed for frequency-domain measurements. Pockel's cell systems are more expensive then AOMs. However, it is possible to use inexpensive incoherent sources (lamps), which also offer wide wavelength ranges, with a Pockel's cell; the main difficulty is getting sufficient intensity through the small aperture of the Pockel's cell.

3.4.4 Fiber optic modulators

These are becoming available for visible light with GHz bandwidth (Khorasani et al., 2002). If enough intensity could be passed through these fibers one would have a very inexpensive light source in a small, convenient package.

3.4.5 Modulated light-emitting diodes

New light-emitting diodes (LEDs) are now available that are very bright and have the bandwidth necessary to operate as frequency-domain sources (ISS, Champaign). They are very inexpensive and small, so it is easy to have several wavelengths. The

LEDs are still too low in intensity for full-frame imaging purposes but should work for other setups.

4. DATA ANALYSIS

4.1 Time-Domain Data

Time-domain data acquisition results in data distributed directly on the time axis following a light pulse. The analysis of fluorescence decay data consists in extracting the exponential time constants and the pre-exponential factors from the data stream. The literature on this problem is extensive, especially for the analysis of fluorescence decay curves. There are many numerical methods and many specific problems, such as deconvoluting the fluorescence decay from the shape of the light pulse (if it is not a delta function; see equation (9) and (10)). These topics have inspired a plethora of publications (Cundall and Dale, 1983; O'Conner and Phillips, 1984; Birch and Imhof, 1991; Enderlein and Erdmann, 1997).

Care must be taken when fitting time-domain data to handle the data correctly, especially when relatively few photons are detected. The correct analysis procedure depends on how the data are acquired. Although this problem is not limited to time-resolved fluorescence data (it is a general matter of numerical data analysis), it is appropriate to discuss this problem here. It is common for time-domain measurements to bin the data (photon counts) into equal time intervals. This is often done regardless of whether the data are initially collected in preselected time intervals (e.g., this is the case when the data are collected as analogue signals) or the data are collected by photon counting.

The most accurate and a fast way to carry out photon counting is to record the time that a single photon arrives at the detector following every repetition of the excitation pulse; thus the data are not distributed evenly on the time axis. Only the first photon is detected for every light pulse repetition; the light intensity must be adjusted so that considerably fewer than one photon per light pulse is detected—approximately fewer than 1 in 100—to avoid pulse pile-up in which photons would be lost (not counted). This methodology can be accomplished with great time resolution and at very high data rates (Becker et al., 2003). An amplitude voltage ramp is started upon the arrival of a photon, which is stopped at a preselected time before the next light pulse. This is called *time amplitude conversion* (TAC), and is a time-honored method of photon-counting data acquisition (Cundall and Dale, 1983). The amplitude of the ramp at the end is digitized to give a very accurate reading of the time of each photon arrival; the repetitive period of and timing between the laser pulses are very accurate and stable. Therefore, the data are not acquired in preset, equally spaced time intervals but exist as a density of arrival times on the time axis at non-preset time points. For convenience and compatibility with fitting routines, it is common to bin the density of counts subsequently into time intervals.

However, it is better if the data are not subsequently binned into artificial time intervals, because the time resolution is lost and the fitting procedures on the binned data may produce false results. With newer instrumentation, the time of arrival of a photon is usually known exactly, much more accurately than the time width of subsequent bins. If one bins the time axis data of a photon-counting experiment so that there is at least one count in each bin (after collecting data over a longer time), the time width of the bins will have to be large; therefore, the resolution of the time axis is reduced considerably. On the other hand, if the data are binned into very short time intervals to retain time resolution, many bins will not have any counts. This latter situation is common in microscopy imaging because the number of counts at any location is in general not large. If this is the case, subsequent fitting of the data to exponentials where the data are equally spaced on the time axis with many commercial least-squares regression algorithms will result in false results. If the data are not binned into equal time intervals before least-square fitting to exponentials, but many of the common fitting algorithms are used, then the fitting procedure will take an enormously long time (if the time intervals between consecutive data points are equal; very efficient algorithms are available to speed up fitting routines).

There are several ways to circumvent this problem. The consecutive data points (times of arrival) can simply be added, and then this sum is divided by the number of acquired data points. This gives the first moment of the distribution of arrival times. Higher moments can similarly be calculated, which gives a more accurate description of the form of the fluorescence decay. This *method of moments* will avoid the loss of accuracy and exceptionally long fitting times. The method is fast because it is noniterative.

There is another simple way to avoid empty time intervals. The arrival times of the original data record can be organized consecutively in the order of photon arrival following the light pulse. Then a new data stream can be created in which the components of the new data stream are partial sums (the integral) of the number of data points collected up to the ith consecutive equally spaced time point, t_i. In this way, there will never be a time interval with no data points, and the time interval between data points can be chosen as small as desired. This new data stream can be fitted to the integral of the model equations describing the fluorescence decay with normal fitting routines. The actual model equations describing the fluorescence decay are the derivative of the expression for the partial-sum data stream.

These considerations are important only when collecting point-wise photon-counting data. If the time between the start of data acquisition and display of the lifetime image is important, then accuracy must be sacrificed in favor of efficiency in data acquisition and analysis. For full-field time-domain imaging (by means of an intensified CCD camera, exciting the full field of the object simultaneously and transferring the full image collected within a preset time interval to the computer), the data are acquired directly in the form of a histogram, where the data within consecutive serial time domains are already binned (summed). Usually, for full-field time-domain measurements, only a few time bins are acquired over the period of the light pulse to reduce the time required for data acquisition and data analysis. These binned histogram data are then fitted to exponentials (often with analytical, noniterative methods). In this case, it is never a problem that there could be less than one count in a bin, and the analysis, although less precise, is much faster.

A general discussion of fitting data densities, rather than preset time interval data, can be found in the literature (Gerschenfeld, 2000).

Another very efficient method to accurately fit time-domain data with exponentials is by fitting the data to a series of Chebyshev polynomials (Hamming, 1973; this is an extremely efficient way to fit data to polynomials) and then extracting the lifetimes according to recursion relations between the coefficients of the consecutive polynomials (Redford and Clegg, in preparation). This method can be used to obtain the lifetimes directly, independent of the amplitudes. Then the amplitudes can be determined through a linear analytical method, without iterations. This method also does not require initial guesses or iterative multiple passes through the data and is therefore very fast (De Boeck et al., 1985).

4.2 Determining the Phase and Modulation of Frequency-Domain Data in Real Time

Data analysis in the frequency domain consists of the following steps: *(1)* fitting the data for phase and modulation, *(2)* calibrating the phase and modulation by means of a standard, *(3)* calculating the lifetime or lifetimes from the corrected phase and modulation, and *(4)* visualizing the data (Lakowicz, 1999). As data are acquired the lifetimes must be calculated and displayed very quickly to provide real-time images. Data analysis for the frequency domain involves fitting the data to a sine function to extract the phase and modulation (this is assuming that only the fundamental frequency is of interest; otherwise, all higher-frequency overtones can also be fitted).

In the following discussion, we derive the equations for least-squares fit of homodyne data and show how the calculations can be carried out rapidly (Clegg and Schneider, 1996). The recorded fundamental frequency of the fluorescence signal over one period can always be expressed in general as a pure sinusoidal function with varying phase: $S_i = A + B \cos\varphi_i$). A and B are constants, and φ_i varies in N steps from an arbitrary starting phase of φ_0 to $[\varphi_0 + (N-1)(2\pi/N)]$; the measurements involve N equally spaced phase settings. That is, $\varphi_i = \varphi_0 + (i-1)2\pi/N$ for $i = 0$ to $N-1$. We define the following χ^2 function, which is the difference squared between the data, Y_i and the fitting function, $A + B \sin(\varphi_I)$. That is,

$$\chi^2 = \sum_{i=0}^{N-1} \left(\left(A + B\sin(\varphi_i) \right) - Y_i \right)^2 \tag{47}$$

Note that we could just as well have used the cosine function for this operation, which would have phase readings differing by $\pi/2$. We can then minimize the χ^2 error function by the usual least-squares methods of varying the three parameters A, B, and φ_0.

$$\frac{\partial \chi^2}{\partial A} = \sum_{i=0}^{N-1} 2\left(A + B\sin(\varphi_i) - Y_i \right) = 0$$

$$\frac{\partial \chi^2}{\partial B} = \sum_{i=0}^{N-1} 2\sin(\varphi_i)\left(A + B\sin(\varphi_i) - Y_i \right) = 0 \tag{48}$$

$$\frac{\partial \chi^2}{\partial \varphi_0} = \sum_{i=0}^{N-1} 2B\cos(\varphi_i)\left(A + B\sin(\varphi_i) - Y_i \right) = 0$$

These are three equations with three unknowns (A, B, and φ_0) that we can solve algebraically to give the phase and modulation of the homodyne signal:

$$\tan\varphi_0 = Y_{\text{sin}}/Y_{\text{cos}}$$
$$M^2 = \left(Y_{\text{sin}}^2 + Y_{\text{cos}}^2\right)/Y_{\text{dc}}^2 \tag{49}$$

We have defined $M = B/A$, and we have defined the following sums:

$$Y_{\text{sin}} = \sum_{i=0}^{N-1} \sin(\varphi_i)Y_i$$
$$Y_{\text{cos}} = \sum_{i=0}^{N-1} \cos(\varphi_i)Y_i \tag{50}$$
$$Y_{\text{DC}} = \sum_{i=0}^{N-1} Y_i$$

The latter sum equations are just lowest-order digital transforms of one period of the data, and we could have derived the expressions for the phase and modulation directly from digital Fourier transform theory.[1] These are not fast Fourier transforms, but the digital transform of the fundamental frequency over one period. This analysis is very rapid (no iterations, just $3N$ multiplications and three summations) and allows us to fit the data quickly using a single pass through the data; no iterative techniques are required.

Because we are particularly interested in minimizing the acquisition and display time, we can speed up the process if we do not wait until all phases have been acquired before calculating the digital transform. We can take advantage of the idle time between each of the images (at different phases) to calculate the components of the summation for the previously acquired data. We calculate and store the different components of the Y_{sin} and Y_{cos} sums for that image so that the remaining part of the fit is then a simple summation of the stored values. After progressing through all phases, the final arithmetic manipulations are carried out for calculating the phase and modulation (Holub et al., 2000). This procedure is especially advantageous if there is a large number of images being used in the fit and reduces the delay for the real-time calculation. One should use optimized code or look-up tables for the trigonometric functions, as these calculations are time consuming.

It is not necessary that the excitation light be modulated as a sinusoid. Any repetitive excitation pulse train will do. For instance, very short repetitive laser pulses, which are approximately repetitive delta functions at rates between 1 and 100 mhz, or repetitive square waves can be used. The frequency-domain analysis discussed above for sinusoidal light modulation can be applied to any repetitive data, regardless of the waveform. This simple, digital Fourier analysis can be used to determine the fundamental frequency parameters of any repetitive waveform. If the excitation pulse contains high-frequency components, the overtones (higher-frequency components) of the fundamental repetitive pulse period are also available for analysis, provided that the detector is modulated with a waveform containing the higher-frequency components. The presence of the higher frequencies does not interfere with calculation of the lower frequencies of the Fourier analysis. The data can be analyzed for the higher frequencies using the higher-order digital transforms (and possibly using the fast Fourier transform [FFT] algorithm, if several higher frequencies are analyzed and sufficient number of phase delays are acquired (Jameson et al., 1984; Clegg and Schneider, 1996).

If exceptionally high-frequency overtones are available in the data acquisition (delta function excitation and very high–frequency data acquisition), then data normally considered to be in the time domain can be fitted using these frequency domain methods (Hamming, 1973; Brigham, 1974; Bracewell, 1978; Hamming, 1989). That is, the time-recorded fluorescence decay can be analyzed in the Fourier domain. Of course, essentially all methods for measuring fluorescence decay signals involve data collection in the time domain. The separation into frequency domain and time domain is somewhat artificial. The essential difference is the form of the excitation modulation, which historically has influenced the type of data analysis, and the artificial division into frequency and time domains. For different excitation modulation forms, different analysis techniques are more convenient and efficient.

4.3 Calibrating and Correcting the Phase and Modulation

The values of $\varphi_0 + \varphi_R$ and M_0 are instrumentation specific and can be determined by measuring calibrated standards. The ordinary procedure is to use a sample with a known lifetime to calculate $\varphi_0 + \varphi_R$ and M_0 and and then use these values for subsequent imaging. For some setups, $\varphi_0 + \varphi_R$ and M_0 can be calibrated using reflected light or another zero lifetime target (second harmonic generation, for example, in a two photon instrument). Other setups measure $\varphi_0 + \varphi_R$ and M_0 directly by measuring a part of the excitation beam. A fluorescence standard with known fluorescence lifetime can also be used (because then we know $\arctan(\omega \tau_{ref})$. If there is drift in the excitation or the detector, $\varphi_0 + \varphi_R$ and M_0 vary slowly over time. An accurate method for instruments with high drift problems is to use self-calibrating images (images with regions of known lifetimes). Regardless of the method used, a simple addition/subtraction and division are needed to correct the phase and modulation for calculating the lifetimes.

4.4 Calculating Lifetimes from the Phase and Modulation

The phase and modulation lifetimes can be calculated directly according to the equations given previously. However, these equations involve very time-consuming trigonometric functions and square roots, which should be optimized with look-up tables or other routines. A library of optimized routines for the particular computer processor used to analyze the data can aid in carrying out the arithmetic operations on large amounts of data. Use of these routines is usually sufficient to ensure real-time processing of the data.

4.5 Data Visualization

The data need to be presented in a manner that can be easily interpreted by the user and in real time (Holub et al., 2000). A single data set has on the order of

800 kB of calculated data that needs to be displayed in less than 40 ms for real-time display. The form of the presentation needs to be such that the user can readily understand and interpret the relevant information quickly. The display of the information needs to be very fast.

A single lifetime image data set consists of three images: an intensity image and modulation and phase images. The phase and modulation can be used to calculate the single exponential effective phase and modulation lifetimes for display, if desired. Because the mapping is nonlinear, using the phase and modulation directly for display can lead to confusion when comparing the phase to the modulation. Therefore, we use the calculated lifetimes for display, because this allows the single lifetime regions to be easily seen (phase and modulation lifetimes are the same in these areas). The software has been written so that we can display the three images separately or side by side with a gray-scale or false-colorization map to create the displayed images. Another technique for providing very informative images is to take advantage of the multiple degrees of freedom in color space to display more than one data image in the same image. For example, one can put one image in the red channel of a red–green–blue (RGB) color-space and put another image in the green or blue channel (see Fig. 11.–4). Although this displays both data images in a single display image, the interpretation of the display colors may be confusing to some.

A better technique, called *bump-mapping*, is to display one data image (typically the intensity) as the total intensity and another data image as the color information (using any uniform-intensity color map). This provides the benefit of emphasizing the areas in the image where the data are valid (have enough intensity), acting like an intensity mask with no hard cutoff. By adding a third dimension to the display image, using an elevation map, for example, one can add in the other data image (as the height of the display image).

All of these image-processing techniques are computationally intensive, which makes it difficult to avoid delaying the data acquisition. However, modern video hardware will handle most of the image processing, freeing the central processor for acquisition tasks. These video display systems were created to handle three-dimensional display of objects, including texture mapping and pixel shading, which are needed for high-performance games. Images and three-dimensional elevation maps can be processed and displayed automatically with the video hardware now found in most personal computers.

4.6 Signal-to-Noise Considerations

One of the drawbacks to using fast-gated intensifiers is that they add appreciable noise to the system (Sturz, 1999). Somewhat larger signals are needed to accurately measure lifetimes than is required for simply measuring an intensity image. However, this difference is not large for the homodyne method described above. Nevertheless, often data with relatively smooth intensity spatial distributions can result in noisy pixel-by-pixel lifetime images. The nonlinearity of the equations given above accentuates the noise in a nonlinear fashion, causing spikes in the calculated parameters for low intensities (for example, background noise is often measured with random phase and extreme or no modulation, therefore regions of background

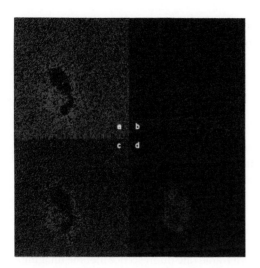

FIGURE 11–4 Lifetime images of protoporphyrin IX in a 3T3 tumor cell after 4 h of incubation with aminolevulinic acid. Lifetime imaging systems provide multiparameter images. Using color channels and bump-mapping, multiple parameters can be displayed in one image. Here frequency-domain data give phase and modulation lifetime images. *a:* The phase lifetime image is shown in a red false-color scale with black to red representing 0 to 8 ns. Note that the image is dominated by noise surrounding the object of interest because noise has a random phase and will result in lifetimes distributed across the range. *b:* The modulation lifetime image is shown in a blue false-color scale with black to blue representing 0 to 8 ns. The two lifetime images have different lifetimes in some regions of the cell (which is hard to visualize in the two separate images), indicating multiple lifetime systems. *c:* The two images are combined using the separate color channels (because red and blue are orthogonal in the color space, the sum can be displayed without loss of information). The resultant image has a false colorization representing both lifetimes. Magenta areas have single lifetimes, whereas blue areas have multiple lifetimes. The background noise surrounding the cell is still present in the image. *d:* The combined lifetime image is bump-mapped (see text) using the intensity information to show the structure of the object being imaged and to remove areas of the image that are primarily noise. The hue of the image still shows the lifetime information as in *c.* Now one can see the boundaries of the object of interest and how the lifetimes are related to the structure.

in the image will have extreme variations in lifetimes). One must be sure that the image-processing software can handle invalid data points (i.e., the noise can lead to negative square-roots and undefined arctangents at pixels with low intensities). Another problem is phase roll over (for instance, the computer will report –1° to be 359°). Also, the noise can cause pixels to have invalid combinations of modulation and phase (combinations that cannot arise from real $P(\tau)$s). All these problems can be solved with intelligent software implementations.

An easy way to reduce the noise in the lifetime images is to average pixels with their nearest neighbors (pixel binning). One can still use the high-resolution intensity image with bump-mapping while using the lower-resolution or smoothed lifetime image as the hue. A rule for avoiding a decrease in accuracy when fitting and a decrease in useful signal-to-noise level is to take the fit of the average, not the average of the fits. This means that one should not use a calculated lifetime image to create the averaged or smoothed image but should average or smooth the raw data before doing the fit. This should also be true of the image analysis software. Often one selects a region of an image to get an average, and the software should calculate the average from the raw data.

NOTE

1. It is well known that the χ^2 minimization for fitting a signal to a sine function and a direct digital Fourier transform analysis give the same result (Hamming, 1973).

REFERENCES

Agronskaia, A.V., L. Tertoolen, and G.H.C. High frame rate fluorescence lifetime imaging. *J. Phys. D Appl. Phys.* 36:1655–1662, 2003.

Bacskai, B.J., J. Skoch, G.A. Hickey, R. Allen, and B.T. Hyman. Fluorescence resonance energy transfer determinations using multiphoton fluorescence lifetime imaging microscopy to characterize amyloid-beta plaques. *J. Biomed. Opt.* 8:368–375, 2003.

Bastiaens, P.I.H. and A. Squire. Fluorescence lifetime imaging microscopy: spatial resolution of biochemical processes in the cell. *Trends Cell Biol.* 9:48–52, 1999.

Becker, R.S. *Theory and Interpretation of Fluorescence and Phosphorescence.* New York: Wiley Interscience, 1969.

Becker, W., A. Bergmann, M.A. Hink, K. König, K. Benndorf and C. Biskup. Fluorescence lifetime imaging by time-correlated single-photon counting. *Microsc. Res. Tech.* 63:58–66, 2003.

Berezovska, O., P. Ramdya, J. Skoch, M.S. Wolfe, B.J. Bacskai, and B.T. Hyman. Amyloid precursor protein associates with a nicastrin-dependent docking site on the presenilin 1-gamma-secretase complex in cells demonstrated by fluorescence lifetime imaging. *J. Neurosci.* 23:4560–4566, 2003.

Birch, D.J.S. and R.E. Imhof. Time-domain fluorescence spectroscopy using time-correlated single-photon counting. In: *Topics in Fluorescence Spectroscopy.* Edited by J. Lackowicz, New York: Plenum Press, 1991, pp. 1–88.

Birks, J.B. *Photophysics of Aromatic Molecules.* London: Wiley, 1970.

Birks, J.B., ed. *Organic Molecular Photophysics.* London: John Wiley & Sons, 1975.

Bracewell, R.N. *The Fourier Transform and Its Applications.* Tokyo: McGraw-Hill, 1978.

Brigham, E.O. *The Fast Fourier Transform.* Englewood Cliffs, NJ: Prentice-Hall, 1974.

Calleja, V., S.M. Ameer-Beg, B. Vojnovic, R. Woscholski, J. Downward, and B. Larijani. Monitoring conformational changes of proteins in cells by fluorescence lifetime imaging microscopy. *Biochem. J.* 372:33–40, 2003.

Chen, S.-H. and M. Kotlarchyk. Interactions of Photons and Neutrons with Matter—An Introduction. NJ: World Scientific, 1997.

Chen, Y., J.D. Mills, and A. Periasamy. Protein localization in living cells and tissues using FRET and FLIM. *Differentiation* 71:528–541, 2003.

Chen, Y. and A. Periasamy. Characterization of two-photon excitation fluorescence lifetime imaging microscopy for protein localization. *Microsc. Res. Tech.* 63:72–80, 2004.

Clayton, A.H.A., Q.S. Hanley, D J. Arndt-Jovin, V. Subramaniam, and T M. Jovin. Dynamic fluorescence anisotropy imaging microscopy in the frequency domain (rFLIM). *Biophys. J.* 83:1631–1649, 2002.

Clegg, R.M. Fluorescence resonance energy transfer. In: *Fluorescence Imaging Spectroscopy and Microscopy.* edited by X.F. Wang and B. Herman. New York: John Wiley & Sons, 1996, pp. 179–252.

Clegg, R.M., T.W.J. Gadella, and T.M. Jovin. Lifetime-resolved fluorescence imaging. *Proc. SPIE* 2137:105–118, 1994.

Clegg, R.M., O. Holub, and C. Gohlke. Fluorescence lifetime-resolved imaging: measuring lifetimes in an image. *Methods Enzymol.* 360:509–542, 2003.

Clegg, R.M. and P.C. Schneider. Fluorescence lifetime-resolved imaging microscopy: a general description of the lifetime-resolved imaging measurements. In: *Fluorescence Microscopy and Fluorescent Probes.* Edited by J. Slavik, New York: Plenum Press, 1996, pp. 15–33.

Clegg, R.M., P.C. Schneider, and T.M. Jovin. Fluorescence lifetime-resolved imaging microscopy. In: *Biomedical Optical Instrumentation and Laser-Assisted Biotechnology.* Edited by A.M. Verga Scheggi, S. Martellucci, A.N. Chester, and R. Pratesi. London: Kluwer Academic Publishers, 1996, pp. 143–156.

Craig, D.P. and T. Thirunamachandran. *Molecular Quantum Electrodynamics. An Introduction to Radiation Molecule Interactions.* Mineola: Dover Publications, 1984, pp. 324–336.

Cubeddu, R., D. Comelli, C. D'Andrea, P. Taroni, and G. Valentini. Time-resolved fluorescence imaging in biology and medicine. *J. Phys. D Appl. Phys.* 35:R61–R76, 2002.

Cundall, R.B. and R.E. Dale, eds. *Time-Resolved Fluorescence Spectroscopy in Biochemistry and Biology.* NATO ASI series. Series A, Life Sciences. New York: Plenum Press, 1983.

De Boeck, H., R.B.J. Macgregor, R.M. Clegg, N. Sharon, and F.G. Loontiens. Binding of *N*-dansylgalactosamine to the lectin from *Erythrina cristagalli* as followed by stopped-flow and pressure-jump relaxation kinetics. *Eur. J. Biochem.* 149:141–145, 1985.

DiCesare, N. and J.R. Lakowicz. Evaluation of two synthetic glucose probes for fluorescence-lifetime-based sensing. *Anal. Biochem.* 294:154–160, 2001.

Dowling, K., M.J. Dayel, P.M.W. French, P. Vourdas, M. J. Lever, A.K.L. Dymoke-Bradshaw, and J.D. Hares. Whole-field fluorescence lifetime imaging with picosecond resolution for biomedicine. *Technical Digest. Summaries of Papers Presented at the Conference on Lasers and Electro-Optics.* 6:308–312, 1998a.

Dowling, K., M.J. Dayel, S.C.W. Hyde, C. Dainty, P.M.W. French, P. Vourdas, M.J. Lever, A.K.L. Dymoke-Bradshaw, J.D. Hares, and P.A. Kellett. Whole-field fluorescence lifetime imaging with picosecond resolution using ultrafast 10–kHz solid-state amplifier technology. *IEEE J. Sel. Top. Quantum Electron.* 4:370–375, 1998b.

Dowling, K., M.J. Dayel, S.C.W. Hyde, P.M.W. French, M.J. Lever, J.D. Hares, and A.K.L. Dymoke-Bradshaw. High resolution time-domain fluorescence lifetime imaging for biomedical applications. *J. Modern Opt.* 46:199–209, 1999.

Duncan, R.R., A. Bergmann, M.A. Cousin, D.K. Apps, and M.J. Shipston. Multi-dimensional time-correlated single photon counting (TCSPC) fluorescence lifetime imaging microscopy (FLIM) to detect FRET in cells. *J. Microsc.* 215:1–12, 2004.

Eliceiri, K.E.W., C.-H. Fan, G.E. Lyons, and J.G. White. Analysis of histology specimens using lifetime multiphoton microscopy. *J. Biomed. Opt.* 8:376–380, 2003.

Enderlein, J. and R. Erdmann. Fast fitting of multi-exponential decay curves. *Opt. Commun.* 134:371–378, 1997.

Förster, T. *Fluoreszenz organischer Verbindungen.* Göttingen: Vandenhoeck & Ruprecht, 1951.

French, T., P.T.C. So, J. Weaver, D.J., T. Coelho-Sampaio, E. Gratton, E.W. Voss, and J. Carrero. Two-photon fluorescence lifetime imaging microscopy of macrophage-mediated antigen processing. *J. Microsc.* 185:339–353, 1997.

Gadella, T.W.J.J., R.M. Clegg, and T.M. Jovin. Fluorescence lifetime imaging microscopy: pixel-by-pixel analysis of phase-modulation data. *Bioimaging* 2:139–159, 2001.

Gadella, W.J.J., T.M. Jovin, and R.M. Clegg. Fluorescence lifetime imaging microscopy (FLIM): spatial resolution of microstructures on the nanosecond time scale. *Biophys. Chem.* 48:221–239, 1993.

Gerritsen, H.C., M.A.H. Asselbergs, A.V. Agronskaia, and W.G.J.H.M. Van Sark. Fluorescence lifetime imaging in scanning microscopes: acquisition speed, photon economy and lifetime resolution. *J. Microsc.* 206:218–224, 2002.

Gerschenfeld, N. *Mathematical Modeling.* Cambridge: Cambridge University Press, 2000.

Gratton, E., S. Breusegem, J. Sutin, R. Qiaoqiao, and N.P. Barry. Fluorescence lifetime imaging for the two-photon microscope: time-domain and frequency-domain methods. *J. Biomed. Opt.* 8:381–390, 2003.

Gratton, E. and M. Limkeman. A continuously variable frequency cross-correlation phase fluorometer with picosecond resolution. *Biophys. J.* 44:315–324, 1983.

Gratton, E., M. Limkerman, J.R. Lakowicz, B.P. Maliwal, H. Cherek, and G. Laczko. Resolution of mixtures of fluorophores using variable-frequency phase and modulation data. *Biophys. J.* 46:479–486, 1984.

Hamming, R.W. *Numerical Methods for Scientists and Engineers.* New York:Dover, 1973.

Hamming, R.W. *Digital Filters.* New York: Dover, 1989.

Hanson, K.M., M.J. Behne, N.P. Barry, T.M. Mauro, E. Gratton and R.M. Clegg. Two-photon fluorescence lifetime imaging of the skin stratum corneum pH gradient. *Biophys. J.* 83:1682–1690, 2002.

Harpur, A.G., F.S. Wouters, and P.I. Bastiaens. Imaging FRET between spectrally similar GFP molecules in single cells. *Nat. Biotechnol.* 19:167–169, 2001.

Harris, D.C. and M.D. Bertolucci. *Symmetry and Spectroscopy: An Introduction to Vibrational and Electronic Spectroscopy.* New York: Dover, 1978.

Hercules, D.M., ed. *Fluorescence and Phosphorescence Analysis.* New York: Wiley Interscience, 1966.

Hink, M.A., T. Bisselin, and A.J.W.G. Visser. Imaging protein–protein interactions in living cells. *Plant Mol. Biol.* 50:871–883, 2002.

Holub, O., M. Seufferheld, C. Gohlke, Govindjee and R.M. Clegg. Fluorescence lifetime-resolved imaging (FLI) in real-time—a new technique in photosynthetic research. *Photosynthetica* 38:581–599, 2000.

Jakobs, S., V. Subramaniam, A. Schonle, T.M. Jovin, and S.W. Hell. EGFP and DsRed expressing cultures of *Erichia coli* imaged by confocal, two-photon and fluorescence lifetime microscopy. *FEBS Lett.* 479:131–135, 2000.

Jameson, D.M., E. Gratton, and R.D. Hall. The measurement and analysis of heterogeneous emissions by multifrequency phase and modulation fluorometry. *Appl. Spectrosc. Rev.* 20:55–106, 1984.

Khorasani, S., A. Nojeh, and B. Rashidian. Design and analysis of the integrated plasma wave micro-optical modulator/switch. *Fiber Integr. Opt.* 21:173–191, 2002.

Koti, A.S.R. and N. Periasamy. Application of time resolved area normalized emission spectroscopy to multicomponent systems. *J. Chem. Phys.* 115:7094–7099, 2003.

Lakowicz, J.R. *Principles of Fluorescence Spectroscopy.* New York: Kluwer Academic/Plenum Publishers, 1999.

Lakowicz, J.R. and H. Szmacinski. Imaging applications of time-resolved fluorescence spectroscopy. In: *Fluorescence Imaging Spectroscopy and Microscopy.* Edited by X.F. Wang and B. Herman, New York: John Wiley & Sons, 1996, pp. 273–311.

Lippert, E. and J.D. Macomber. *Dynamics of Spectroscopic Transitions. Basic Concepts.* Berlin: Springer Verlag, 1995.

Marriott, G., R.M. Clegg, D.J. Arndt-Jovin, and T.M. Jovin. Time-resolved imaging microscopy. Phosphorescence and delayed fluorescence imaging. *Biophys. J.* 60:1374–1387, 1991.

McLoskey, D., D.J.S. Birch, A. Sanderson, K. Suhling, E. Welch, and P.J. Hicks. Multiplexed single-photon counting. I. A time-correlated fluorescence lifetime camera. *Rev. Sci. Instrum.* 67:2228–2233, 1996.

Mitchell, A.C., J.E. Wall, J.G. Murray, and C.G. Morgan. Measurement of nanosecond time-resolved fluorescence with a directly gated interline CCD camera. *J. Microsc.* 206:233–238, 2002.

O'Conner, D.V. and D. Phillips. *Time-Corellated Single Photon Counting.* London: Academic Press, 1984.

Oida, T., Y. Sako, and A. Kusumi. Fluorescence lifetime imaging microscopy (flimscopy). Methodology development and application to studies of endosome fusion in single cells. *Biophys. J.* 64:676–685, 1993.

Parker, C. A. *Photoluminescence of Solutions*. Amsterdam: Elsevier, 1968.

Periasamy, A., M. Elangovan, E. Elliott, and D.L. Brautigan. Fluorescence lifetime imaging (FLIM) of green fluorescent fusion proteins in living cells. *Methods Mol. Biol.* 183:89–100, 2002.

Periasamy, A., P. Wodnicki, X.F. Wang, S. Kwon, G.W. Gordon, and B. Herman. Time-resolved fluorescence lifetime imaging microscopy using a picosecond pulsed tunable dye laser system. *Rev. Sci. Instrum.* 67:3722–3731, 1996.

Peter, M. and S.M. Ameer-Beg. Imaging molecular interactions by multiphoton FLIM. *Biol. Cell* 96:231–236, 2004.

Piston, D.W., G. Marriott, T. Radivoyevich, R.M. Clegg, T.M. Jovin, and E. Gratton. Wideband acousto-optic light modulator for frequency domain fluorometry and phosphorimetry. *Rev. Sci. Instrum.* 60:2596–2600, 1989.

Rabinovich, E., M.J. O'Brien, S.R.J. Brueck, and G.P. Lopez. Phase-sensitive multichannel detection system for chemical and biosensor arrays and fluorescence lifetime-based imaging. *Rev. Sci. Instrum.* 71:522–529, 2000.

Rolinski, O.J. and D.J.S. Birch. Determination of acceptor distribution from fluorescence resonance energy transfer: theory and simulation. *J. Chem. Phys.* 112:8923–8933, 2000.

Rolinski, O.J. and D.J.S. Birch. Structural sensing using fluorescence nanotomography. *J. Chem. Phys.* 116:10411–10418, 2002.

Saleh, B.E.A. and M.C. Teich. *Fundamentals of Photonics*. New York: Wiley-Interscience, 1991.

Sanders, R., A. Draaijer, H.C. Gerritsen, P.M. Houpt, and Y.K. Levine. Quantitative pH Imaging in cells using confocal fluorescence lifetime imaging microscopy. *Anal. Biochem.* 227:302–308, 1995.

Schneckenburger, H., K. Stock, M. Lyttek, W.S. Strauss, and R. Sailer. Fluorescence lifetime imaging (FLIM) of rhodamine 123 in living cells. *Photochem. Photobiol. Sci.* 3:127–131, 2004.

Schneider, P.C. and R.M. Clegg. Rapid acquisition, analysis, and display of fluorescence lifetime-resolved images for real-time applications. *Rev. Sci. Instrum.* 68:4107–4119, 1997.

So, P.T.C., T. French, W.M. Yu, K.M. Berland, C.Y. Dong and E. Gratton. Two-photon fluorescence microscopy: time-resolved and intensity imaging. In: *Fluorescence Imaging Spectroscopy and Microscopy*. Edited by X.F. Wang and B. Herman, New York: John Wiley & Sons, 1996, pp. 351–373.

So, P.T.C., K. Konig, K. Berland, C.Y. Dong, T. French, C. Buhler, T. Ragan, and E. Gratton. New time-resolved techniques in two-photon microscopy. *Cell. Mol. Biol.* 44:771–793, 1998.

Steigmiller, S., B. Zimmermann, M. Diez, M. Börsch, and P. Gräber. Binding of single nucleotides to H+-ATP synthases observed by fluorescence resonance energy transfer. *Bioelectrochemistry* 63:79–85, 2004.

Stepanov, B.I. and V.P. Gribkovskii. *Theory of Luminescence*. London: ILIFFF Books LTD, 1968.

Strickler, S.J. and R.A. Berg. Relationship between absorption intensity and fluorescence lifetime of molecules. *J. Chem. Phys.* 37:814–820, 1962.

Sturz, R.A. Evolving image intensifier technology. *Proc. SPIE Int. Soc. Opt. Eng. USA* 3751:114–125, 1999.

Sytsma, J., J.M. Vroom, C.J. de Grauw, and H.C. Gerritson. Time-gated fluorescence lifetime imaging and microvolume spectroscopy using two-photon excitation. *J. Microsc.* 191:39–51, 1998.

Szmacinski, H. and J.R. Lakowicz. Optical measurements of pH using fluorescence lifetimes and phase-modulation fluorometry. *Anal. Chem.* 65:1668–1674, 1993.

Turro, N.J. *Modern Molecular Photochemistry*. Sausalito: University Science Books, 1991.

Valentini, G., C. D'Andrea, D. Comelli, A. Pifferi, P. Taroni, A. Torricelli, R. Cubeddu, C. Battaglia, C. Consolandi, G. Salani, L. Rossi-Bernardi, and G. De Bellis. Time-resolved DNA-microarray reading by an intensified CCD for ultimate sensitivity. *Opt. Lett.* 25:1648–1650, 2000.

Valeur, B. *Molecular Fluorescence: Principles and Applications*. Weinheim: Wiley-VCH, 2002.

vande Ven, M. and E. Gratton. Time-resolved fluorescence lifetime imaging. In: *Optical Microscopy: Emerging Methods and Applications*, edited by B. Herman and J.J. Lemasters, New York: Academic Press, 1992, pp. 373–402.

Vermeer, J.E., E.B. Van Munster, N.O. Vischer, and T.W.J. Gadella. Probing plasma membrane microdomains in cowpea protoplasts using lipidated GFP-fusion proteins and multimode FRET microscopy. *J. Microsc.* 214:190–200, 2004.

Wang, X.F., A. Periasamy, and B. Herman. Fluorescence lifetime imaging microscopy (FLIM): instrumentation and applications. *Crit. Rev. Anal. Chem.* 23:369–395, 1992.

Webb, S.E.D., Y. Gu, S. Lévêque-Fort, J. Siegel, M.J. Cole, K. Dowling, R. Jones, P.M.W. French, M.A.A. Neil, R. Juškaitis, L.O.D. Sucharov, T. Wilson, and M.J. Lever. A wide-field time-domain fluorescence lifetime imaging microscope with optical sectioning. *Rev. Sci. Instrum.* 73:1898–1907, 2002.

12

Streak Fluorescence Lifetime Imaging Microscopy: A Novel Technology for Quantitative FRET Imaging

V. Krishnan Ramanujan, Jian-Hua Zhang,
Victoria E. Centonze, and Brian Herman

1. Introduction

Fluorescence lifetime imaging microscopy (FLIM) methods are becoming an established technology for quantitative live-cell imaging. When coupled with the high temporal and spatial resolution of optical microscopy, lifetime imaging can provide novel information from intact cellular environments (Wang et al., 1996; Bastiaens et al., 1999; Hanley et al., 2001). This becomes particularly advantageous in situations where the imaged fluorescent probes are spectrally similar. The FLIM methods are relatively insensitive to artifacts that can affect spectroscopic characteristics of fluorophores and can provide distinct lifetimes characteristic of the individual fluorophores attached to each protein (e.g., donor and acceptor).

A variety of FLIM methods have been developed in the past decade for measuring various intracellular parameters by monitoring changes in fluorescence excited-state lifetime (Chapters 2, 11–13). Conventional time-domain FLIM methods such as multigate detection and single-photon counting have a typical temporal resolution of a few hundred picoseconds (ps)—a limitation imposed primarily by the detectors used in these systems. A new multiphoton FLIM system with good spatiotemporal resolution and better photon detection efficiency suitable for rapid lifetime determinations in live-cell applications has recently been developed (Krishnan et al., 2003a, 2003c).

This chapter provides a comprehensive description of this streak camera-based FLIM imaging (StreakFLIM) system and contrasts its performance with that of other currently available time-domain FLIM systems. The applicability of the present system in reliable FRET measurements is exemplified with substrates having cyan (CFP) and yellow (YFP) fluorescent proteins linked by short peptide sequence.

2. TIME-DOMAIN FLUORESCENCE LIFETIME IMAGING MICROSCOPY

2.1 General Principles

Radiative decay of fluorescence emission from a fluorescent molecule in its excited state can be represented by a monoexponential function:

$$I(t) = I_0 \exp(-t/\tau) \tag{1}$$

where $I(t)$ is the measured intensity, I_0 is the intensity at time $t = 0$, and τ is the fluorescence lifetime, characteristic of the molecule. This expression has to be modified for multiexponential fluorescence decays of the fluorophore.

The FLIM methods currently available can be broadly classified into two groups: (1) time-domain and (2) frequency-domain FLIM. In frequency-domain methods, the specimen is excited by a sinusoidally modulated wave of a certain frequency (such as radio frequencies for nanosecond lifetimes). The consequent fluorescence emission is also sinusoidally modulated but with different phase and amplitude relative to the excitation because of the time lag induced by "residence time," or the excited-state lifetime of the fluorophores. Generally, a gain-modulated detector is used to determine the phase shift and amplitude demodulation from which the lifetimes of the fluorescent species are calculated (French et al., 1997; Verveer et al., 2001). It has been found that the efficiency of gain-modulated detectors in frequency–domain methods is usually ~50%, which makes it necessary to obtain several sets of data at different excitation modulation frequencies for separating two or more lifetime components. Conversely, in the case of time-domain methods, the specimen is illuminated with pulsed light and fast detectors record the subsequent fluorescence emission decay. The fluorescence lifetimes of the different fluorescent species within the specimen are extracted by fitting the observed emission decay to an appropriate mathematical model. For a detailed discussion of these methods and a comparative survey of relative merits and shortcomings of these methods, the reader is referred to other chapters in this book and earlier review articles (Wang et al., 1996; Bastiaens et al., 1999). Regardless of the method adopted, either single-photon or multiphoton excitation can be used for measuring lifetimes.

The availability of characteristic laser lines (for exciting most of the commonly used fluorophores) and high-quality excitation and emission filters make single-photon excitation the most straightforward option in all FLIM methods. Conversion of a standard epifluorescence imaging (intensity) system to a lifetime imaging system is simpler if single-photon excitation is chosen. This conversion requires only some additional fast detection electronics.

Alternately, multiphoton excitation methods provide additional advantages that include an increase in spatial resolution, compared with that of wide-field single-photon excitation, deep-tissue imaging, a decrease in photodamage, substantial reductions in spectroscopic cross talk, and better reliability of the information obtained. In this chapter, we will limit our discussion to multiphoton time-domain FLIM methods and we refer the reader to excellent reviews in FLIM methods employing single-photon excitation (Morgan and Mitchell., 1996; French et al., 1998; Emptage, 2001). Laser diodes (visible wavelengths) are an alternate option for single-photon excitation with FLIM systems and provide better stability, but at the expense of lower output power (Mitchell et al., 2002; Mendez et al., 2004).

2.2 Streak Fluorescence Lifetime Imaging Microscopy

The heart of the StreakFLIM system is the *streak camera* (also called *streakscope*), a device that can measure ultrafast light phenomena with a temporal resolution as low as 2 ps. Figure 12–1a depicts the operating principle of the streakscope. It consists of a photocathode surface, a pair of sweep electrodes, a microchannel plate (MCP) to amplify photoelectrons coming off the photocathode, and a phosphor screen to detect the amplified output of MCP. Individual optical pulses (fluorescence emission) are collected from every point along a single line as the beam scanner scans across the *x*-axis. When these optical pulses (originating at different points in space and time) strike the photocathode surface, photoelectrons are emitted. As the photoelectrons pass between the pair of sweep electrodes a high voltage, synchronized with the excitation pulse, is applied to these sweep electrodes. This sweep voltage steers the electron paths away from the horizontal direction at different angles depending on their arrival time at the electrodes. The photoelectrons then get amplified in the MCP and reach the phosphor screen, forming an image of optical pulses arranged in vertical direction according to the time of their arrival at the sweep electrodes. The earliest pulse is arranged in the uppermost position, and the latest pulse is in the bottom-most portion of the phosphor image. The resulting streak image has space as the *x*-axis and time as the *y*-axis.

In addition to intensity and time information, the streakscope can be used to obtain either spatial or wavelength information. To measure fast optical phenomena using a streak camera, a crucial requirement is the trigger section that controls the timing of the streak sweep. Adjustment of the trigger section has to be such that the initialization of the streak sweep synchronizes with the arrival of the light at the streak camera. Figure 12–1b shows the operation timing of the streak camera at the time of sweep.

2.3 Comparison of Streak Approach with Other Time-Domain Fluorescence Lifetime Imaging Methods

In conventional time-domain FLIM, multigate detection with a gated microchannel plate image intensifier and a charge-coupled device (CCD) imaging camera is used (Straub and Hell, 1998; Gerritsen et al., 2002). By gating the image intensifier at different intervals (with a specified time window) along the exponential fluorescence decay profile, one can obtain a set of images of the actual decay of fluorescence during the excited-state lifetime (Krishnan et al., 2003b). In the multigate detection approach, the lifetime is extracted by measuring the fluorescence signal in at least two different time-gated windows. In streak imaging, there is no "gating" of fluorescence emission, so the complete fluorescence decay curves can be collected and then used to extract lifetime information at individual pixels. Photon

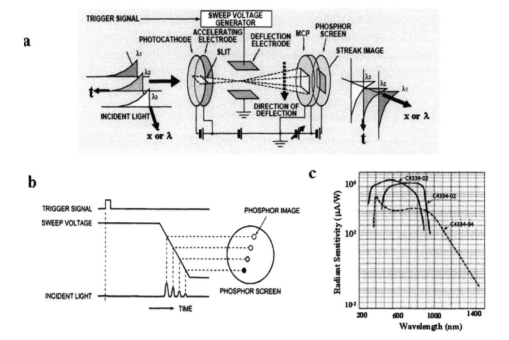

FIGURE 12–1 *a:* Schematic of the principle of streak imaging. An optical image with intensity information at every pixel (or wavelength) is converted to a streak image with spatial information as the horizontal axis and time as the vertical axis. Every point in the streak image provides three-dimensional information—intensity, space, and time, corresponding to every pixel in *a*. In this schematic, three optical pulses arrive at the slit with varying intensity and vary slightly in terms of time and space (or wavelength). As the corresponding photoelectrons from the photocathode pass between the pair of sweep electrodes, the applied sweep voltage steers the electron paths away from the horizontal direction at different angles, depending on their arrival time at the electrodes. The amplified electrons reach the phosphor screen, forming an image of three optical pulses arranged in a vertical direction according to the time of their arrival at the sweep electrodes. The earliest pulse is arranged in the uppermost position and the latest pulse is in the bottom-most portion of the phosphor image. The resulting streak image has space (or wavelength) as the *x*-axis and time as the *y*-axis. *b:* Operation timing in the streakscope at the time of sweep. *c:* Plot of radiant sensitivity of the streakscope (C4334) as a function of wavelength.

use and time resolution of the multigate detection approach are still limited by detector performance (Gerritsen et al., 2002). One also can not avoid background noise arising from scattered fluorescence emission when an imaging (area) detector is used in multigate detection. Regardless of these limitations, the multigate detection FLIM method has been very successful with single-photon excitation in the past (Sytsma et al., 1998).

For multiphoton applications, it is imperative to use point-scanning detectors to eliminate background noise. Streak imaging essentially gives four-dimensional (intensity, time, *x*- and *y*-axes) information with high photon efficiency. The streak camera has good radiant sensitivity over a wide bandwidth of wavelengths, making it suitable for both single-photon and multiphoton imaging applications without compromising detection sensitivity (Fig. 12–1c). To date, streak camera is the fastest photon detector available, with a time resolution of ~2 ps.

Another time-domain approach is *time-correlated single-photon counting* (TCSPC), which involves detection of single photons of a periodic light signal, measurement of their detection times, and construction of a histogram of detection times of the single emitted photons from these individual time measurements (Becker et al., 2004; Chapter 13). Photon-counting detectors are well suited for low-light detection, providing quantitized pulses for every photon event and thus making the measurement of lifetimes more accurate (Schonle et al., 2000; Becker et al., 2002). These detectors suffer from a poor dynamic range, compared with that of their area-detector counter parts. However, improved fast detectors, compatibility with laser scanning microscope modules, and advanced data analysis algorithms make TCSPC FLIM systems a reliable option.

The limiting factor in achieving good temporal resolution is the transition-time spread (TTS) of fast photomultiplier tubes used in these systems, which are minimal in streak camera detection. Although MCP versions yield a higher time resolution than that of fast photomultiplier tubes, they suffer from limitations in photon count rate. Furthermore, in low-light situations, the histogramming approach used in TCSPC systems can lead to significantly longer data acquisition times, making live-cell FLIM imaging difficult. For a similar signal-to-noise ratio, the streak camera has better photon detection efficiency and, hence, shorter data acquisition times under similar imaging conditions (Krishnan et al., 2003c). Table 12–1 summarizes the performance features of different ultrafast photon detectors with the streak camera.

3. Experimental System

A detailed technical description of the StreakFLIM system can be found elsewhere (Krishnan et al., 2003a). A concise description of the system configuration is provided below. Figure 12–2 shows a block diagram of the complete StreakFLIM system.

3.1 Light Sources

The laser system (Model Mira 900, Coherent Inc.) consists of a titanium:sapphire gain medium, providing mode-locked, ultrafast, femtosecond pulses with a fundamental frequency of 76 mHz. The tunability between 700 and 1000 nm allows a wide range of fluorophores to be studied using multiphoton excitation. A pulse

TABLE 12–1 Comparison of Time Resolution of Various Fast Photon Detector Types

Detector Type	Time Resolution
Conventional Photomultiplier tubes	~0.6 – 1.0 ns
Fast photomultiplier tubes	~140 – 220 ps
Microchannel plate photomultipliers	~25 – 30 ps
Avalanche photodiodes	~60 – 500 ps
Streak camera	~1.5 – 50 ps

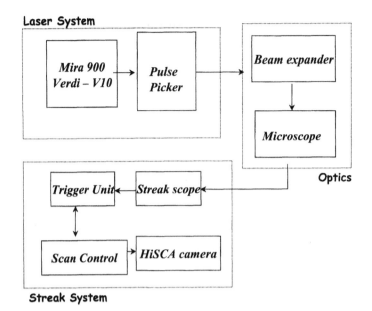

Laser System

Optics

Streak System

FIGURE 12–2 The streak FLIM imaging system is composed of a laser system (Coherent Mira 900 and Coherent pulse picker 9200), optics (Olympus IX70 inverted microscope, FLIM optics), and streak system (trigger unit, streakscope, HiSCA Argus camera).

picker (Model 9200, Coherent Inc.) is used to produce variable repetition rates (146 kHz to 4.75 mHz) and derives its synchronization signal from the 76 mHz photodiode output of the Mira laser. By reducing the repetition rate, which decreases the number of excitation pulses, the specimen is not exposed to unwanted excitation pulses and is thus protected from undue photobleaching. In optimum laser operation conditions, the output pulses of the pulse picker have a repetition rate of 500 kHz and power typically around 10 mW. Alternate possibilities for excitation include pulsed single-photon sources, laser diode modules, or continuous-wave laser sources with acousto-optical modulators.

3.2. Fluorescence Lifetime Imaging Microscopy Optics

In the FLIM optical path, the output of the pulse picker (typically 500 kHz–1 MHz) passes through a beam expander ($\times 3$) to achieve a large numerical aperture (NA) at the back focal plane of the microscope objective. The entire optical alignment of the FLIM elements is performed with respect to the optical axis of microscope objective. A beam scanner (Galvometer mirror) is used to scan the laser spot in the horizontal direction (x-axis) and is enabled by a triangular wave from a pulse generator (Hewlett Packard 8112A). By varying the peak-to-peak voltage amplitude of the triangular wave, one can change the scanning length. A raster scanning mechanism is employed for both the x- and y-axes. A focusing relay lens is positioned to sharpen the focus of scanned beam onto the dichroic D1 (with a cutoff at 750 nm), which also transmits the fluorescence emission to the streakscope. The imaging lens prior to the streakscope photocathode critically influences collection efficiency.

3.3 Microscope

An inverted Olympus IX70 microscope with a specially designed 63× (1.2 NA, IR) water-immersion objective is used in the system. The high NA provides for spatial confinement of the excitation power in a small focal volume, thus increasing the two-photon excitation probability. Experiments can be done with cover glasses (thickness of 0.17 mm) and specimens can be either solutions sandwiched between two coverslips or fixed cellular specimens mounted on standard microscope slides or live cells grown in appropriate medium.

3.4 Streak Detection and Readout

A specific trigger electronics module and a digital readout module are required to measure ultrafast optical phenomena with a streak camera. The trigger module controls the timing of the streak sweep (Fig. 12–1b) and initializes the streak sweep at the time of arrival of optical pulses at the streak camera. The streakscope (C4334, Hamamatsu Photonics, Japan) used in the present system has a temporal resolution ~20 ps and has a photocathode dark current three orders of magnitude smaller than that in the photomultiplier tubes used in single-photon counting systems, thus offering a high signal-to-noise ratio for measuring even weak fluorescence signals.

The electrical output of MCP is converted to an optical output (streak image) on the phosphor screen. This image is then read out by a fast CCD camera (Argus/HiSCA, Hamamatsu Photonics, Japan), which is lens-coupled to the streakscope, and then undergoes analog-to-digital conversion. An important feature of this camera is that it has a CCD that offers exceptionally high-speed and high-sensitivity detection. A maximum sampling rate of ~500 frames/s can be achieved with this camera for single-wavelength fluorescence measurements, which is more than 10 times higher than the normal video rate. This feature makes the system unique for rapid data acquisition with a better signal-to-noise ratio than that expected from conventional CCD cameras. Faster frame acquisition can be achieved by binning the pixels at the expense of loss of spatiotemporal resolution; thus a balance between the speed of data acquisition and required resolution has to be determined in every measurement. The camera's maximum number of effective pixels is 658 × 494 in the unbinned condition; a maximum of 32 × 32 binning (in both space and time axes of the streak image) can be achieved. The system can be operated in two scanning modes: slow scan with a better linearity (12 bits) or fast scan (10 bits).

3.5 Data Acquisition Scheme

Figure 12–3 schematically documents the data acquisition flow chart using a homogeneous fluorescence specimen. As the x-axis beam scanner scans a single excitation spot along a single line across the x-axis (rate of scanning ~1 ms per line, i.e., 658 points), the streak camera detects fluorescence emission decay at each individual pixel (Streak Image). When a synchronous y-scanning is carried out on the region of interest, the streak-imaging process gives a complete stack of (x,t) streak images. This stack contains the complete information of optical intensity as well as

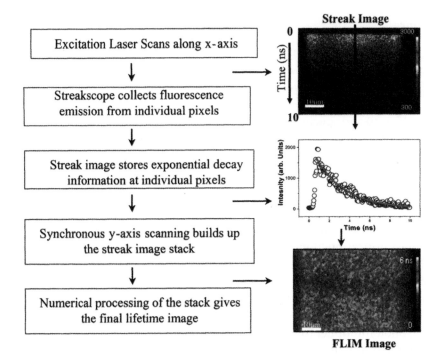

FIGURE 12–3 Data acquisition flowchart with a simplified representation of the steps involved in streak imaging and calculation of lifetime image. A fluorescent plastic specimen was used to demonstrate these steps (840 nm excitation; 1 MHz pulse repetition rate). The right panel shows the streak image, a representative exponential decay profile (arrow shown in streak image), and the calculated lifetime image after the (x,y) scan.

the spatial and temporal information from the optical image. Numerical processing of all these streak images (i.e., the exponential decay profiles at every pixel of the raw streak image) gives the final FLIM image. Every pixel in the FLIM image now contains the lifetime information (in contrast to intensity information in the optical image). The acquisition of FLIM data is controlled by use of AquaCosmos software (version 2.5, Hamamatsu Photonics, Japan).

4. STREAK FLUORESCENCE IMAGING MICROSCOPY–FRET APPLICATION

The StreakFLIM system can be employed to measure reliably Förster resonance energy transfer (FRET) in living cells, as exemplified in Figure 12–4. FRET measurements via lifetime imaging involve measuring the changes in lifetime of the donor molecule in the absence and presence of the acceptor. Spatial proximity (<10 nm) and sufficient overlap between the donor emission and acceptor absorption spectra determine the extent of energy transfer. The FLIM-FRET method does not suffer from the spectral cross talk and artifacts arising from protein expression variability, as are often encountered in intensity-based FRET measurements. One

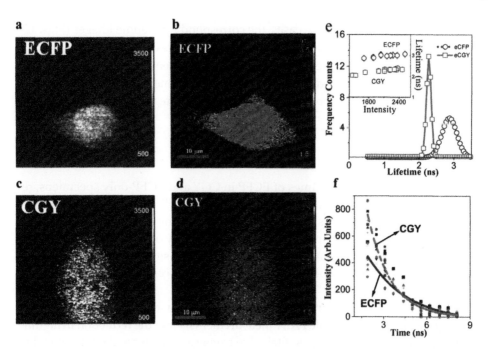

FIGURE 12–4 FRET imaging in cells. Fluorescence intensity (*a, c*) and lifetime (*b, d*) images of baby hamster kidney cells transfected with DNA constructs encoding ECFP and CGY (see text for details) are shown. *e:* Lifetime histograms were calculated from whole-cell analysis and individual regions (inset). *f:* Exponential decay profiles were obtained from individual pixels randomly selected in (*a*) and (*c*). Solid lines are the mean exponential fit to these pixels and serve as guides to the eye. Laser output from Mira 900 (76 mHz rep rate; 840 nm wavelength; ~200 fs pulse width) was rendered to the pulse picker, which stepped down the repetition rate to 500 kHz optimized for the streakscope triggering. The average power at the entrance of the FLIM optics was 8 mW. The 12-bit streak images were obtained in the 1 × 32 binning condition. Exposure times for imaging (~98 ms per single line) were optimized to minimize photobleaching during data acquisition. The lifetime images were calculated from the raw streak images with AquaCosmos software (Hamamatsu Photonics, Japan). An intensity threshold criterion of about 10% of the maximum dynamic range was used in calculation of the lifetime images and the final images were median filtered. The calculated mean lifetimes of the ECFP and CGY specimens were 2.9 ± 0.18 ns and 2.44 ± 0.27 ns, respectively, which implies an energy transfer efficiency of ~18%.

can therefore expect to obtain a more reliable estimate of energy transfer efficiency in FLIM-based FRET methods.

A variety of cell biological applications have been reported in which enhanced cyan (ECFP) and yellow (EYFP) fluorescent proteins were employed as the donor–acceptor pair, owing to their favorable spectral criteria for FRET detection. In the following example, FRET measurements are demonstrated for a FRET substrate (CGY) engineered in our laboratory by introducing a polyglycine peptide linker between ECFP and EYFP. This substrate was designed in such a way that the donor–acceptor fluorophores would be close enough to allow energy transfer between them.

Figures 12–4a and 12–4c show intensity images of baby hamster kidney (BHK) cells expressing ECFP only and CGY, respectively, in the cytosol. Although the intensity is similar in both these cases, the corresponding lifetime images (Figs. 12–4b, d) reveal a drastic reduction in ECFP lifetime in cells expressing CGY, thereby demonstrating significant energy transfer in this specimen. The efficiency (E) of energy transfer can be calculated from the mean ECFP lifetimes, as shown in the whole-cell histogram (Fig. 12–4e), and/or from the individual regions (inset of Fig. 12–4e) and by the following equation:

$$E = 1 - (\tau_{DA}/\tau_D)$$

where τ_{DA} and τ_D are the lifetimes of ECFP in CGY and ECFP-only specimens, respectively. In the above example, the lifetime of ECFP alone was 2.9 ± 0.18 ns (Fig. 12–4b). For the CGY fusion protein, the ECFP lifetime was reduced to 2.44 ± 0.27 ns (Fig. 12–4d), giving an energy transfer efficiency of ~18%. The lifetime values reported in this chapter are obtained by assuming a single-component lifetime, although there are recent reports suggesting that ECFP can have two lifetime components (Rizzo et al., 2004).

There are two interesting features that are relevant to sensitivity of the StreakFLIM system for reliable FRET measurements. First, the reduction in lifetime observed in the CGY specimen is solely due to energy transfer and not to any other intensity artifacts. This is illustrated in the inset of Figure 12–4e, where the observed lifetime was constant over a wide range of intensities in both the specimens. This feature makes it easier to implement FRET measurements in situations where the protein expression levels can vary significantly. Regardless of this feature, it is important in practice to minimize photobleaching effects, as these can directly affect the excited-state reactions significantly. Second, it is possible to carry out reliable FRET measurements in individual pixels without compromising the spatiotemporal resolution. For example, Figure 12–4f shows the intensity decay in individual pixels randomly selected from both Figures 12–4a and 12–4c. The solid lines are the mean exponential decay fits to these pixels. Clearly the CGY substrate exhibits a sharper decay than the ECFP-only specimen, exemplifying reliable FRET measurements even in a single pixel. This sensitive feature can be exploited in FRET determinations in intracellular organelles such as mitochondria, Golgi neurons, and endosomes whose sizes are comparable to a few pixels. Finally, the entire set of measurements can be carried out rapidly (~1–30 s, depending on the size of the region) so that live-cell FRET measurements via the StreakFLIM system are possible.

5. Perspectives

In this chapter, the examples of biomedical application of the multiphoton StreakFLIM system are limited to FRET measurements in living cells. It is possible, however, to extend the scope of the streak imaging to tissues and small animals because multiphoton excitation enables one to perform deep-tissue imaging comfortably. It is imperative that one resolves the potential autofluorescence contributions from different sections of the tissues while imaging because tissue autofluorescence itself has characteristic lifetimes. Alternatively, there is a potential application of the StreakFLIM system in the study of cancer biology by imaging the modifications in tissue autofluorescence lifetimes as evidenced in tumor growth (Zint et al., 2003). Another potential medical application of streak imaging is in the design of endoscopic FLIM probes for noninvasive detection of intrinsic fluorescence from liver and stomach (Siegel et al., 2003). The medical applications of streak imaging are not straightforward and should be accompanied by concomitant development of advanced computational procedures for three-dimensional reconstruction of lifetime data and for analysis of lifetime images on a per-pixel basis. With the current advancements in laser excitation sources, fiber optic systems, and computational software, it is reasonable to envisage that the present StreakFLIM system will find realistic applications in biomedical research and medical imaging technology.

We thank Hamamatsu Photonics K.K., Japan, for technical support, for many useful discussions with colleagues, and for some of the figures used in this chapter.

REFERENCES

Bastiaens, P.I. and A. Squire. Fluorescence lifetime imaging microscopy: spatial resolution of biochemical processes in the cell. *Trends. Curr. Biol.* 9:48–52, 1999.

Becker, W., A. Bergmann, M.A. Hink, K. König, and C. Biskup. Fluorescence lifetime imaging by time-correlated single-photon counting. *Microsc. Res. Tech.* 63:58–66, 2004.

Becker, W., A. Bergmann, and G. Weiss. Lifetime imaging with the Zeiss LSM- 510. *Proc. SPIE* 4620:30–35, 2002.

Emptage, N.J. Fluorescent imaging in living systems. *Curr. Opin. Pharmacol.* 1:521–525, 2001.

French, T., P.T. So, C.Y. Dong, K.M. Berland, and E. Gratton. Fluorescence lifetime imaging techniques for microscopy. *Methods Cell Biol.* 56:277–304, 1998.

French, T., P.T. So, D.J. Weaver, T. Coelho-Sampaio, E. Gratton, E.W. Voss, and J. Carrero. Two-photon fluorescence lifetime imaging microscopy of macrophage-mediated antigen processing. *J. Microsc.* 185:339–353, 1997.

Gerritsen, H.C., M.A.H. Asselbergs, A.V. Agronskaia, and W.G.J.A.H.M. Van Sark. Fluorescence lifetime imaging in scanning microscopes: acquisition speed, photon economy and lifetime resolution, *J. Microsc.* 206:218–224, 2002.

Hanley, Q.S, V. Subramaniam, D.J. Arndt-Jovin, and T.M. Jovin. Fluorescence lifetime imaging: multi-point calibration, minimum resolvable differences, and artifact suppression. *Cytometry* 43:248–260, 2001.

Krishnan, R.V., H. Saitoh, H. Terada, V.E. Centonze, and B. Herman. Development of a multiphoton fluorescence lifetime imaging microscopy system using a streak camera. *Rev. Sci. Instrum.* 74:2714–2721, 2003a.

Krishnan, R.V., A. Masuda, V.E. Centonze, and B. Herman. Quantitative imaging of protein–protein interactions by multiphoton fluorescence lifetime imaging microscopy using as streak camera. *J. Biomed. Opt.* 8:362–368, 2003b.

Krishnan, R.V., E. Biener, J.H. Zhang, R. Heckel, and B. Herman. Probing subtle flurescence dynamics in cellular proteins by streak camera based fluorescence lifetime imaging microscopy. *Appl. Phys. Lett.* 83:4658–4660, 2003c.

Mendez, E., D.S. Elson, M. Koeberg, C. Dunsby, D.D.C. Bradley, and P.M.W. French. Fluorescence lifetime imaging using a compact, low-cost, diode-based all-solid-state regenerative amplifier. *Rev. Sci. Instrum.* 75:1264–1267, 2004.

Mitchell, A.C., J.E. Wall, J.G. Murray, and C.G. Morgan. Direct modulation of the effective sensitivity of a CCD detector: a new approach to time-resolved fluorescence imaging. *J. Microsc.* 206:225–232, 2002.

Morgan, C.G. and A.C. Mitchell. Fluorescence lifetime imaging: an emerging technique in fluorescence microscopy. *Chromosome Res.* 4:261–263, 1996.

Rizzo, M.A., G.H. Springer, B. Granada, and D.W. Piston. An improved cyan fluorescent protein variant useful for FRET. *Nat. Biotechnol.* 22:445–449, 2004.

Schonle, A., M. Glatz, and S.W. Hell. Four-dimensional multiphoton microscopy with time-correlated single-photon counting. *Appl. Opt.* 39:6308–6311, 2000.

Siegel, J., D.S. Elson, S.E.D. Webb, K.C. Benny Lee, A. Vlandas, S. Leveque-Fort, M.J. Lever, P.J. Tadrous, G.W.H. Stamp, A. Wallace, A. Sandison, T.F. Watson, F. Alvarez, and P.M.W. French. Studying biological tissue with fluorescence lifetime imaging: microscopy, endoscopy and complex decay profiles. *Appl. Opt.* 42:2995–3004, 2003.

Straub, M. and S.W. Hell. Fluorescence lifetime three-dimensional microscopy with picosecond precision using a multifocal multiphoton microscope. *Appl. Phys. Lett.* 73:1769–1771, 1998.

Sytsma, J., J.M. Vroom, C.J. de Grauw, H.C. Gerritsen. Time-gated lifetime imaging and micro-volume spectroscopy using two-photon excitation. *J. Microsc.* 191:39–51, 1998.

Verveer, P.J., A. Squire, and P.I. Bastiaens. Improved spatial discrimination of protein reaction states in cells by global analysis and deconvolution of fluorescence lifetime imaging microscopy data. *J. Microsc.* 202:451–456, 2001.

Wang, X.F., A. Periasamy, P. Wodnicki, G.W. Gordon, and B. Herman. Fluorescence lifetime imaging microscopy. In: *Fluorescence Imaging Spectroscopy and Microscopy*, edited by X.F. Wang and B. Herman, New York: Wiley Interscience, 1996, pp. 313–321.

Zint, C.V., W. Uhring, M. Torregrossa, B. Cunin, and P. Poulet. Streak camera: a multi-detector for diffuse optical tomography. *Appl. Opt.* 42 pp. 3313–3320, 2003.

13

Time-Correlated Single-Photon Counting Fluorescence Lifetime Imaging–FRET Microscopy for Protein Localization

YE CHEN AND AMMMASI PERIASAMY

1. INTRODUCTION

A fluorophore molecule is characterized not only by its emission spectrum but also by its characteristic fluorescence lifetime. Therefore, measurements of fluorescence lifetime can provide information about a molecule's environment in biological systems. Typical examples are the mapping of cell parameters such as pH, ion concentrations, or oxygen saturation by fluorescence quenching, and probing protein or DNA structures by lifetime-sensitive dyes (Eftink, 1992; Gerritsen et al., 1997; Knemeyer et al., 2000; Periasamy, 2001; van Zandvoort et al., 2002).

Methods for measuring fluorescence lifetimes are divided into two major categories based on the instruments used: frequency domain (Lakowicz, 1999) and time domain (O'Connor and Phillips, 1984). Frequency-domain fluorometers excite fluorescence with sinusoidal light that is modulated at radio frequencies (for nanosecond decays). The phase shift and amplitude attenuation of the fluorescence emission are then measured relative to the phase and amplitude of the exciting light. Different lifetime values will cause specific phase shifts and attenuation at a given frequency. Lifetime imaging can be achieved by modulating the image intensifiers (Squire et al., 2000; Clayton et al., 2002; Chapters 2, 11) or by modulating single-channel detectors (Carlsson and Liljeborg, 1998).

In time-domain methods, pulsed light is used as the excitation source, and fluorescence lifetimes are measured from the fluorescence signal directly or by using time-correlated single-photon counting (TCSPC). Lifetime imaging is achieved by gated image intensifiers (Elangovan et al., 2002), by streak camera (Chapter 10), by counting the photons in several parallel time gates (Gerritsen et al., 1997, 2001), or by a time-correlated photon counting (Bacskai et al., 2003; Becker et al., 2004).

The success of two-photon TCSPC lifetime imaging in biological specimens has paralleled the development of ultrafast (femtosecond) tunable lasers and sensitive photomultiplier tubes (PMTs). The TCSPC provides high time resolution, which depends on the transit time spread of the detector such as fast conventional PMTs (150 ps) and microchannel plate (MCP) PMTs (30 ps) (Becker et al., 2004). Instead of using curve-fitting methodology based on fluorescence decay, the time-resolved fluorescence microscopic technique allows measurement of dynamic events at very high temporal resolution and monitoring of interactions between cellular components with very high spatial resolution. The fluorescence lifetime imaging microscopy (FLIM) system is used effectively to measure the fluorescence anisotropy decays of the fluorophore molecule to obtain the degree of molecular orientation. This helps to study the dynamics and structure of molecules in biological systems (Lakowicz, 1999; Volkmer et al., 2000; Clayton et al., 2002; Chapter 2). Moreover, by combining FLIM with Förster resonance energy transfer (FRET) we provide a means for depicting the dynamic events of protein associations in four dimensions (three-dimensional space and time).

In this chapter, we describe the development and characterization of a two-photon excitation TCSPC FLIM-FRET microscopy system that uses TCSPC imaging mode hardware from Becker & Hickl, Biorad Radiance 2100 confocal/multiphoton microscopy, and high-speed Coherent ti:sapphire laser systems. We have used TCSPC FRET-FLIM to visualize the localization and distribution of interactions of the transcription factor CAATT/enhancer binding protein alpha (C/EBPα) in living pituitary cells.

2. TIME-CORRELATED SINGLE-PHOTON COUNTING FLIM-FRET MICROSCOPY

2.1 FRET Microscopy

The various chapters in this book, as well as the literature cited in those chapters, describe the basis for FRET microscopy. Briefly, FRET is a distance-dependent physical process by which energy is transferred nonradiatively from an excited molecular fluorophore (the donor) to another fluorophore (the acceptor) by means of intermolecular long-range dipole–dipole coupling. FRET can be an accurate measurement of molecular proximity at Ångstrom distances (10–100 Å) and is highly efficient if the donor and acceptor are positioned within the Förster radius (the distance at which half the excitation energy of the donor is transferred to the acceptor, typically 3–6 nm). The efficiency of FRET is dependent on the inverse sixth power of intermolecular separation, making it a sensitive technique for investigating a variety of biological phenomena that produce changes in molecular proximity (Förster, 1948, 1965; dos Remedios et al., 1987; Clegg et al., 1994; Clegg, 1996; Lakowicz, 1999; Sekar and Periasamy, 2003; Chapter 2).

2.2 Two-Photon Excitation Microscopy

A significant improvement over confocal FRET can be achieved if the out-of-focus signal is eliminated altogether by limiting excitation to only the fluorophore at the focal plane (Chapter 5). This is precisely what two-photon FRET (2p-FRET) microscopy does. Because two-photon excitation occurs only in the focal volume, the detected emission signal is exclusively in-focus light. Furthermore, because two-photon excitation uses a longer-wavelength, high-speed pulsed light source, it is less damaging to living cells, thus limiting problems associated with fluorophore photobleaching and photodamage in the specimen, as well as intrinsic fluorescence of cellular components. Since the selected donor excitation wavelength can also excite (about <10%) the acceptor fluorophore molecule present, the technique requires correction to remove unwanted fluorescence signal in the FRET image (Elangovan et al., 2003; Chapters 6, 7). It is important to note that appropriate average power should be used to reduce photobleaching.

The infrared illumination in two-photon excitation penetrates deeper into the specimen than visible light excitation because of its higher energy, making it ideal for many applications involving depth penetration through thick sections of tissue (see Chapter 6). Since pulsed lasers are used as an excitation source, this configuration is an ideal system for FLIM.

2.3 Fluorescent Lifetime Imaging Microscopy

The FLIM microscopic technique allows the measurement of dynamic events at very high temporal resolution (nanoseconds). An important advantage of FLIM measurements is that they are independent of change in probe concentration, photobleaching, and other factors that limit intensity-based steady-state measurements. When combined with FRET, this approach can provide direct evidence for the physical interactions between proteins with very high spatial and temporal resolution. Because only one protein partner, the donor, is monitored, it is unnecessary to use spectral bleedthrough (SBT) correction in FRET-FLIM images. The one- and two-component analysis of the donor molecule lifetime in the presence of acceptor demonstrates the distance distribution among interacting proteins (Elangovan et al., 2002; Chen and Periasamy, 2004).

The energy transfer efficiency (E), the rate of energy transfer (k_T), and the distance between donor and acceptor molecule (r) are calculated using the following equations (Lakowicz, 1999; Elangovan et al., 2002):

$$E = 1 - (\tau_{DA}/\tau_D) \tag{1}$$

$$k_T = (1/\tau_D)\,(R_0/r)^6 \tag{2}$$

$$r = R_0\{(1/E) - 1\}^{1/6} \tag{3}$$

$$R_0 = 0.211\{k^2\,n^{-4}\,Q_D\,J(\lambda)\}^{1/6} \tag{4}$$

where τ_D and τ_{DA} is the donor excited state lifetime in the absence and presence of the acceptor; R_0 is the Förster distance—that is, the distance between the donor and acceptor at which half the excitation energy of the donor is transferred to the acceptor while the other half is dissipated by all other processes, including light emission; n is the refractive index; Q_D is the quantum yield of the donor; and κ^2 is a factor describing the relative dipole orientation (normally assumed to be 2/3; Lacowicz, 1999).

2.4 Instrumentation

The system used for the particular experiments described here consists of a Nikon TE300 epifluorescent microscope with a 100W Hg arc lamp. A Plan Fluor 60× numerical aperture (NA) 1.4 oil infrared (IR) and 60× W objective lens were used for 2p-FRET and 2p-FRET-FLIM image acquisition (Chen and Periasamy, 2004). The TE300 was coupled to the Biorad Radiance 2100 confocal/multiphoton system (www.zeiss.de/cellscience). A 10W Verdi pumped, tunable (model 900 Mira, www.coherentinc.com), mode-locked, ultrafast (78 mHz) pulsed (<150 fs) laser was coupled to the laser port of a Radiance 2100 (Fig. 13–1). This laser is equipped with x-wave optics for an easy tunable range of the entire wavelength (700–1,000 nm). The system was equipped with a laser spectrum analyzer (Model E201; www. istcorp.com) to monitor the excitation wavelength and a power meter to measure the laser power at the specimen plane (Model SSIM-VIS & IR; www.coherentinc. com). The Radiance system was equipped with an external detector (DDS) and four internal detectors (not shown) for fluorescence imaging. The transmission detector was used for transmission imaging and also for second harmonic generation (SHG) imaging. LaserSharp 2000 software was used to acquire 2p-FRET with both the internal and direct (external) detectors. FLIM images were acquired using lifetime photomultiplier tubes (LPMT) and the LaserSharp 2000 and Becker and Hickl SPCM acquisition software.

The FLIM PMT (LPMT; model# PMH-100) was coupled in the middle of the arm connected to the direct detector as shown in Figure 13–1. The coupler was removed in the arm and a flip mirror [FM] was inserted to direct the emission fluorescence signals from the specimen to the FLIM detector (LPMT) or to the direct

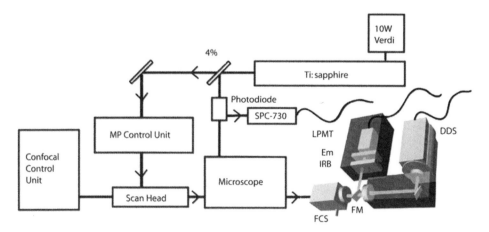

FIGURE 13–1 Schematic illustration of time-correlated single-photon counting FRET-FLIM microscopy. DDS, direct detector system (external detectors); Em, emission filter wheel; FCS, filter cube slider; FM, flip mirror; IRB, infrared laser light-blocking filter; LPMT-Lifetime photomultiplier tube; MP, multiphoton.

detector (DDS). It is also possible to detect simultaneously both 2p-FRET and FLIM signal through this configuration. The six-position dual filter wheel is installed between the LPMT and the VSM to select appropriate emission filters, depending on the fluorophore used for protein molecular imaging. In another filter wheel we used BG-36 glass filter, which blocks the excitation IR laser light and also transmits the visible spectrum, about 70% at 500 nm (www.chromatech.com). The whole system including the microscope was covered with a black box to reduce the background counts to as low a level as possible.

The Becker & Hickl GmbH (http://www.becker-hickl.de) photon-counting module (TCSPSC, SPC-730) is widely used to acquire FLIM images for various biological applications (Bacskai et al., 2003; Eliceiri et al., 2003; König and Riemann, 2003; Becker et al., 2004; Chen and Periasamy, 2004). The x- and y-scan synchronizing pulses, together with a pixel clock signal from the Radiance 2100 control unit, are used to synchronize data collection in the SPC-730 board (Fig. 13–2). This

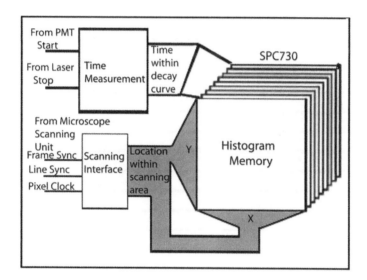

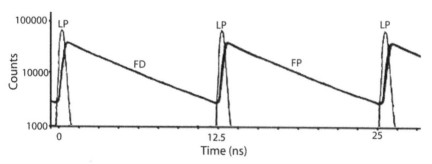

FIGURE 13–2 Synchronization of various signals. *a:* Schematic illustration of synchronization of various sync signals (excitation laser pulse, pixel clock, and x, y scanning signals) with the SPC-730 board. PMT, photomultiplier. *b:* Illustration of laser pulse (LP) excitation vs. fluorescence decay (FD). This demonstrates that one can measure lifetime reliably up to 10 ns. [Adapted from Becker & Hickl technical manual, www.becker-hickl.de]

allows pixel-by-pixel registration of the accumulated photons during laser scanning. Laser pulses are detected by a high-speed PIN photodiode and used by the SPC-730 board to determine the detection time of a photon (anode pulse from the PMT) relative to the laser pulses. This measurement system requires that the timer (a time-to-amplitude converter [TAC]) be activated only on receipt of a detected photon rather than at every laser pulse. The SPC-730 system starts timing at the receipt of a detected photon and measures the time interval until the next laser pulse. A fluorescence decay histogram of photon emission times relative to the laser excitation pulse is generated from the distribution of interpulse intervals at each pixel of the image (Fig. 13–3). The detector is a fast PMT (LPMT), with a full width half maximum (FWHM) response time of ~150 ps (PMH-100, Becker and Hickl). This

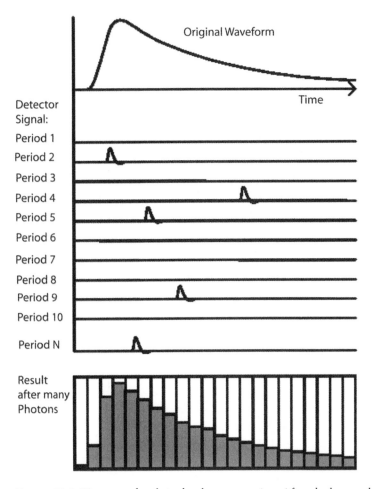

FIGURE 13–3 Time-correlated single-photon counting. After the laser pulse excitation, the photon is detected and the time of the detector pulse events are collected in a memory location with an address proportional to the detection time as shown in the figure. After collecting many photons, a histogram of the detection times is stored in the memory (as shown at the bottom of the figure) for further data analysis to estimate the lifetime of the molecule. [Adapted from Becker & Hickl technical manual, www.becker-hickl.de]

detector is fast enough to resolve lifetimes due to environmental changes in a biological system or protein–protein interactions.

2.5 Cell Preparation

For the studies described here, the sequence encoding the DNA binding and dimerization domain of the transcription factor C/EBPα was fused in-frame to the commercially available enhanced yellow fluorescent protein (ECFP) or enhanced yellow fluorescent protein (EYFP) or enhanced green fluorescent protein (EGFP) color variants (www.clontech.com) to generate CFP-C/EBPΔ244 and YFP-C/EBPΔ244 (Day et al., 2003; Chapter 4). For transfections, mouse pituitary GHFT1–5 cells were harvested and transfected with the indicated plasmid DNA(s) by electroporation (Day, 1998). The total input DNA was kept constant with empty vector DNA. Cell extracts from transfected cells were analyzed by western blot to verify that the tagged proteins were of the appropriate size, as described previously (Day, 1998). For imaging, the cells were inoculated drop-wise onto a sterile cover glass in 35 mm culture dishes, allowed to attach prior to gently flooding the culture dish with media, and maintained for 18 to 36 h prior to imaging. The cover glass with cells attached was inserted into a chamber containing the appropriate medium and the chamber was then placed on the microscope stage.

2.6 Selection of Excitation Spectra

It is important to determine the absorption cross-section or the excitation spectra of the fluorophore molecules under investigation (Xu and Webb, 1997; Chen and Periasamy, 2004). To obtain the two-photon excitation spectra for enhanced blue fluorescent protein (EBFP), EGFP, ECFP, EYFP, and dsRED1 we fused these proteins to C/EBPα in a living GHFT1–5 cell nucleus and tuned the ti:sapphire laser from 700 to 1000 nm. The measured intensity at various excitation wavelengths was normalized for the excitation intensity and then plotted as shown in Figure 13–4. The peak wavelength was used to excite the respective fluorophore molecules.

2.7 Selection of Average Excitation Power

Photobleaching is an inherent phenomenon in fluorescence imaging. To reduce the photobleaching it is important to select appropriate average excitation power at the specimen plane. This is particularly important for FLIM-FRET data acquisition. Some of the components of lifetime measurements will be lost from photobleaching and it is difficult to determine the correct level of energy transfer. The other alternative is to use a correction methodology for photobleaching. The selection of average excitation power varies according to the experimental setup and cell preparation. Excitation average power and acquisition time should be optimized to obtain reasonable photon counts (>500) to achieve a good statistical fit.

As shown in Figure 13–5, we used single expressed CFP-C/EBPΔ244 live cells and collected the images via time-lapse configuration at about 1 mW average excitation power at the specimen plane. Figure 13–5a shows negligible photobleaching

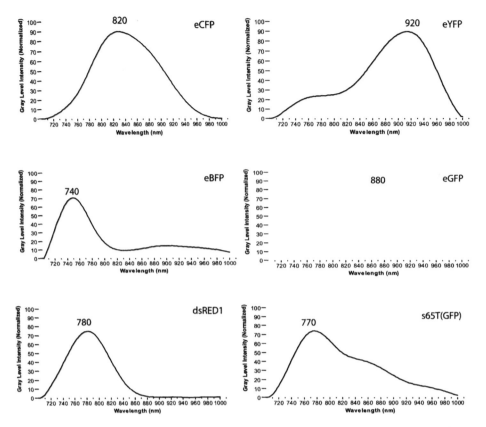

Figure 13–4 Two-photon excitation spectra for eCFP, eYFP, eBFP, eGFP, dsRED1, and s65T. [Adapted from Chen and Periasamy 2004.]

and the mean lifetime remains the same at 2.5 ns for continuous data collection at 1 mW (Figs. 13–5b–d). If the average excitation power is increased to 4 mW at the specimen plane, the mean lifetime appears to decrease (Figs. 13–5e–j), which suggests that some components of the lifetime distribution were removed by photobleaching (see Fig. 13–5 for more details).

3. Acquisition of Fluorescence Lifetime Imaging Microscopy–FRET Images and Analysis

In lifetime imaging there is always concern about an autofluorescence component, which is about 100 ps (Kelbauskas and Dietel, 2002). The lifetime due-to-dye aggregation and dyes connected to metallic nanoparticles are about 50 ps or shorter (Geddes et al., 2003). In contrast, the fluorophores used for cellular imaging are typically <10 ns. In FLIM-FRET imaging the lifetimes are usually multicomponent due to the nonuniform local environment and binding state, different dipole–dipole orientation, and distance distribution between donor and acceptor molecules (Lakowicz, 1999; Squire et al., 2000; Elangovan et al., 2002; Chen and Periasamy, 2004).

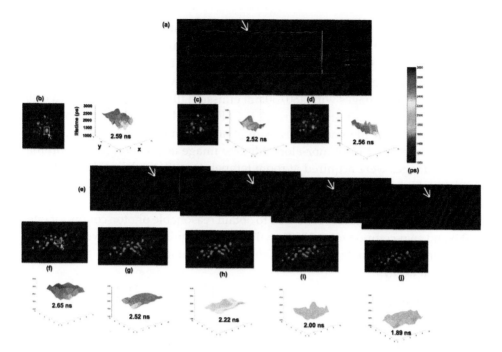

FIGURE 13–5 Selection of average excitation power. The average power of the excitation intensity was determined using the Radiance 2100 confocal/multiphoton microscopy system before collecting the lifetime images with the Becker & Hickl SPC-730 board. In our experimental observation 30 s was required for data collection for ECFP-C/EBPα protein expressed in single GHFT1-5 pituitary living cell nucleus. The three-dimensional lifetime distribution graph was obtained from a single protein complex. Excitation power was about 1 mW at the specimen plane (*a*). Negligible photobleaching was observed after 120 s scanning, shown as a horizontal line in *a*. Then the image was collected with the FLIM detector for ~30 s and the processed mean lifetime value (within the box) was 2.59 ns (*b*). At the same average power, the images were collected repeatedly, and the processed mean lifetime value (same box area) remained the same—2.52 ns (*c*) and 2.56 ns, (*d*). Another cell was selected and used the same average power (1 mW) excitation. The processed lifetime value did not change, as shown in *f* (within the box). Higher average power (4 mW) excitation was then used to collect the images. The lifetime image in *g* remains closer to the lifetime value in *f*. The lifetime value started decreasing when the image was colleted again at 4 mW (*h–j*). The intensity graph in *e* indicates the photobleaching trend of the molecule at 4 mW average powers and demonstrates that an appropriate average power excitation is required to reduce the photobleaching so that the correct lifetime values can be measured. The lifetime values remain the same if the photobleaching level is within 30 gray-level intensity.

Here we describe the acquisition and analysis of TCSPC FLIM-FRET images for unknown distances, positive and negative controls, and various mutant forms of green fluorescent proteins (EBFP, EGFP, ECFP, and EYFP) FLIM images.

3.1 Acquisition

Cells expressed with CFP-C/EBPΔ244 (donor) were placed on the microscope stage and identified using an arc lamp light source. As described above, the 820 nm laser line was used as an excitation wavelength to illuminate the cell with appropriate excitation average power as described above. LaserSharp 2000 software was used to display the two-photon image on the monitor to adjust the focusing of the protein molecule. While the laser was scanning the cell the filter cube slider (FCS) was switched toward the LPMT as shown in Figure 13–1. The SPCM software was used to acquire lifetime images with 480/30 nm emission filters. We next imaged cells expressing both CFP-C/EBPΔ244 and YFP-C/EBPΔ244 (i.e., donor in the presence of the acceptor). The lifetime images were acquired using the same excitation and emission filters used for donor-alone lifetime image acquisition (www.chroma. com). All images were acquired at room temperature (74°F). We did not observe any detectable autofluorescence signals when we used the unlabeled cells in the same media used for FRET-FLIM imaging.

The accumulation time was 30–60 s to obtain reasonable photon counts for one- (>500) and two-component (>1000) analysis. The acquired images were saved to the disk for further analysis to obtain the lifetime values for single- and double-exponential decay.

3.2 Analysis

The data analysis software (SPCImage; Becker & Hickl) allows multiexponential curve fitting of the acquired data on a pixel-by-pixel basis using a weighted least-squares numerical approach. The sum of all time bins is equivalent to the intensity image, and this is displayed as an image that is pseudocolored according to the curve fit results. Therefore, each image can be easily displayed in a meaningful way to compare lifetimes within or between other images.

3.3 Positive and Negative Controls

It is important to characterize the microscope system for detection of FRET signals using negative and positive controls. For a positive FRET control in the data presented here, images of cells expressing ECFP coupled directly to EYFP through a flexible 15–amino acid linker (ECFP-15aa-EYFP) were used. Positive and negative controls have been used in wide-field FRET microscopy (Chen et al., 2003); here we demonstrate their use in TCSPC FLIM-FRET microscopy.

Using the donor filter set, we selected an appropriate cell to collect the lifetime image as described in the previous section and then processed the cell to identify the quenched lifetime in the presence of acceptor. As shown in Figure 13–6, the lifetime of the donor molecule in the presence of acceptor was about 1.87 ns, which demonstrates the occurrence of dimerization of the protein molecules. The estimated average distance separating the linked ECFP and EYFP, determined with the donor lifetime (2.6 ns) in the absence of acceptor, was 52.442 Å. The shortest distance from the fluorophore to the outside of the fluorescent protein β barrel is 11 Å (Rekas et al., 2002), which limits the minimum distance that can separate the donor and acceptor fluorophores to approximately 22 Å. The 15-amino acid linker separating the fluorescent proteins has a maximum extended chain length of approximately 50 Å, but flexibility in its conformation should allow a range of distances separating the fluorophores at any given time. When expressed in the living

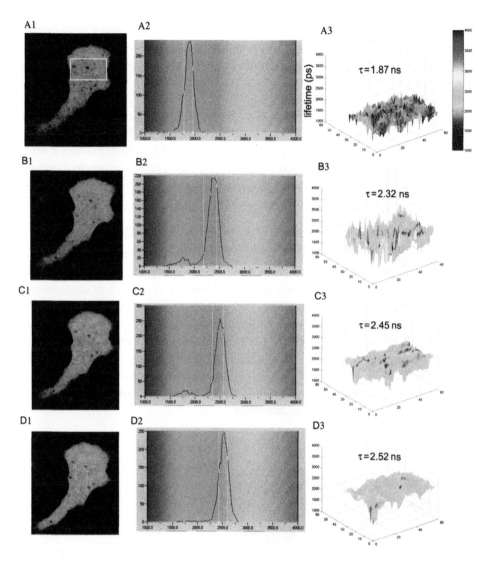

FIGURE 13–6 Positive FRET-FLIM control. Images of cells expressing ECFP coupled directly to EYFP through a 15–amino acid linker (ECFP-15aa-EYFP) were acquired and analyzed using time-correlated single-photon counting FRET–fluorescence lifetime imaging microscopy. The mean lifetime of the selected region of interest (the box) was 1.87 ns in the presence of acceptor (A1 and A2). Photobleaching of the acceptor molecule showed the occurrence of FRET and the lifetime returned to its natural lifetime, 2.52 ns (D1 and D2). The lifetime of the donor molecule was increased by increasing the photobleaching of the acceptor molecule as illustrated in two- (A2–D2) and three-dimensional (A3–D3) lifetime distribution. (Adapted from Wallrabe and Periasamy, 2005)

cell, it is difficult to predict the conformation that this fusion protein would adopt, but the FRET results provide an estimate of the spatial relationship between the linked fluorescent proteins. We then photobleached the acceptor molecules step by step, using the 514 laser line to demonstrate that the donor molecule returns to its natural lifetime of 2.6 ns (see Fig. 13–6).

We repeated FLIM-FRET imaging for negative control. In this case, in cells co-expressing the unlinked CFP and YFP-tagged 4-hep C/EBPα, which were co-localized but did not interact, no (or negligible) change in lifetime of the donor molecule was observed, as demonstrated in Figure 13–7. As seen elsewhere in dynamic biological processes, there is a heterogeneous distribution of efficiency and distances, demonstrated in data presented in this chapter.

3.4 Localization of Protein (C/EBPα) Dimerization

The C/EBP proteins bind to specific DNA elements as obligate dimers and preferentially bind to satellite DNA repeat sequences located in regions of centromeric heterochromatin (Tang and Lane, 1999). In this study we used TCSPC FRET-FLIM microscopy to characterize intranuclear dimer formation for the transcription factor C/EBPα in living pituitary GHFT1–5 cells. Members of the C/EBP family of transcription factors are critical determinants of cell differentiation. The b-zip region of C/EBPα is fused to GFP, which contains the dimerization domain (Day et al., 2003). We sought to determine whether the expressed fusion proteins were associated as dimers in this subnuclear sites.

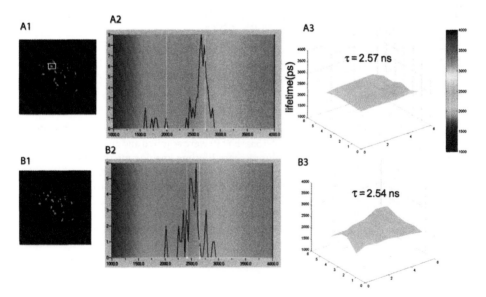

FIGURE 13–7 Negative FRET-FLIM control. Cells coexpressing unlinked ECFP and EYFP, which were co-localized but did not interact, served as a negative FRET-FLIM control (A1, B1). There was no change in donor lifetime by photobleaching the acceptor molecule, as shown in two- (A2, B2) and three-dimensional (A3, B3) lifetime distribution. The mean lifetime value for A1 and B1 was calculated within the region of interest (the box) shown in A1.

As shown in Figure 13–6, the C/EBPα protein dimers can be localized by means of FLIM-FRET methodology. Identification of the protein dimers is determined by following the lifetime in the presence and absence of acceptor. In the absence of acceptor, the CFP-C/EBPα is $\tau_D = 2.6$ ns ($\chi^2 = 1.0$; see Fig. 13–9) and the lifetime decreases to $\tau_{DA} = 2.14$ ns ($\chi^2 = 1.2$) for a particular pixel in the presence of acceptor. The rate of energy transfer efficiency (E) and the distance between donor and acceptor molecule can be determined using equations (1) and (3):

$$E = 1-\tau_{DA}/\tau_D = 1-(2.14/2.6) = 1-0.82 = 18\%$$

$$r = R_0 * (1-1/E)^{1/6} = 52.767 \, (1-1/18)^{1/6} = 5.2 \text{ nm}$$

The other alternative for verifying the results is by acceptor photobleaching (see Chapters 4 and 8). We used the 514 nm laser line to photobleach the YFP-C/EBPΔ244 molecules. The distribution of the lifetime shifted toward higher lifetime after photobleaching the acceptor molecule (Fig. 13–8), indicating that the donor molecule returns to its natural lifetime of 2.6 ns. The three-dimensional plot of a region of interest (ROI) shows change in color to the higher lifetime distributions.

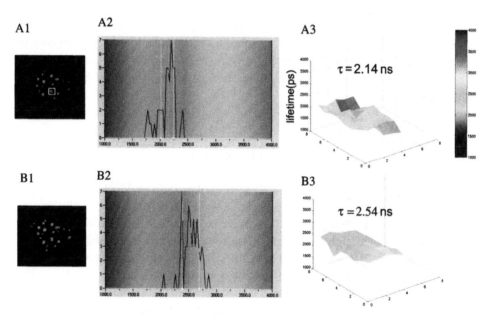

FIGURE 13–8 Demonstration of lifetime distribution for ECFP-EYFP-C/EBPΔ244 protein dimerization in GHFT1–5 cell nucleus. The mean lifetime of the donor in the absence of the acceptor (τ_D) was 2.6 ns (see Fig. 13–10) and in the presence of acceptor (τ_{DA}) was 2.14 ns (mean lifetime of the selected region of interest [ROI], A1). In the presence of acceptor, not all the proteins participated in the energy transfer process. When the acceptor molecule was photobleached at 514 nm, the donor molecule lifetime returned to close to the natural lifetime, 2.54 ns, as shown in B1 and B2. A3 and B3 show the three-dimensional distribution of the selected ROI.

3.5 Estimation of Donor Molecules

As mentioned in the literature, intensity-based or steady-state protein–protein interaction imaging techniques introduce errors in estimating the distance between the donor and acceptor molecules. Moreover, because of the SBT and cross talk, it is difficult to determine whether one is observing true sensitized emission, bleed-through signals, or a combination of both, unless accurate SBT correction methodology is applied (Gordon et al., 1998; Elangovan et al., 2003; Mills et al., 2003; Wallrabe et al., 2003). In contrast, FLIM measurements are sensitive to competing environmental and physical processes, such as resonance energy transfer and quenching, which can alter the fluorescence lifetime. Thus measurements of fluorescence lifetimes provide a very accurate reflection of the fluorophore's local environment (Elangovan et al., 2002; Chapters 2, 11, 12).

As shown in Figure 13–9a, the dimerization reduces donor lifetime ($\tau_1 = 0.83$ ns) at the occurrence of FRET when energy transfer quenches the donor, resulting in different lifetime distributions compared with non-FRET/unquenched donors ($\tau_2 = 2.54$ ns). Non-FRET/unquenched donors in the nucleus are those that did not dimerize. It should be noted that for the intensity-based method we speak of "apparent" energy transfer efficiency, as that calculation is based on all donor molecules, including those that do not participate in FRET. In FLIM, we can separate the FRET and non-FRET donors at a single pixel or voxel on the basis of lifetime distributions. Thus we believe that this is a more realistic measurement.

Moreover, it is possible to estimate the percentage of donor molecules involved in the energy transfer process by following the a1 and a2 values in the presence of acceptor molecules at a single pixel (Fig. 13–9b). In future investigations this could be calibrated to obtain a real number of protein molecules involved in the energy transfer process.

4. RATIONALE IN FLIM-FRET DATA ANALYSIS

There is great interest in studying the dynamics of protein molecules under physiological condition because protein–protein interactions mediate many cellular processes. Identification of the interacting protein partners is critical to understanding its function and place in the biochemical pathway, thereby establishing its role in important disease processes. The FLIM-FRET microscopy techniques described above can provide more detailed information about molecular interactions in living cells. For instance, measurement of fluorescence lifetimes of the protein population within the living cell may expose several different molecular species, each with characteristic lifetimes. Repeated measurements may reveal dynamic changes in fluorescence lifetimes within the protein population and can be used to evaluate temporal changes in the extent of resonance energy transfer (Chapter 2).

Fluorescence lifetime imaging-FRET microscopy relies on the ability to capture weak and transient fluorescent signals efficiently and rapidly from the interactions of labeled molecules in single living or fixed cells. FRET microscopy has the advantage that the spatial distribution of FRET efficiency can be visualized through-

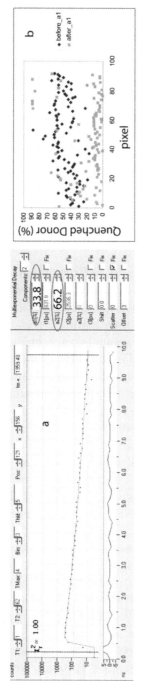

FIGURE 13–9 Demonstration of estimation of donor molecules involved in the energy transfer process. *a*: Double-exponential decay at one pixel from the FLIM image in Figure 13–8. a1 = 33.8% represents the percentage of quenched donor molecules, τ_1(t1) = 831.9 ps, quenched donor lifetime; a2 = 66.2% represents the percentage of unquenched donor molecules, τ_2(t2) = 2536 ps, unquenched donor lifetime. *b*: Comparison of percentage of quenched donor molecule before and after photobleaching.

out the image, rather than registering only an average over the entire cell or population. Because energy transfer occurs over distances of 1–10 nm, a FRET signal corresponding to a particular location within a microscope image provides an additional magnification surpassing the optical resolution (~0.2 μm) of the light microscope. Thus, within a voxel of microscopic resolution, FRET resolves average donor–acceptor distances beyond the microscopic limit down to the molecular scale (0.001–0.01 μm). This is one of the principal and unique benefits of FRET microscopic imaging: not only can co-localization of the donor- and acceptor-labeled probes be seen, but intimate interactions of molecules labeled with donor and acceptor can be demonstrated. Below we describe the hurdles we have experienced in quantifying the distance between donor and acceptor molecule using TCSPC FLIM-FRET microscopy systems.

4.1 Enhanced Cyan Fluorescent Protein: One or Two Components?

As has been described in the literature, FLIM-FRET is the ideal technique to study protein associations in living specimens. It has also been noted that different techniques have different advantages and disadvantages. When fluorophore molecules are used alone (not conjugated), the lifetime distribution is a very narrow histogram. If the same fluorophore were conjugated to an antibody, we would get a wider lifetime distribution. As explained in this and other chapters, the distributions were due to the competitive environmental changes in the biological systems. Sometimes this results in a two-component lifetime distribution.

It has been reported that ECFP has two-component lifetimes (Tramier et al., 2002; Rizzo et al., 2004), which may be responsible for the presence of multiple fluorescent states in the ECFP molecules. This may be due to the two different conformations found in the crystal structure of ECFP (Hyun Bae et al., 2003). We also observed that there is no two-component lifetime distribution in other fluorescent proteins, such as EBFP, EGFP, and EYFP, as shown in Figure 13–10. The two-component distribution creates a problem for FLIM-FRET microscopy because the fluorescence lifetime decreases with FRET.

In our experimental system, even though the ECFP-alone expressed cell provides a good fit for single exponential decay, it can also provide a reasonable fit for two-component analysis (Fig. 13–11c). We have observed a very clear separation of one and two components for ECFP in the presence of an acceptor molecule, as shown in Figure 13–11e. We have also tested the newly generated cerulean (Rizzo et al., 2004) and have observed a very clear single-exponential distribution (Fig. 13–11b) for the unfused compared with the enhanced CFP (Fig. 13–11a). When the cerulean was expressed with C/EBPα protein, we observed some fast components as shown in Figure 13–11d. In the presence of acceptor molecules, the cerulean provides clear two-component distribution, demonstrating the occurrence of FRET (Fig. 13–11f).

In this situation one would compare the two-component distribution in the presence and absence of acceptor molecules. As shown in our data (Fig. 13–11e), the separation between one and two components is wider in the presence of acceptor than that in the absence of acceptor (Fig. 13–11c). We have also shown in

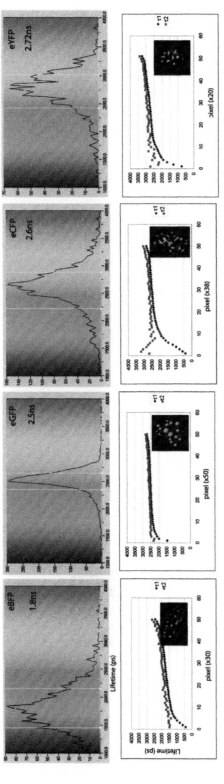

FIGURE 13–10 One- and two-component analysis of EBFP, EGFP, ECFP, and EYFP. It appears that, except for ECFP, all the fluorescent variants provide single-exponential decay (one-component lifetime distribution).

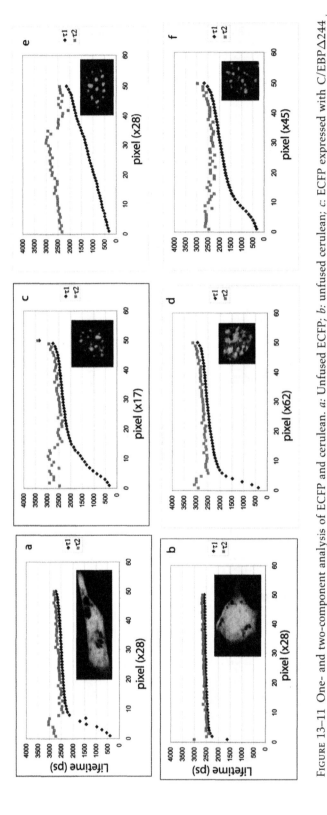

FIGURE 13–11 One- and two-component analysis of ECFP and cerulean. *a:* Unfused ECFP; *b:* unfused cerulean; *c:* ECFP expressed with C/EBPΔ244 protein; *d:* cerulean-EYFP-C/EBPΔ244; *e:* ECFP-EYFP-C/EBP-C/EBPΔ244; *f:* cerulean-EYFP-C/EBPΔ244 (see text for more detailed discussion).

Figure 13–9 the percentage of quenched and unquenched donor molecules at a single pixel. In Figures 13–11c and 13–11e, for example, the percentage of the number of pixels is about 14%, compared to 7% in the absence of acceptor at 1.9–2.0 ns range. This demonstrates that it is possible to differentiate the quenched donor molecules in the presence of acceptor when ECFP and EYFP are used as the FRET pair.

5. CONCLUSION

We have demonstrated the integration of FLIM hardware with existing laser scanning two-photon microscopy to measure the fluorescence lifetime for protein molecules in living cells. The advantage of using FLIM is that lifetime measurements are independent of change in concentration and excitation intensity. The TCSPC FRET-FLIM imaging system has the added advantage of tracking dynamic protein–protein interactions and providing more precise determinations of the distance between donor and acceptor molecules compared to intensity-based (wide-field, confocal, and multiphoton) methods. Spectral imaging FRET is yet another technique that can be used to study protein associations in living specimens (Chapter 10). Although the techniques described here are very promising, they are limited by sensitivity that is not yet comparable to that of the intensity-based FRET imaging systems. Because each technique has its own advantages and disadvantages, one should determine the appropriate technique or methodology to be used according to the experimental conditions at hand.

We wish to thank Prof. Richard Day for providing the cells and for his critical comments. We wish to thank Drs. Axel Bergmann and Wolfgang Becker for their help and support. We also thank Mr. Horst Wallrabe and Ms. Erica Caruso for their valuable help in preparing the chapter. We acknowledge financial support from the University of Virginia.

REFERENCES

Bacskai, J.B., J. Skoch, G.A. Hickey, R. Allen, and B.T. Hyman. Fluorescence resonance energy transfer determinations using multiphoton fluorescence lifetime imaging microscopy to characterize amyloid-β plaques. *J. Biomed. Opt.* 8:368–375, 2003.

Becker, W., A. Bergmann, M.A. Hink, K. König, K. Benndorf, and C. Biskup. Fluorescence lifetime imaging by time-correlated single-photon counting. *Microsc. Res. Tech.* 63:58–66, 2004.

Carlsson, K. and A. Liljeborg. Simultaneous confocal lifetime imaging of multiple fluorophores using the intensity modulated multiple wavelength scanning technique. *J. Microsc.* 191:119–127, 1998.

Chen, Y., J.D. Mills, and A. Periasamy. Protein localization in living cells and tissues using FRET and FLIM. *Differentiation* 71:528–541, 2003.

Chen, Y. and A. Periasamy. Characterization of two-photon excitation fluorescence lifetime imaging microscopy for protein localization. *Microsc. Res. Tech.* 63:72–80, 2004.

Clayton, A.H.A., Q.S. Hanley, D.J. Arndt-Jovin, V. Subramaniam, and T.M. Jovin. Dynamic fluorescence anisotropy imaging microscopy in the frequency domain (rFLIM). *Biophys. J.* 83:1631–1649, 2002.

Clegg, R.M. Fluorescence resonance energy transfer. In: *Fluorescence Imaging Spectroscopy and Microscopy*, edited by X.F. Wang and B. Herman, New York: John Wiley & Sons, 1996, pp. 179–251.

Clegg, R.M., A.H. Murchie, and D.M.J. Lilley. The solution structure of the four-way DNA junction at low-salt conditions: a fluorescence resonance energy transfer analysis. *Biophys. J.* 66:99–109, 1994.

Day, R.N. Visualization of Pit-1 transcription factor interactions in the living cell nucleus by fluorescence resonance energy transfer microscopy. *Mol. Endocrinal.* 12:1410–1419, 1998.

Day, R.N., T.C. Voss, J.F. Enwright III, C.F. Booker, A. Periasamy, and F. Schaufels. Imaging the localized protein interactions between Pit-1 and the CCAAT/enhancer binding protein alpha (C/EBPα) in the living pituitary cell nucleus. *Mol. Endocrinal.* 17:333–345, 2003.

dos Remedios, C.G., M. Miki, and J.A. Barden. Fluorescence resonance energy transfer measurements of distances in actin and myosin: a critical evaluation. *J. Muscle Res. Cell Motil.* 8:97–117, 1987.

Eftink, M.R. Fluorescence quenching. In: Topics in Fluorescence Spectroscopy, Vol. 2., edited by J.R. Lakowicz, New York: Kluwer Academic Publishers, 1992, pp. 53–126.

Elangovan, M., R.N. Day, and A. Periasamy. A novel nanaosecond FRET-FLIM microscopy to quantitate the protein interactions in a single living cell. *J. Microsc.* 1:2–14, 2002.

Elangovan, M., H. Wallrabe, Y. Chen, R.N. Day, M. Barroso, and A. Periasamy. Characterization of one- and two-photon excitation resonance energy transfer microscopy. *Methods* 29:58–73, 2003.

Eliceiri, K.W., C.-H. Fan, G.E. Lyons, and J.G. White. Analysis of histology specimens using lifetime multiphoton microscopy. *J. Biomed. Opt.* 8:376–380, 2003.

Förster T. Intermolecular energy migration and fluorescence. *Ann. Phys.* (Leipzig) 2:55–75, 1948.

Förster T. Delocalized excitation and excitation transfer. In: *Modern Quantum Chemistry*, Vol. 3, edited by O. Sinanoglu, New York: Academic Press, 1965, pp. 93–137.

Geddes, C.D., H. Cao, I. Gryczynski, Z. Gryczynski, J. Fang, and J.R. Lakowicz. Metal-enhanced fluorescence (MEF) due to silver colloids on a planar surface: potential applications of indocyanine green to in vivo imaging. *J. Phys. Chem. A* 107:3443–3449, 2003.

Gerritsen, H.C. and K. de Grauw. One- and two-photon confocal fluorescence lifetime imaging and its applications. In: *Methods in Cellular Imaging,* edited by A. Periasamy, New York: Oxford University Press, 2001, pp. 309–323.

Gerritsen, H.C., R. Sannders, A. Draaijer, and Y.K. Levine. Fluorescence lifetime imaging of oxygen in cells. *J. Fluoresc.* 7:11–16, 1997.

Gordon, G.W., G. Berry, X.H. Liang, B. Levine, and B. Herman. Quantitative fluorescence resonance energy transfer measurements using fluorescence microscopy. *Biophys. J.* 74:2702–2713, 1998.

Hyun Bae, J., M. Rubini, G. Jung, G. Wiegand, M.H. Seifert, M.K. Azim, J.S. Kim, A. Zumbusch, T.A. Holak, L. Moroder, R. Huber, and N. Budisa. Expansion of the genetic code enables design of a novel "gold" class of green fluorescent proteins. *J. Mol. Biol.* 328:1071–1081, 2003.

Kelbauskas, L. and W. Dietel. Internalization of aggregated photosensitizers by tumor cells: subcellular time-resolved fluorescence spectroscopy on derivatives of pyropheophorbide-a ethers and chlorine6 under femtosecond one- and two-photon excitation. *Photochem. Photobiol.* 76:686–694, 2002.

Knemeyer, J-P., N. Marmé, and M. Sauer. Probe for detection of specific DNA sequences at the single-molecule level. *Anal. Chem.* 72:3717–3724, 2000.

König, K. and I. Riemann. High-resolution multiphoton tomography of human skin with subcellular spatial resolution and picosecond time resolution. *J. Biomed. Opt.* 8:432–439, 2003.

Lakowicz, J.R. *Principles of Fluorescence Spectroscopy*, 2nd ed. New York: Plenum Press, 1999.

Mills, J.D., J.R. Stone, J.D. Rubin, D.E. Melon, D.O. Okonkwo, A. Periasamy, and G.A. Helm. Illuminating protein interactions in tissue using confocal and two-photon excitation fluorescent resonance energy transfer (FRET) Microscopy. *J. Biomed. Opt.* 8:347–356, 2003.

O'Connor, D.V. and D. Phillips. *Time-Correlated Single Photon Counting.* London: Academic Press. 1984.

Periasamy A. *Methods in Cellular Imaging.* New York: Oxford University Press, 2001.

Rekas, A., J.R. Alattia, T. Nagai, A. Miyawaki, and M. Ikura. Crystal structure of venus, a yellow fluorescent protein with improved maturation and reduced environmental sensitivity. *J. Biol. Chem.* 277:50573–5058, 2002.

Rizzo, M.A., G.H. Springer, B. Granada, and D.W. Piston. An improved cyan fluorescent protein variant useful for FRET. *Nat. Biotech.* 22:445–449, 2004.

Sekar, R.B. and A. Periasamy. Fluorescence resonance energy transfer (FRET) microscopy imaging of live cell protein localization. *J. Cell Biol.* 160:629–633, 2003.

Squire, A., P.J. Verveer, and P.I.H. Bastiaens. Multiple frequency fluorescence lifetime imaging microscopy. *J. Microsc.* 197:136–149, 2000.

Tang, Q.Q. and M.D. Lane (1999). Activation and centromeric localization of CCAAT/enhancer-binding proteins during the mitotic clonal expansion of adipocyte differentiation. *Genes Dev.* 13:2231–2241.

Tramier, M., I. Gautier, T. Piolot, S. Ravalet, K. Kemnitz, J. Coppey, C. Durieux, V. Mignotte, and M. Coppey-Moisan. Picosecond-hetero-FRET microscopy to probe protein–protein interactions in live cells. *Biophys. J.* 83:3570–3577, 2002.

van Zandvoort, M.A.M.J., C.J. de Grauw, H.C. Gerritsen, J.L.V. Broers, M.G.A. Egbrink, F.C.S. Ramaekers, and D.W. Slaaf. Discrimination of DNA and RNA in cells by a vital fluorescent probe: lifetime imaging of SYTO13 in healthy and apoptotic cells. *Cytometry* 47:226–232, 2002.

Volkmer, A., V. Subramaniam, D.J. Birch, and T.M. Jovin. One- and two-photon excited fluorescence lifetimes and anisotropy decays of green fluorescent proteins. *Biophys. J.* 78:1589–1598, 2000.

Wallrabe, H., M. Elangovan, A. Burchard, A. Periasamy, and M. Barroso. Confocal FRET microscopy to measure clustering of receptor–ligand complexes in endocytic membranes. *Biophys. J.* 85:559–571, 2003.

Wallrabe, H. and A. Periasamy. Imaging protein molecules using FRET and FLIM microscopy. *Curr. Opin. Biotechnol.* 16:In Press, 2005.

Xu, C. and W.W. Webb. Multiphoton excitation of molecular fluorophores and nonlinear laser microscopy. In: *Topics in Fluorescence Spectroscopy: Nonlinear and Two-Photon-Induced Fluorescence*, Vol. 5., edited by J.R. Lakowicz, New York: Plenum Publishing Corp., 1997, pp. 471–540.

14

Bioluminescence Resonance Energy Transfer: Techniques and Potential

Mohammed Soutto, Yao Xu, and Carl H. Johnson

1. Introduction

This volume is about the applications of resonance energy transfer (RET) techniques to biological problems. Resonance energy transfer techniques are based on the nonradiative transfer of energy between suitable donor and acceptor species via the Förster mechanism (Milligan, 2004). This chapter will describe a RET technique that uses bioluminescence rather than fluorescence, namely, Bioluminescence Resonance Energy Transfer (BRET). Most of the genetically encoded fluorescence proteins (e.g., jellyfish green fluorescent protein [GFP]) are part of bioluminescence systems in the organisms from which they are derived. In the jellyfish, *Aequorea victoria*, GFP functions to accept energy transferred from the luminescent photoprotein aequorin, converting aequorin's blue light emission to green light (Morin and Hastings, 1971). In other words, GFP was "designed" by evolution to participate in BRET. There are many different bioluminescent organisms, from bacteria and fungi to mollusks, crustaceans, insects, fishes, and others. The different organisms display a similar diversity of mechanisms, meaning that the substrates (luciferins) and enzymes (luciferases), as well as the relevant gene sequences, are also different (Hastings and Johnson, 2003). Many or most of the different systems apparently originated independently during evolution, so bioluminescence is an excellent case of convergent evolution (Hastings, 1983). Bioluminescence is an enzyme-catalyzed chemiluminescence reaction in which the energy released (~50 kcal) is used to produce an intermediate or product in an electronically excited state, P^*, which then emits a photon. The emission does not come from or depend on light absorbed, as in fluorescence or phosphorescence, but the excited state produced is indistinguishable from that produced in fluorescence after the absorption of a photon by the ground state of the molecule concerned (Hastings and Johnson, 2003).

Most of the fluorescence proteins used for FRET (e.g., GFP and its mutant variants) and the luciferases used for BRET come from the organisms known as Cnidaria (Coelenterates). The sea pansy *Renilla*, the jellyfish *Aequorea*, and the hydroid *Obelia* are the best studied of many bioluminescent species in this group. These organisms all emit light as brief, bright flashes, typically caused by a mechanical stimulation that elicits the calcium ion fluxes that directly activate bioluminescence. These flashes are believed to deter predation. The bioluminescence requires oxygen and a luciferin substrate, coelenterazine, which is an imidazolopyrazine. Its luciferase-catalyzed reaction with oxygen gives a dioxetanone intermediate, which breaks down to give CO_2 and oxidized luciferin (coelenteramide) in the excited state (Shimomura, 1982; Hastings and Johnson, 2003). In the *Aequorea* system, two proteins are involved, the Ca^{2+}-dependent luciferase aequorin (~20 kDa) and its resonance energy partner, GFP. Bioluminescence emission of isolated aequorin is blue (λ_{max} ~486 nm) when the reaction is carried out in extracts, but green (λ_{max} ~508 nm) from the living organism (Shimomura and Johnson, 1962). We now know this spectral shift is due to RET with GFP (Morin and Hastings, 1971; Hastings and Johnson, 2003). In the case of aequorin, the enzyme is directly sensitive to calcium ions, but in the *Renilla* system, coelenterazine binds to a nonenzymatic protein (18.5 kDa, with 3 Ca^{2+} binding sites); calcium triggers the release of coelenterazine and light emission is catalyzed by *Renilla* luciferase (RLUC; 35 kDa). Therefore, RLUC is not directly sensitive to Ca^{2+}. As with aequorin, the luminescence from the RLUC-catalyzed reaction is blue (λ_{max} ~480 nm) in vitro, but green in vivo (λ_{max} ~509 nm) because of the RET to *Renilla* GFP (Hastings and Johnson, 2003).

As in the case of FRET, BRET can be used as a "molecular yardstick" to determine whether molecules are within 50–100 Å of each other. As such, BRET has many of the same strengths and limitations as FRET. However, because FRET demands that the donor fluorophore be excited by external illumination, the practical usefulness of FRET can be limited because of the concomitant results of excitation: photobleaching, autofluorescence, and direct excitation of the acceptor fluorophore (see Comparing Bioluminescence Resonance Energy Transfer and FRET, below). Furthermore, some tissues might be easily damaged by the excitation light or might be photoresponsive (e.g., retina and most plant tissues). Because BRET depends upon an enzyme-catalyzed reaction, it is measured in the absence of any incident excitation.

2. A BIOLUMINESCENCE RESONANCE ENERGY TRANSFER SYSTEM TO ASSAY PROTEIN INTERACTIONS

Renilla luciferase was originally chosen as the donor luciferase for BRET because its emission spectrum is similar to the cyan mutant of GFP (λ_{max} ~480 nm) that had been shown to exhibit FRET with the acceptor fluorophore YFP, which is a yellow-emitting GFP mutant (Miyawaki et al., 1997). The excitation peak of Enhanced YFP (EYFP, λ_{max} ~513 nm) does not perfectly match the emission peak of RLUC, but the emission spectrum of RLUC is sufficiently broad that it provides good excitation of EYFP. The spectral overlap between RLUC and EYFP is similar to that between EYFP and the enhanced cyan mutant of GFP, ECFP, which yields a critical Förster radius (R_0) for FRET of ~50 Å (Mahajan et al., 1998). Thus, we would expect significant BRET between RLUC and EYFP, with a R_0 of ~50 Å. The fluorescence emission of EYFP is yellow, peaking at 527 nm, which is distinct from the RLUC emission peak. Furthermore, RLUC and EYFP do not naturally interact with each other. Fortunately, the substrate for RLUC, coelenterazine, is a hydrophobic molecule that easily permeates cell membranes.

As depicted in Figure 14–1A, in the BRET assay of protein interactions RLUC is genetically fused to one candidate protein, and EYFP is fused to another protein of interest that might interact with the first protein. If RLUC and EYFP are brought close enough for RET to occur, the bioluminescence energy generated by RLUC can be transferred to EYFP, resulting in the enhanced emission of yellow light (Fig. 14–1A). If there is no interaction between the two proteins of interest, RLUC and EYFP will be too far apart for significant transfer and only the blue-emitting spectrum of RLUC will be detected. Thus, protein–protein interactions can be monitored both in vivo and in vitro by detecting the emission spectrum and quantifying the emission ratio at 530:480 nm.

The BRET between RLUC and EYFP was first demonstrated in control experiments in which RLUC was fused directly to EYFP through a linker of 11 amino acids (Xu et al., 1999). The luminescence profile of the *Escherichia coli* cells expressing this RLUC:EYFP fusion construct yielded a bimodal spectrum, with one peak centered at 480 nm (RLUC luminescence) and a new peak centered at 527 nm (equivalent to EYFP fluorescence) (Xu et al., 1999). This result suggests that a significant propor-

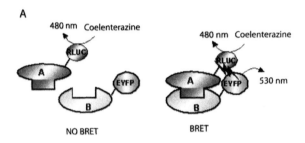

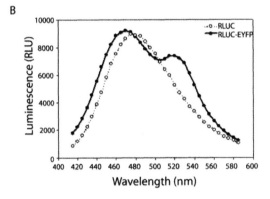

FIGURE 14–1 Diagram of bioluminescence resonance energy transfer (BRET) used for a protein–protein interaction assay. *A:* One protein of interest (A) is genetically fused to the donor *Renilla* luciferase (RLUC) and the other candidate protein (B) is fused to the acceptor fluorophore enhanced yellow fluorescent protein (EYFP). In the absence of interaction between A and B (NO BRET) and in the presence of the substrate, coelenterazine, RLUC emits luminescence with a peak at 480 nm. If interaction between the two fusion proteins brings RLUC and EYFP into close proximity (between 10 and 100 Å), BRET can occur, with an additional emission at a longer wavelength (e.g., peak at 530 nm for the EYFP partner). *B:* Luminescence emission spectra from HEK=293 cells transfected with constructs expressing either RLUC or RLUC::EYFP. RLU,

tion of the energy from RLUC is transferred to EYFP and emitted at the character-istic wavelength of EYFP. Similar results can be obtained in mammalian cells, as shown in Figure 14–1B. BRET was first applied in mammalian cells in an assay of β_2-adrenergic receptor interaction (Angers et al., 2000). Therefore, RLUC–EYFP is an effective combination to apply in a protein–protein interaction assay.

3. COMPARING BIOLUMINESCENCE RESONANCE ENERGY TRANSFER AND FRET

Bioluminescence resonance energy transfer has potential advantages over FRET because it does not require the use of excitation illumination; BRET should be supe-rior for cells that are either photoresponsive (e.g., retina or any photoreceptive tissue) or damaged by the wavelength of light used to excite FRET. Moreover, photo-bleaching of the fluorophores can be a serious limitation of FRET, but it is irrelevant to BRET. Cells that have significant autofluorescence would also be better assayed by BRET than by FRET. This is particularly true for highly autofluorescent tissue, but all cells are autofluorescent to a degree because of ubiquitous fluorescent mol-ecules such as NADH, collagen, and flavins. In plant cells BRET is particularly promis-ing because the highly fluorescent photosynthetic pigments and cell wall compounds prevalent in plants can interfere with FRET-based assays. The first application of BRET to plant cells has demonstrated the interaction of the phototransduction molecule COP1 in onion and *Arabidopsis* (Subramanian et al., 2004a, 2004b).

In addition, FRET may be prone to complications due to simultaneous excita-tion of both donor and acceptor fluorophores. Specifically, even with monochro-matic laser excitation, it is impossible with the current generation of fluorescent proteins to excite only the donor without exciting the acceptor fluorophore to some degree (see Chapters 4–7). In contrast, because BRET does not involve optical exci-tation, all the light emitted by the fluorophore must result from resonance trans-fer. Therefore, BRET is theoretically superior to FRET for quantifying resonance transfer. One of the most important advantages of BRET over FRET is that the rela-tive levels of expression of the donor and acceptor partners can be quantified in-dependently: the donor by luminescence and the acceptor by fluorescence. This is difficult with FRET because the acceptor is generally excited to some extent by the excitation wavelength used to excite the donor. With BRET, measuring the system's fluorescence gives the relative level of the acceptor (YFP–GFP fusion partner) and when coelenterazine is added, the total luminescence of the system measured in darkness gives the relative level of the donor (luciferase fusion partner). Knowl-edge of the relative levels of the fusion partners is crucial for comparing results from one experiment to the next.

No technique is perfect, however, and there are characteristics of BRET that limit its use. One of these limitations concerns the substrate for RLUC, coelenterazine. Coelenterazine is hydrophobic and can permeate all the different cell types that we have tested, including bacteria (*E. coli* (Xu et al., 1999) and cyanobacteria, yeast, the green alga *Chlamydomonas* (Minko et al., 1999), plants (Subramanian et al., 2004a, 2004b), and animal cells in culture (Fig. 14–1; Angers et al., 2000). But native coelenterazine is easily oxidized in aqueous medium and can also autoluminesce in complex medium, especially media containing lipids (e.g., serum) or detergents. This autoluminescence can cause a background that can obscure the true signal. Fortunately, new substrates are available that can overcome this limitation (see Section 4, below). Another major limitation of BRET is that the luminescence may sometimes be too dim to accurately measure without a very sensitive light-measuring apparatus. With FRET, dim signals can be amplified by simply increasing the intensity or duration of excitation (possibly at the cost of light-induced damage to the cells), whereas with BRET, the only option to improve low signal levels is to integrate the signal for a longer time and/or use a more sensitive apparatus. New plate-reading luminometers have been designed with BRET in mind so that two wavelengths of luminescence can be measured and the ratio computed. These luminometers include the Fusion and Victor (Perkin-Elmer), the Mithras (Berthold), and the FluoStar and PolarStar (BMG). Other instruments are capable of BRET measurements, such as the single-channel Turner TD20/20 luminometer with BRET accessory (used in the studies of Subramanian et al., 2004a, 2004b). Manufacturers are continuously developing improved instrumentation for measuring low-light levels, and these improvements in technology will undoubtably aid the further development of BRET assays of real-time protein–protein interactions in living organisms.

4. NEW TOOLS FOR BIOLUMINESCENCE RESONANCE ENERGY TRANSFER

4.1 Substrates

Very recently, several new tools for BRET have appeared that may prove useful. The first tool is new substrates for RLUC introduced by Promega that may solve the problems of coelenterazine instability and autoluminescence, even in complex medium. The Promega substrate, EnduRen™, is an analog of coelenterazine in which the site of oxygenation is protected (EnduRen = 6-(4-acetyoxyphenyl)-2,8-dibenzylimidazo[1,2-a]pyrazin-3-pivalyloxymethyl ester). This protection prevents EnduRen from acting as a substrate for RLUC-catalyzed luminescence or from autoluminescence. However, the protecting group can be readily cleaved by the intracellular esterases found in mammalian cells to generate native coelenterazine. EnduRen is permeable to cells, freely moving from the extracellular medium into cells, where its protecting group is cleaved off preventing export and converting it to the active form. It can then act as a substrate for RLUC. In the extracellular medium, EnduRen is stable for days and has minimal autoluminescence, even in the presence of serum (Fig. 14–2). Also, the BRET ratio obtained with EnduRen is stable over prolonged incubations in both serum-containing and serum-free medium (Fig. 14–2). The luminescence generated from EnduRen is dimmer than that obtained with native coelenterazine, but other protected coelenterazine analogs are being developed that are as bright (or brighter!) as native coelenterazine (personal communication from Drs. Erika Hawkins and Keith Wood, Promega Corporation).

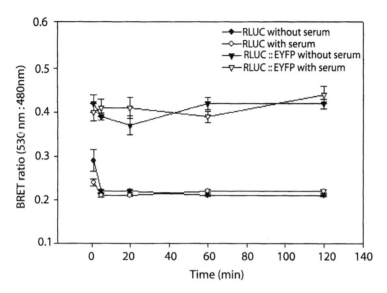

FIGURE 14–2 Stability of BRET ratio obtained with EnduRen™. HEK=293 cells were transfected with either RLUC or RLUC::EYFP. Using the Mithras plate-reading luminometer (Berthold), the BRET ratio was measured after adding 10 μM of EnduRen with or without 10% fetal bovine serum during a time course of 2 h. The data shown represent the mean ± SEM of four independent readings.

PerkinElmer BioSignal, Inc. markets another coelenterazine analog, DeepBlueC™, in which the spectrum of emission as catalyzed by RLUC is shifted to shorter wavelengths (λ_{max} ~400 nm). When used with a GFP mutant adapted to this emission wavelength (GFP^2), this BRET pair results in a higher signal-to-noise ratio. This system is now available under the trademark BRET$^{2™}$. The intensity of luminescence obtained with DeepBlueC is significantly lower than that obtained with native coelenterazine; in our laboratory, we rarely obtain BRET^2 signals of sufficient strength to be quantifiable. However, other laboratories have used BRET^2 successfully and effectively (e.g., Bertrand et al., 2002; Dionne et al., 2002; Milligan, 2004).

4.2 Luciferases

Two RLUCs have been codon optimized for mammalian expression. One is available from PerkinElmer BioSignal, Inc. (Rluc(h)). Transfection of the Rluc(h) construct into mammalian cells results in significantly higher luminescence levels than with a comparable construct encoding wild-type Rluc (Xu et al., 2002a). The other optimized Rluc (available from Promega as hRL) was reported to be more highly expressed than the native RLUC in mammalian cells. A new luciferase isolated from *Gaussia* has an emission spectrum similar to that of RLUC but has a molecular weight of only 20 kDa (available from Prolume Ltd.). Like RLUC, this luciferase uses coelenterazine as a substrate. By virtue of its smaller size, GLUC may better allow native interactions without steric hindrance in fusion proteins. Transfection of the humanized version of *Gaussia* luciferase (hGluc) also allows strong luminescence signals in mammalian cells that are somewhat more stable over time than with hRluc(h) (Xu et al., 2002a).

4.3 Fluorescent Acceptor Proteins

YFP is an effective RET partner for RLUC. However, an even more red-shifted acceptor fluorophore that could receive RET from RLUC would be tremendously useful to increase the spectral separation between RLUC's emission and the BRET emission, thereby improving the signal-to–noise level. Therefore, we eagerly await the development of a red fluorescent protein that is excellent for BRET (e.g., Campbell et al., 2002).

5. BIOLOGICAL APPLICATIONS OF BIOLUMINESCENCE RESONANCE ENERGY TRANSFER: PAST, PRESENT, AND FUTURE

5.1 Application to Various Organisms

The BRET technique was originally used to test the interaction of cyanobacterial circadian clock proteins in *E. coli* (Xu et al., 1999). BRET has now been successfully extended to other cell types, from plants (Subramanian et al., 2004a, 2004b) to mammalian cells. Figure 14–1 shows spectra of RLUC and the RLUC::EYFP fusion protein expressed in mammalian cells (HEK-293 cells). There are many publications in which BRET has been used in mammalian cells, where it has been particularly successful in studies involving dimer and/or oligomer formation among receptors in vivo (Eidne et al., 2002; Kroeger and Eidne, 2004; Milligan, 2004). The first such study was that of Angers et al. (2000), who used BRET to demonstrate that human β-2-adrenergic receptors form constitutive homodimers in HEK-293 cells. Treatment with the agonist isoproterenol increased the BRET signal, indicating that the agonist interacted with receptor dimers at the cell surface. Since that ground-breaking publication, BRET studies of receptors have been particularly useful as applied to a variety of G protein–coupled receptors (Bertrand et al., 2002; Kroeger and Eidne, 2004; Milligan, 2004), but have also been successful in studying arrestins, dopamine receptors, NMDA receptors, neuropeptide Y receptors, serotonin receptors, substance P/μ-opioid receptors, calmodulin, calcium-sensing receptors, transcription factors, and others (Bertrand et al., 2002; Jensen et al., 2002; Berglund et al., 2003; Charest and Bouvier, 2003; Fiorentini et al., 2003; Germain-Desprez et al., 2003; Pfeiffer et al., 2003; Turner et al., 2004).

5.2 High-Throughput Screening with Bioluminescence Resonance Energy Transfer

In our original publication on BRET, we proposed that it might be easily adapted to high-throughput screening (HTS) for protein interactions and drug effects (Xu et al., 1999). We have repeated this suggestion in subsequent publications (Xu et al., 2002b, 2002c, 2003), and other researchers have taken up the trumpet call (Boute et al., 2002; Roda et al., 2003; Milligan, 2004). High-throughput screening is an application for which luminometry is superior to fluorimetry, and therefore one in which BRET should outshine FRET. Our first suggestion was to use BRET for a protein interaction screen in *E. Coli* or yeast by expressing a "bait" protein fused to RLUC and a library of "prey" molecules fused to YFP (or vice versa). By measuring the light emission collected through interference filters, the 530:480 nm luminescence ratio could be measured from colonies of bacteria or yeast on agar plates using a camera imaging system or a photomultiplier-based instrument designed to measure luminescence of liquid cultures in titer plates. Colonies (or wells) that show higher light intensity at 530 nm or exhibit an above-background ratio of the 530:480 nm ratio signal could be selected and the "prey" DNA sequence further characterized. Proof of principle for this idea was demonstrated in the original BRET publication using *E. coli* colonies (Xu et al., 1999). This protein interaction assay could be used in any transformable or transfectable cell type (e.g., mammalian cell cultures). The advantage of this type of screen over the yeast two-hybrid methodology is that the interacting proteins could be tested in the original cell type and in the native cellular compartments rather than in the foreign environment of the yeast nucleus.

Another HTS application for BRET would be to establish a particular BRET interacting partnership and then use that signal to screen for drugs that enhance or disrupt the interaction in vivo (Boute et al., 2002; Roda et al., 2003; Milligan, 2004). In this context, the application of BRET to discovering drugs that modulate the interaction of G protein–coupled receptors has been discussed by Milligan (2004). Therefore, BRET should be of great usefulness in HTS drug discovery. The aforementioned plate-reading luminometers (Fusion, Victor, Mithras, FluoStar, and PolarStar) are designed with BRET HTS capability. Moreover, with the new protected coelenterazine substrates from Promega, the assays can be performed in serum-containing or another type of complex medium. Thus, all the tools are in place for an efficient BRET HTS system.

5.3 Bioluminescence Resonance Energy Transfer Imaging

As evidenced by contributions in this volume, FRET has been most useful for imaging the subcellular localization(s) of protein interaction. Because (1) the luminescence signal from BRET is dim and (2) the microscopic resolution of bioluminescence is generally poorer than that for fluorescence, some investigators have decided that BRET might not be amenable to imaging (Boute et al., 2002; Milligan, 2004). We do not share this view. With continual improvements in imaging dim signals, we believe that BRET will be useful for imaging both macroscopic and microscopic samples. Regarding macroscopic imaging, we have shown in the original BRET publication that bacterial colonies can be imaged for BRET ratios (Xu et al., 1999). In addition, we have shown that localized BRET ratio signals can be visualized along the various parts of plant seedlings (Subramanian et al., 2004b).

In terms of microscopic imaging, the dimness of BRET signals and the large loss of signal in microscopes are significant hurdles. We are currently engaged in establishing feasibility for subcellular BRET imaging. The dimness of the signal requires maximum throughput in the microscopic system. This means we need to use high numerical aperture (NA) objectives and a camera port on the microscope that allows maximum photon throughput. In addition, the long exposure times required (2–30 min) mean that switching between 480 nm and 530 nm bandpass filters is not optimal. Within a long exposure time, the total signal intensity might have changed, thereby invalidating any quantitative comparisons of the 480 nm and 530 nm images. We have therefore opted to use a Dual-View™ (Optical Insights) with filters at 480 nm and 530 nm. The Dual-View allows two images to be collected simultaneously and projected to separate halves of the charge-coupled device (CCD) chip. Although the use of the Dual-View requires an even longer exposure time (because the amount of light reaching each pixel is reduced by the Dual-View's dichroic and filters), the simultaneous acquisition of two wavelengths allows the collection of valid BRET ratios even when there are alterations in the overall luminescence intensity. Finally, it is necessary to collect the images with a camera that has high sensitivity, low dark counts, a broad dynamic range, linear response, and the ability to integrate images for a long time. Based on our comparisons, the camera that has the best compromise of these characteristics is the cooled (–25°C) EB-CCD camera from Hamamatsu. The electron bombardment characteristics of this camera cause a type of shot noise, which causes a few very bright pixels in long integrations. While annoying, this type of noise is easily filtered out as long as the exposure time for the real image is not so long that it causes saturation (which we don't want anyway because it prevents quantification).

Figure 14–3 depicts our preliminary attempts to establish a BRET imaging system. The figure shows HEK-293 cells transfected with a construct expressing RLUC (Fig. 14–3A–D) or with a construct expressing RLUC::EYFP (Fig. 14–3E–I). The cells were imaged with the Hamamatsu EB-CCD through the Dual-View mounted on the bottom port of an Olympus IX71 inverted microscope (40× objective, 1.30 NA). Figures 14–3A and 14–3E are bright-field images (exposure time of 120 ms). Figures 14–3B and 14–3F are the BRET images at 480 nm (panel B exposed for 3 min,

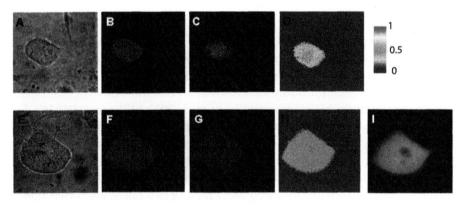

Figure 14–3 Bioluminescence microscopy and BRET signal with Remilla luciferase (RLUC) and enhanced yellow fluorescent protein (EYFP). HEK=293 cells were transiently transfected with either RLUC (*A–D*) or RLUC::EYFP (*E–I*), *A, E*: Bright-field image from a CCD camera and an inverted microscope. *B, F*: Luminescence image with interference filter transmitting light of 480 nm. *C, G*: Luminescence image with interference filter transmitting light of 530 nm. *D, H*: Comparison of BRET ratio (530:480 nm) with the pseudocolor scale shown immediately to the right of panel *D*. *I*: Fluorescence image of RLUC::EYFP-expressing cells (EX 500 BP20, EM 515 LP). Panels *B, C, F*, and *G* were acquired simultaneously through the Dual-View™.

panel F for 2 min), and Figures 14–3C and 14–3G are the BRET images at 530 nm (panel C for 3 min, G for 2 min). Figures 14–3B, 14–3C 14–3F, and 14–3G were acquired simultaneously through the Dual-View. Note the weaker signal at 530 nm than that at 480 nm for the RLUC-expressing cell (compare panels B with C), whereas the 480 nm and the 530 nm signals are similar for the RLUC::EYFP-expressing cell (compare panels F and G). Figures 14–3D and 14–3H show a comparison of the BRET ratio over the entire cell via the pseudocolor scale to the right of panel D. The average BRET ratio over the RLUC-expressing cell is approximately 0.25 (panel D), whereas for the RLUC::EYFP-expressing cell, it is approximately 0.60 (panel H). Figure 14–3I illustrates the fluorescence of the RLUC::EYFP-expressing cell (EX 500 plus or minus 10 nm bandpass filter, EM 515 nm longpass filter, exposure of 120 ms).

In summary, BRET ratio imaging is clearly possible over the entire cell for constructs that show a large enough difference in BRET ratio. Will it be possible to detect subcellular differences in BRET signals? We predict that BRET ratio differences between the nuclear and cytoplasmic compartments will be detectable. Whether it will be possible to make finer structure determinations of BRET ratio imaging will depend upon the continued development of sophisticated imaging devices. There is every reason to believe that technological advances in this capability will continue, and therefore that BRET imaging will be taken to finer levels. Therefore, for applications in which fluorescence measurements are problematic (e.g., because of autofluorescence, bleaching, phototoxicity, and sensitivity of the tissue to light), our advice is: don't FRET, do BRET!

We thank Dr. David Piston for advice relating to BRET imaging and Mark Hobson and Tadashi Maruno of Hamamatsu Inc. for the use of the EB-CCD camera. This research was supported by the National Institutes of Health (GM 65467), the National Institute of Mental Health (MH 43836 and MH 01179), and the National Science Foundation (MCB 0114653 to C.H.J. and Dr. Albrecht von Arnim).

References

Angers, S., A. Salahpour, E. Joly, S. Hilairet, D. Chelsky, M. Dennis, and M. Bouvier. Detection of beta 2-adrenergic receptor dimerization in living cells using bioluminescence resonance energy transfer (BRET). *Proc. Natl. Acad. Sci. U.S.A.* 97:3684–3689, 2000.

Berglund, M.M., D.A. Schober, M.A. Esterman, and D.R. Gehlert. Neuropeptide Y Y4 receptor homodimers dissociate upon agonist stimulation. *J. Pharmacol. Exp. Ther.* 307: 1120–1126, 2003.

Bertrand, L., S. Parent, M. Caron, M. Legault, E. Joly, S. Angers, M. Bouvier, M. Brown, B. Houle, and L.J. Menard. The BRET^2/arrestin assay in stable recombinant cells: a platform to screen for compounds that interact with G protein–coupled receptors (GPCRs). *Receptor Signal Transduct. Res.* 22:533–541, 2002.

Boute N., R. Jockers, and T. Issad. The use of resonance energy transfer in high-throughput screening: BRET versus FRET. *Trends Pharmacol. Sci.* 23:351–354, 2002.

Campbell, R.E., O. Tour, A.E. Palmer, P.A. Steinbach, G.S. Baird, D.A. Zacharias, and R.Y. Tsien. A monomeric red fluorescent protein. *Proc. Natl. Acad. Sci. U.S.A.* 99:7877–7882, 2002.

Charest, P.G. and M. Bouvier. Palmitoylation of the V2 vasopressin receptor carboxyl tail enhances beta-arrestin recruitment leading to efficient receptor endocytosis and ERK ½ activation. *J. Biol. Chem.* 278:41541–41551, 2003.

Dionne, P., M. Caron, A. Labonté, K. Carter-Allen, B. Houle, E. Joly, S.C. Taylor, and L. Menard. BRET^2: efficient energy transfer from *Renilla* luciferase to GFP^2 to measure protein–protein interactions and intracellular signaling events in live cells. In: *Luminescence BioTechnology*, edited by K. Van Dyke, C. Van Dyke, and K. Woodfork, New York: CRC Press, 2002, pp. 539–555.

Eidne, K.A., K.M. Kroeger, and A.C. Hanyaloglu. Applications of novel resonance energy transfer techniques to study dynamic hormone receptor interactions in living cells. *Trends Endocrinol. Metab.* 13:415–421, 2002.

Fiorentini, C., F. Gardoni, P. Spano, M. Di Luca, and C. Missale. Regulation of dopamine D1 receptor trafficking and desensitization by oligomerization with glutamate N-methyl-D-asparate receptors. *J. Biol. Chem.* 278:20196–20202, 2003.

Germain-Desprez, D., M. Bazinet, M. Bouvier, and M. Aubry. Oligomerization of transcriptional intermediary factor 1 regulators and interaction with ZNF74 nuclear matrix protein revealed by bioluminescence resonance energy transfer in living cells. *J. Biol. Chem.* 278:22367–22373, 2003.

Hastings, J.W. Biological diversity, chemical mechanisms and evolutionary origins of bioluminescent systems. *J. Mol. Evol.* 19:309–321, 1983.

Hastings, J.W. and C.H. Johnson. Bioluminescence and chemiluminescence. In: *Methods in Enzymology*, edited by A.G. Mariott and I. Parker, San Diego, CA: Biophotonics, 2003, pp. 75–104.

Jensen, A.A., J.L. Hansen, S.P. Sheikh, and H. Brauner-Osborne. Probing intermolecular protein–protein interactions in the calcium-sensing receptor homodimer using bioluminescence resonance energy transfer (BRET). *Eur. J. Biochem.* 269:5076–5087, 2002.

Kroeger, K.M., and K.A. Eidne. Study of G-protein coupled receptor interactions by bioluminescence resonance energy transfer. *Methods Mol. Biol.* 259:323–334, 2004.

Mahajan, N.P., Linder, G. Berry, G.W. Gordon, R. Heim, and B. Herman. Bcl-2 and Bax interactions in mitochondria probed with green fluorescent protein and fluorescence resonance energy transfer. *Nat. Biotechnol.* 16:547–552, 1998.

Milligan, G. Applications of bioluminescence and fluorescence resonance energy transfer to drug discovery at G protein–coupled receptors. *Eur. J. Pharm. Sci.* 21:397–405, 2004.

Minko I., S.P. Holloway, S. Nikaido, O.W. Odom, M. Carter, C.H. Johnson, and D.L. Herrin. Renilla luciferase as a vital reporter for chloroplast gene expression in *Chlamydomonas*. *Mol. Gen. Genet.* 262:421–425, 1999.

Miyawaki, A., J. Llopis, R. Heim, J.M. McCaffery, J.A. Adams, M. Ikura, and R.Y. Tsien. Fluorescent indicators for Ca^{2+} based on green fluorescent proteins and calmodulin. *Nature* 388:882–887, 1997.

Morin, J.G. and J.W. Hastings. Energy transfer in a bioluminescent system. *J. Cell. Physiol.* 77:313–318, 1971.

Pfeiffer, M., S. Kirscht, R. Stumm, T. Koch, D. Wu, M. Laugsch, H. Schroder, V. Hollt, and S. Schulz. Heterodimerization of substance P and mu-opioid receptors regulates receptor trafficking and resensitization. *J. Biol. Chem.* 278:51630–51637, 2003.

Roda A., M. Guardigli, P. Pasini, and M. Mirasoli. Bioluminescence and chemiluminescence in drug screening. *Anal. Bioanal. Chem.* 377:826–833, 2003.

Shimomura, O. Mechanism of bioluminescence. In: *Chemical and Biological Generation of Excited States*, edited by W. Adam and G. Cilento. New York: Academic Press, 1982, pp. 249–276.

Shimomura, O., F.H. Johnson, and Y. Saiga. Extraction, purification and properties of Aequorin, a bioluminescent protein from the luminous hydromedusan, *Aequorea*. *J. Cell. Comp. Physiol.* 59:223–239, 1962.

Subramanian, C., B.-H. Kim, N.N. Lyssenko, X. Xu, C.H. Johnson, and A.G. von Arnim. The *Arabidopsis* repressor of light signaling, COP1, is regulated by nuclear exclusion: mutational analysis by bioluminescence resonance energy transfer. *Proc. Natl. Acad. Sci. U.S.A.* 2004a, in press.

Subramanian, C., Y. Xu, C.H. Johnson, and A.G. von Arnim. In vivo detection of protein–

protein interaction in plant cells using BRET. In: *Signal Transduction Protocols* (*Methods in Molecular Biology* series), edited by R.C. Dickson and M.D. Mendenhall, Totowa, NJ: Humana Press, 2004b, pp. 271–286.

Turner, J.H., A.K. Gelasco, and J.R. Raymond. Calmodulin interacts with the third intracellular loop of the serotonin 5–HT (sub1A) receptor at two distinct sites. Putative role in receptor phosphorylation by protein kinase C. *J. Biol. Chem.* 279:17027–17037, 2004.

Xu Y., C.H. Johnson, and D. Piston. Bioluminescence resonance energy transfer assays for protein–protein interactions in living cells. *Methods Mol. Biol.* 183:212–233, 2002a.

Xu, Y., A. Kanauchi, D.W. Piston, and C.H. Johnson. Resonance energy transfer as an emerging technique for monitoring protein–protein interactions *in vivo*: BRET *vs.* FRET. In: *Luminescence BioTechnology*, edited by K. Van Dyke, C. Van Dyke, and K. Woodfork, New York: CRC Press, 2002b, pp. 529–538.

Xu, Y., A. Kanauchi, A.G. von Arnim, D.W. Piston, and C.H. Johnson. Bioluminescence resonance energy transfer: monitoring protein–protein interactions in living cells. In *Methods in Enzymology*, edited by A.G. Mariott and I. Parker, San Diego, CA: Biophotonics, 2003, pp. 289–301.

Xu, Y., D.W. Piston, and C.H. Johnson. A bioluminescence resonance energy transfer (BRET) system: application to interacting circadian clock proteins. *Proc. Natl. Acad. Sci. U.S.A.* 96:151–156, 1999.

Xu, Y., D. Piston, and C.H. Johnson. BRET assays for protein–protein interactions in living cells. In: *Green Fluorescent Protein*: *Applications and Protocols* (*Methods in Molecular Biology* series). edited by B.W. Hicks, (Totowa, NJ: Humana Press) 2002c, pp. 121–133.

15

Quantifying Molecular Interactions with Fluorescence Correlation Spectroscopy

Keith M. Berland

1. Introduction

Förster resonance energy transfer (FRET) microscopy has become a powerful tool for quantifying molecular interactions in living systems. For various practical and technical reasons, however, FRET methods cannot always be successfully applied to a given biological system of interest. Fluorescence correlation spectroscopy (FCS) (Magde et al., 1972, 1974; Elson and Magde, 1974; Eigen and Rigler, 1994; Rigler and Elson, 2001; Schwille, 2001; Hess et al., 2002) and related fluctuation spectros- copy methods (Petersen et al., 1993; Chen et al., 1999, 2000, 2002; Muller et al., 2000) are emerging as powerful tools for detecting specific biomolecular interactions, and FCS measurements can provide an important complement to FRET methods. Fluo- rescence correlation spectroscopy also provides a versatile and highly sensitive means of characterizing chemical and physical dynamics in biological systems.

In this chapter, we focus primarily on the capability of quantifying protein– protein and other molecular interactions by means of FCS. We begin with a gen- eral introduction to FCS and then discuss how to apply it to quantify molecular interactions, including the use of both autocorrelation and cross-correlation mea- surements. Our goal is to introduce the fundamental concepts behind FCS detec- tion of molecular interactions so that readers can better determine when FCS measurements can be usefully applied in their particular experimental system and how these measurements may complement FRET measurements.

FRET signals result from the excited-state interaction between two spectrally distinct fluorescent entities, the donor and acceptor molecules (see Chapters 2, 4– 13). The dynamics of the excited-state dipole–dipole coupling (the FRET process) is highly distance dependent (with Ångstrom-scale resolution), which is the basis for quantifying specific molecular interactions. Thus FRET applications require carefully engineered fluorescence-labeling geometries to maximize the FRET in- teraction. In contrast, FCS has no specific requirements on fluorescence-labeling geometries or fluorescent probe spacing. Instead, interactions are determined by analyzing the hydrodynamic mobility, concentration, and molecular brightness of fluorescent molecules. Thus FCS has an important advantage in being able to quantify interactions even for molecular assemblies in which it is not possible to label the molecules to yield good FRET efficiencies. It is important to keep in mind that FCS analysis requires observation of molecular dynamics. Thus traditional FCS measurements are not easily applied to investigate interactions between stationery molecules. There are variations of FCS that offer similar advantages to those discussed below for samples with substantial immobile or slow-moving populations (see Chapter 16).

2. FLUORESCENCE CORRELATION SPECTROSCOPY THEORY

The methods of FCS entail essentially counting experiments, in which the fluorescence signal from within a small optically defined volume serves as a measure of the number of molecules (N) within that volume. This number fluctuates as molecules diffuse into and out of the observation volume, as shown schematically in Figure 15–1. There are two key sources of information in the fluctuation signal and each can be used to quantify molecular interactions. First, the duration of the fluc-

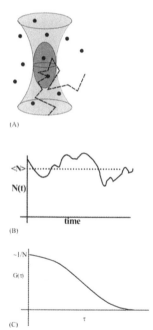

(A)

(B)

(C)

FIGURE 15–1 *A:* Fluorescence correlation spectroscopy observation volume is optically defined. The light gray cone represents the excitation laser profile, and the dark gray ellipsoid, the observation volume. The volume is defined by the laser and fluorescence detection optics. Brownian motion of individual molecules will cause them to diffuse into and out of the volume. The dashed line represents the trajectory of a single particle. *B:* As molecules diffuse in and out of the volume, the number in the volume will fluctuate, as described in the text. *C:* Correlation functions calculated from the fluctuation data are analyzed to determine diffusion coefficients and molecular concentrations.

tuations is related to the average time individual molecules reside within volume. This time, directly measured with FCS, can be used to recover a diffusion coefficient (D) for the fluorescent species in the volume. Since the diffusion coefficient depends on the molecular mass, changes in diffusion rate following molecular association and dissociation can be quantified with FCS. Second, the amplitude of the fluctuations is determined by the number concentration of independently diffusing molecules occupying the volume. Unlike most experimental measures such as absorption, weight concentration (mg/ml), or total fluorescence, the number concentration changes when molecules associate. For example, there are one-quarter as many independent diffusers following a monomer–tetramer association reaction. Thus the number concentration as measured by FCS can be used to directly quantify the extent of molecular interactions. This concentration-based approach is further complemented by considering the molecular brightness of fluorescent species, another quantity that can be determined from FCS measurements and the fluorescence intensity. Molecular brightness has dimensions of fluorescent counts per unit time per independently diffusing molecule. If two fluorescent molecules associate, the resulting complex will be twice as bright as the component molecules, providing further evidence of the molecular interaction.

To analyze fluorescence fluctuation signals, temporal correlation functions are calculated from the fluctuation data (Fig. 15–1C). A correlation function compares a signal with either itself (autocorrelation) or a related signal (cross-correlation). The correlation function amplitude is inversely proportional to the average number of molecules in the volume ($\sim 1/N$), and the temporal relaxation of the correlation function is governed by the residency time of molecules diffusing through the volume. Physical models have been developed that permit accurate recovery of experimental parameters through curve fitting of FCS data (Thompson, 1991; Rigler and Elson, 2001).

In a typical FCS experiment, the average fluorescence signal, $\langle F_i \rangle$, is proportional to the concentration of fluorescent molecules and the size of the volume, and can be written simply in terms of the number of molecules that occupy the volume:

$$\langle F_i \rangle = \psi_i \langle C_i \rangle V_{\text{eff}} = \psi_i \langle N_i \rangle \tag{1}$$

Here C_i, $V_{\textit{eff}}$, and N_i represent the molecular concentration, observation volume, and number of independent molecules within the volume, respectively. Angular brackets represent time averages. The parameter ψ can be identified as the molecular brightness of a fluorescent species (with units of photons per molecule per second) and is determined by both intrinsic molecular properties (e.g., fluorescence lifetime, quantum yield, absorption cross-section), and the measurement instrumentation. ψ is one of the most critical parameters in determining the fidelity of FCS measurements (Koppel, 1974), and generally speaking, instrumentation with single-molecule sensitivity is required for robust applications of FCS. For the following discussion, it is useful to generalize the brightness parameter to the format $\psi_{i,n}$, where the double index represents the brightness of a particular molecular species, i, in a particular detection channel n (e.g., a "red" or "green" detection channel), since cross-correlation measurements require multiple detector channels.

The general theoretical form of the correlation function, $G(\tau)$, can be written as

$$G_{mn}(\tau) = \gamma \frac{\sum_i \psi_{i,m} \psi_{i,n} \langle N_i \rangle A_i(\tau)}{\left[\sum_i \psi_{i,m} \langle N_i \rangle\right]\left[\sum_i \psi_{i,n} \langle N_i \rangle\right]} \cdot \qquad (2)$$

The summation is over all molecular species, i, in the sample. For autocorrelation measurements, detectors m and n are the same, and for cross-correlation they are two different detection channels. The constant geometrical prefactor, γ, is determined by the shape of the observation volume. The temporal dependence of $G(\tau)$ is contained in the functions $A_i(\tau)$. The volume is typically assumed to have a "three-dimensional Gaussian" profile, for which $A_i(\tau)$ can be written explicitly as

$$A_i(\tau) = (1 + a\varphi_i)^{-1}\left(1 + a\varphi_i \frac{\omega_0^2}{z_0^2}\right)^{-\frac{1}{2}} \qquad (3)$$

The dimensionless parameter $\varphi_i = \frac{4D_i\tau}{\omega_0^2}$ and D_i, ω_0, and z_0 are the diffusion coefficient of species i and the radial and axial dimensions of the observation volume, respectively. In the above expression, the parameter a has the value 1 for one-photon excited fluorescence and the value 2 for two-photon excited fluorescence. This model is applied to fit experimental data and recover information about dynamics and molecular interactions. It is useful to note that A_i has unit value for short correlation times $\left(\tau \ll \frac{\omega_0^2}{D}\right)$. Thus the amplitude of the correlation functions, $G(0)$, can be written explicitly as a sum over molecular species, weighted by the molecular brightness of each species. For example, the $G(0)$ for the autocorrelation function from two molecular species, A and B, can be written as

$$G(0) = \gamma \frac{\psi_A^2 \langle N_A \rangle + \psi_B^2 \langle N_B \rangle}{\left[\psi_A \langle N_A \rangle + \psi_B \langle N_B \rangle\right]^2} \qquad (4)$$

Note that the contribution of each molecular species to the $G(0)$ value is weighted by the square of its molecular brightness. Readers interested in a more detailed derivation of basic FCS theory are referred to Thompson (1991).

3. DATA ACQUISITION

3.1. Instrumentation

Measurements with FCS are performed using either two-photon or confocal microscopes to measure time-dependent fluctuations in the fluorescence signal from minute observation volumes (Berland et al., 1995; Schwille et al., 1999). Details on the differences between confocal and two-photon FCS methods are beyond the scope of this chapter, although we will note that two-photon methods are often preferred for intracellular applications, sharing many of the advantages of two-photon excitation for biological imaging (Denk et al., 1990; So et al., 2000; Zipfel et al., 2003). Users of one-photon FCS instrumentation should be aware of potential artifacts associated with certain optical configurations (Hess and Webb, 2003).

Both confocal and two-photon fluorescence detection methods monitor sub-femtoliter observation volumes. For nanomolar sample concentrations, the total number of molecules present in the observation volume at any one time is thus quite small (~1–1000), which ensures that fluctuations will be resolvable relative to the average fluorescence signal levels. Observation of such small molecular populations requires appropriately designed optical systems that have single-molecule sensitivity. Commercial confocal microscopes and associated photodetectors are often not sufficiently sensitive for optimal FCS measurements, although high-performance FCS instrumentation has recently become commercially available. A schematic diagram of the two-photon FCS microscope used in the author's lab is shown in Figure 15–2 (Berland, 2004).

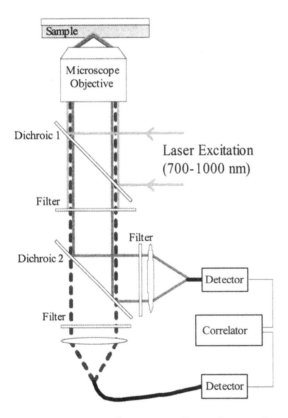

FIGURE 15–2 A typical experimental setup for two-photon fluorescence correlation spectroscopy: specialized hardware with single-molecule sensitivity is required. In this instrument, minimal optical elements and fiber-coupled avalanche photodiodes are used to achieve high sensitivity detection.

3.2 Sample Preparation

Specific details regarding fluorescence labeling of biomolecules are also beyond the scope of this chapter, but there are several general considerations that are important in preparing samples for FCS measurements. First, clean samples are essential. Only very small sample concentrations are used, thus impurities in the sample are more detrimental to FCS measurements than to the successful use of many other techniques. Each fluorescent species contributes to the overall FCS measurement as the square of its molecular brightness (see equation (4)), and bright fluorescent "dust" or bright aggregates in samples can easily overwhelm the signal of interest. Second, one should strive to have uniform fluorescent labeling of the proteins. The quality of an FCS signal will be far better with uniformly labeled molecular populations (e.g., a single fluorophore per protein monomer). For this reason, genetically encoded fusion proteins with GFP variants expressed in living cells can be particularly appealing for use in FCS measurements. In contrast to other techniques described in this book, one particular challenge when using over-expressed proteins in live cells is to select cells expressing sufficiently low concentrations to get good FCS results.

4. METHODS FOR QUANTIFYING INTERACTIONS

Here two distinct methods are described to quantify molecular interactions with FCS. The first, diffusion analysis, is based on the molecular weight dependence of diffusion coefficients. The second method is based on using FCS to quantify the concentrations of independently diffusing molecular species. As noted above, number concentrations are altered by molecular association. In addition, measurements of concentration and fluorescence signal levels can be used together to determine the molecular brightness of sample species. This brightness can also serve as a direct reporter of interactions between fluorescent species.

4.1 Diffusion Analysis

The FCS diffusion-based approach resembles fluorescence polarization assays in that one monitors a shift in diffusion coefficients following molecular association. However, fluorescence polarization (anisotropy) assays measure rotational diffusion and FCS methods typically monitor translational diffusion. This gives FCS an important advantage in that there is no restriction on the total molecular weight of the molecular complex[1] being investigated, and FCS can thus be applied for even very large molecules.

Diffusion coefficients, however, are not particularly sensitive measures of molecular weight. For spherical particles the diffusion coefficient is given by

$$D = \frac{kT}{6\pi\eta r}$$

where D represents the diffusion coefficient and r the hydrodynamic radius of the molecules. Parameters k, T, and η represent the Boltzman constant, temperature, and solution viscosity, respectively. Since the hydrodynamic radius of a molecule varies approximately as the cube root of its molecular weight, the diffusion coefficient varies only as the inverse cube root of the molecular weight. FCS diffusion-based analysis is thus best applied to quantify interaction between molecules that have substantially different molecular weights. In particular, changes in diffusion are easily resolved with FCS when a small fluorescent molecule interacts with a much larger molecule. On the other hand, it is difficult to detect accurately the binding of a large fluorescent protein to a small protein or ligand using diffusion-based analysis. The change in a diffusion coefficient required to clearly resolve distinct interactions is approximately a factor of two, corresponding to an eightfold change in molecular weight (Meseth et al., 1999).

Figure 15–3 illustrates the use of the diffusion-based approach for measuring the interaction of fluorescein-tagged SV40 nuclear localization signal (NLS) peptides with the nuclear import receptor importin-α. The importin-α protein was unlabeled. Importin-α has a molecular weight near 60 kD[a], which is much larger than the NLS peptide. Upon binding, the NLS–importin-α complex diffuses more slowly than the free NLS molecule. A series of FCS measurements were performed with fixed NLS concentration and serial dilutions of importin-α. The measurements were then fitted to recover the fraction of NLS molecules bound at various importin-α sample concentrations. The resulting binding curves are also shown in the Figure 15–3. This type of measurement provides very accurate quantitation of the protein–protein interactions, and has many useful applications.

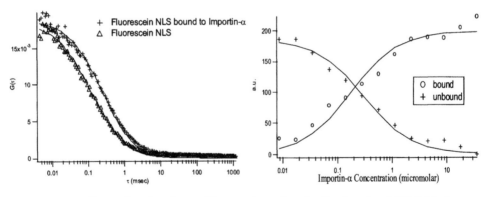

FIGURE 15–3 Diffusion-based analysis of fluorescein nuclear localization signal (NLS) binding to importin-α. *Left:* Fluorescence correlation spectroscopy (FCS) curves for free NLS and NLS fully bound to importin-α. The importin-α bound form diffuses more slowly and the FCS curve is thus shifted to the right (longer correlation times). *Right:* A series of similar measurements were performed for fixed concentrations of NLS and serial dilutions of importin-α. The FCS curves were then fitted, using the known diffusion times for free and bound NLS molecules, to determine the fraction bound and unbound at various importin-α concentrations. The full set of measurements was then fitted to a binding curve, yielding dissociation constant for the interaction of 180 nM.

One major drawback of the diffusion-based approach is that the diffusion rates are dependent on both the molecular weight and the local viscosity. Confined environments and transient interactions with nearby structures can also influence apparent diffusion rates. It can thus be very difficult to use diffusion analysis to quantify interactions in living cells because it can be a challenge to decouple molecular weight changes from variations in the local environment or other more complex interactions.

4.2 Molecular Counting and Brightness

The amplitude of the measured correlation functions, $G(0)$, is a measure of the number of independently diffusing molecules in the volume. Coupled with the average fluorescence intensity, it is also a reporter of molecular brightness. When molecules associate there is a corresponding change in both number concentration and molecular brightness. For example, the association of a population of fluorescent monomers to form dimers will result in a change in the number of independently diffusing molecules by a factor of two, relative to the monomer population. In addition, the molecular brightness, i.e., fluorescence signal from each independently diffusing molecular species, will increase by two times. These two changes together are sufficient to quantify the extent of molecular interactions with FCS by analyzing the amplitude of the correlation function and the overall fluorescence intensity. The accuracy of these measurements is not dependent on the molecular weight and diffusion coefficient of the observed species. Therefore, the molecular counting approach is especially suitable for applications in live cells and can also be used to monitor interactions of molecules, even if diffusion rates do not change appreciably following association events. This represents one of the more powerful advantages of FCS. While beyond the scope of this chapter, it is useful to note that when the sample is not monodisperse, implementation of this strategy is somewhat more complex. Although the basic concept remains unchanged, other analysis procedures such as the photon-counting histogram may be required to resolve various different populations in samples with complex distributions of interactions (Chen et al., 2000, 2002; Muller et al., 2000).

When using this concentration–brightness approach, all molecular species of interest should be fluorescent (unlike diffusion analysis). They can be labeled with either the same color fluorophore for autocorrelation measurements or spectrally distinct fluorophores for use in dual-color cross-correlation measurements. In cross-correlation measurements, one labels the molecular species that are thought to interact with spectrally distinct fluorescent probes, e.g., red and green dyes. If these molecules associate, the red and green fluorescence signals will have correlated fluctuations (i.e., the red and green channels will go up and down together as dual-color molecules diffuse into or out of the observation volume). If they do not interact, the two channels will fluctuate independently. There is thus no cross-correlation signal unless the sample contains a molecular species that produces a fluorescent signal in both detector channels, i.e., a positive interaction between the two species. This is the basis of the dual-color FCS capability to "tune in" to particular interacting molecular components even within an otherwise heterogeneous sample population. In using dual-color cross-correlation, carefully chosen fluorophores and filter sets must be used to ensure that cross talk between detection channels does not degrade the signal quality (Schwille et al., 1997).

To highlight the strength of molecular counting–based applications of FCS, a comparison of the diffusion analysis and dual-color cross-correlation methods was made in quantifying the hybridization of nucleic acids (Berland, 2004). The experimental system shown in Figure 15–4 consists of two 18-mer single-stranded DNA molecules, denoted A and B, labeled with red (Cye 3.5) or green (Oregon green) dyes, respectively. Each of these oligonucleotides is complementary to adjacent regions on a nonfluorescent 40-mer DNA strand (X). The hybridization reaction observed is $A + B + X \leftrightarrow ABX$, i.e., all three species are bound together to form a single complex. In most dual-color FCS applications, interactions between two different molecular species labeled with spectrally distinct chromophores are studied. The altered experimental configuration shown here does not change the details of the FCS measurement, yet highlights the possibility of monitoring the formation of molecular complexes even when a particular component of the complex is not fluorescent.

Both autocorrelation and dual-color cross-correlation measurements were performed on samples with and without the target sequence, X. Figure 15–5 shows the autocorrelation curves for each detection channel. The measured diffusion coefficient for the Oregon green single-stranded molecule is $D = 1.3 \times 10^{-6}$ cm²/s. This value decreases to $D = 9.2 \times 10^{-7}$ cm²/s upon binding to its target sequence. A less pronounced shift is observable in the red channel for the Cy3.5 oligo. The simi-

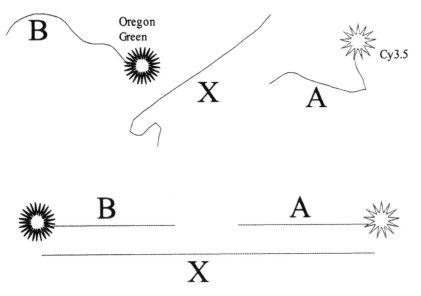

FIGURE 15–4 The three molecular components of the oligonucleotide hybridization reaction; the top of the figure shows the individual strands, and the bottom the geometry of the trimeric complex. Species A is labeled with Cy3.5, B with Oregon green 488, and X is unlabeled. The sequences, selected from the gamma tubulin gene, were chosen to minimize self-dimerization and hairpin formation. [Adapted from Berland, 2004.]

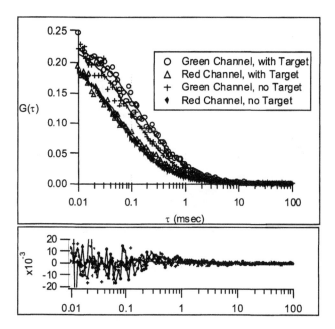

FIGURE 15–5 Autocorrelation traces for mixtures of species *A* and *B*, both with and without the target sequence *X*. A small shift in the diffusion coefficient is apparent following the hybridization reaction. [Adapted from Berland, 2004.]

larity of the bound (with target) and unbound (no target) curves demonstrates the difficulty in recovering information about molecular interactions based on diffusion analysis when the interacting species have similar size. On the other hand, quantifying the interactions works extremely well using the amplitude of the correlation functions, as shown in Figure 15–6. This result highlights the strength of dual-color cross-correlation methods for resolving molecular interactions—in this case, the formation of the *ABX* (Cy3.5–Oregon green–target) double-stranded complex. The time-dependent population of double-stranded DNA steadily increases as the hybridization reaction proceeds, and the concentration of the trimolecular DNA complex (*ABX*) can be monitored directly through the amplitude of the cross-correlation curves. It is thus quite clear that molecular counting is the preferred approach to quantifying interactions for molecules of similar molecular weight. Moreover, the counting approach is more robust for intracellular applications. A detailed analysis of these data can be found in Berland (2004).

5. CONCLUSIONS

Fluorescence correlation spectroscopy is just beginning to be widely applied as a tool for studying intracellular dynamics and interactions, and both experimental methods and analysis tools are continuing to evolve rapidly. Here we have attempted to introduce the basic concepts behind FCS measurements so that the reader may begin to understand the kinds of experimental systems for which FCS may be useful.

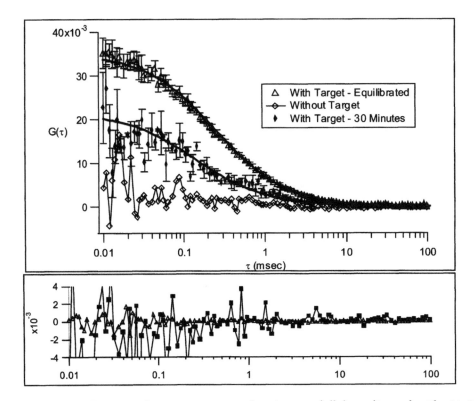

FIGURE 15–6 Cross-correlation measurements for mixtures of all three oligonucleotides (*A, B, X*) that hybridize to form double-stranded DNA. Equal concentrations of the three single-stranded species (*A, B, X*) were mixed together prior to performing FCS measurements. The increasing population of double-stranded DNA, reflecting the progress of the hybridization reaction, is readily apparent from the growing amplitude of these dual-color cross-correlation curves. This clearly demonstrates the effectiveness of the dual-color FCS method for quantifying the extent of this molecular interaction. [Adapted from Berland, 2004.]

In many cases, FCS can provide an important complement to FRET for quantifying molecular interactions. Moreover, for certain experiments in which FRET cannot be practically applied, variations of FCS may often prove to be a superior alternative.

NOTE

1. Polarization measurements are not particularly effective for larger proteins because they rotate too slowly relative to the fluorescence lifetime of standard fluorophores.

REFERENCES

Berland, K.M. Detection of specific DNA sequences using dual-color two-photon fluorescence correlation spectroscopy. *J. Biotech.* 108:127–136, 2004.
Berland, K.M., P.T.C. So, and E. Gratton. Two-photon fluorescence correlation spectroscopy: method and application to the intracellular environment. *Biophys. J.* 68:694–701, 1995.

Chen, Y., J.D. Muller, Q. Ruan, and E. Gratton. Molecular brightness characterization of EGFP in vivo by fluorescence fluctuation spectroscopy. *Biophys. J.* 82:133–144, 2002.

Chen, Y., J.D. Muller, P.T.C. So, and E. Gratton. The photon counting histogram in fluorescence fluctuation spectroscopy. *Biophys. J.* 77:553–567, 1999.

Chen, Y., J.D. Muller, S.Y. Tetin, J.D. Tyner, and E. Gratton. Probing ligand protein binding equilibria with fluorescence fluctuation spectroscopy. *Biophys. J.* 79:1074–1084, 2000.

Denk, W., J.H. Strickler, and W.W. Webb. Two-photon laser scanning fluorescence microscopy. *Science* 248:73–76, 1990.

Eigen, M. and R. Rigler. Sorting single molecules: application to diagnostics and evolutionary biotechnology. *Proc. Natl. Acad. Sci. U.S.A.* 91:5740–5747, 1994.

Elson, E.L. and D. Magde. Fluorescence correlation spectroscopy I. Conceptual basis and theory. *Biopolymers* 13:1–27, 1974.

Hess, S.T., S. Huang, A.A. Heikal, and W.W. Webb. Biological and chemical applications of fluorescence correlation spectroscopy: a review. *Biochemistry* 41:697–705, 2002.

Hess, S.T. and W.W. Webb. Focal volume optics and experimental artifacts in confocal fluorescence correlation spectroscopy. *Biophys. J.* 83:2300–2317, 2003.

Koppel, D.E. Statistical accuracy in fluorescence correlation spectroscopy. *Phys. Rev. A* 10:1938–1945, 1974.

Magde, D., E. Elson, and W.W. Webb. Thermodynamic fluctuations in a reacting system. Measurement by fluorescence correlation spectroscopy. *Phys. Rev. Lett.* 29:705–708, 1972.

Magde, D., E.L. Elson, and W.W. Webb. Fluorescence correlation spectroscopy. II. Experimental realization. *Biopolymers* 13:29–61, 1974.

Meseth, U., T. Wohland, R. Rigler, and H. Vogel. Resolution of fluorescence correlation measurements. *Biophys. J.* 76:1619–1631, 1999.

Muller, J.D., Y. Chen, and E. Gratton. Resolving heterogeneity on the single molecular level with the photon-counting histogram. *Biophys. J.* 78:474–486, 2000.

Petersen, N.O., P.L. Hoeddelius, P.W. Wiseman, O. Seger, and K.E. Magnusson. Quantitation of membrane receptor distributions by image correlation spectroscopy: concept and application. *Biophys. J.* 63:1135–1146, 1993.

Rigler, R. and E.S. Elson. *Fluorescence Correlation Spectroscopy Theory and Applications.* Vol. 65. New York: Springer, 2001.

Schwille, P. Fluorescence correlation spectroscopy and its potential for intracellular applications. *Cell Biochem. Biophys.* 34:383–408, 2001.

Schwille, P., U. Haupts, S. Maiti, and W.W. Webb. Molecular dynamics in living cells observed by fluorescence correlation spectroscopy with one- and two-photon excitation. *Biophys. J.* 77:2251–2265, 1999.

Schwille, P., F.J. Meyer-Almes, and R. Rigler. Dual-color fluorescence cross-correlation spectroscopy for multicomponent diffusional analysis in solution. *Biophys. J.* 72:1878–1886, 1997.

So, P.T.C., C.Y. Dong, B.R. Masters, and K.M. Berland. Two-photon excitation fluorescence microscopy. In: *Annual Review of Biomedical Engineering*, edited by M.L. Yarmush. Palo Alto: Annual Reviews, 2000, pp. 399–429.

Thompson, N.L. Fluorescence correlation spectroscopy. In: *Topics in Fluorescence Spectroscopy*, edited by J.R. Lakowicz, New York: Plenum Press, 1991, pp. 337–378.

Zipfel, W.R., R.M. Williams, and W.W. Webb. Nonlinear magic: multiphoton microscopy in the biosciences. *Nat. Biotech.* 21:1369–1377, 2003.

16

Mapping Molecular Interactions and Transport in Cell Membranes by Image Correlation Spectroscopy

ELEONORA KEATING, CLAIRE M. BROWN,
AND NILS O. PETERSEN

1. INTRODUCTION

Protein–protein interactions control most events at the plasma membrane, such as signal transduction, endocytosis, attachment, and locomotion. It is evident from the theme of this book that measurement of intermolecular interactions in cellular systems is critical to our understanding of mechanisms of many cellular processes. It is equally evident that determining the nature of the interactions and their dynamics is challenging and requires multiple approaches that take advantage of fundamental physical and chemical tools to inform the biochemistry and the cell biology (Webb et al., 2003). In this chapter we present a family of image processing tools that are complementary to the Förster resonance energy transfer (FRET) measurements of intermolecular interactions and discuss their strengths and weaknesses.

1.1 Molecular Interactions

Intermolecular interactions of proteins in a cell membrane can take many forms characterized by the energetics, the dynamics, and the length scale over which the interactions take place.

Strong interactions will form complexes that withstand solubilization and interrogation by biochemical techniques such as immunoprecipitation and chromatography or electrophoresis. The presence of these complexes can be ascertained without ambiguity, but it is virtually impossible to provide quantitative estimates of the number of such interactions within a complex or the fraction of a particular species that is involved. *Weak interactions* can also lead to complexes, particularly if there are many such interactions, but these cannot withstand isolation procedures. They must be measured directly in the membrane in the living cell.

Persistent interactions will last for long periods of time because the kinetics of dissociation is slow. These will usually also correspond to the strong interactions and they can therefore be detected following isolation procedures. However, *transient interactions* will last for short periods of time only because they are governed by rapid exchange between the free and complexed forms of the molecules. During isolation, the equilibrium may shift so that these interactions are best determined in situ.

Whether intermolecular interactions are strong, weak, persistent or transient, measurement in living cells is a benefit, particularly if it is possible to determine the stoichiometry, kinetics, and spatial distribution of the interactions and hence get insight into the mechanisms underlying the formation of functional protein complexes.

1.2 Organization of Complexes, Clusters, and Aggregates

Intermolecular interactions can lead to formation of assemblies of molecules that have properties distinct from their surroundings. There are a number of terms that have been used to describe such assemblies in cell membranes and these include complexes, clusters, aggregates, and domains. For the purpose of this chapter, we will use the following definitions:

> *Complex:* an assembly of similar or dissimilar molecules created by specific intermolecular interactions that determine the stoichiometry and structure.
>
> *Cluster:* a complex of proteins or lipids or both that persists for a period of time long enough to be detected as a unit by microscopic techniques.
>
> *Aggregate:* a complex of identical molecules (or a cluster in which only a single component is detected or studied).
>
> *Association complex:* a complex of different molecules.
>
> *Domain:* a region of the membrane that is structurally and functionally distinct from the rest of the membrane. There are both lipid and protein domains.

Large complexes have dimensions an order of magnitude or more greater than the dimensions of the individual proteins in the complex, yet they may be smaller than the resolution of most optical tools used for their detection (~300 nm). *Small complexes* contain only a few of the proteins of interest and each would be in molecular contact with the others.

Fluorescence resonance energy transfer is an excellent tool for measuring intermolecular interactions among pairs of molecules in small complexes and aggregates in particular because the distance between pairs of molecules needs to be relatively small (~10 nm or less; see Chapters 2, 4–13). In large complexes and domains, the density of proteins may be high, but the distance between the proteins of interest may be greater than the optimal energy transfer distance. In those circumstances, FRET will fail to detect the interactions. FRET is also a very sensitive tool for measurement of the distances of the interactions among a small number of fluorescent molecules (or sections of molecules), but because of the sensitivity to distance and orientation, interpretation of FRET is difficult when there are many chromophores, as in large complexes or aggregates. This makes it difficult to assess the number of proteins participating in the formation of the complex. Further, it can be difficult to prepare and introduce good probes, and the absence of energy transfer does not always preclude that proteins are indeed interacting. Nevertheless, FRET imaging shows tremendous potential for mapping the nature and extent of intermolecular interactions across the surface of living cells. We hope that image correlation spectroscopy, as described here, will complement FRET imaging by providing a means to measure protein interactions in larger complexes, even when these proteins may not be in close proximity or bound directly to one another.

2. Fluctuation Spectroscopy

Fluctuation spectroscopy refers to a number of techniques in which the variation in a signal is measured and analyzed to obtain information about the average properties of the system (Magde et al., 1972). Examples include noise analysis in electronics, quasielastic light scattering in liquids, and fluorescence correlation spectroscopy in solutions. In each case, a signal is monitored with time and the temporal fluctuations from the mean value are analyzed by spectral analysis, autocorrelation of the function with itself, or cross-correlation of the function with a reference signal. The outcome is information about the dynamic properties of the system at a characteristic length scale (defined by the measurement tool) and in some cases information about the number of entities that give rise to the fluctuations in the signal.

In the early 1970s, Magde, Elson, and Webb (1974) introduced fluorescence correlation spectroscopy (FCS) as a tool to measure the diffusion of proteins in solution and in cell membranes. They demonstrated that the autocorrelation function, $g(\tau)$, of the relative fluctuation in intensity of fluorescence, $(i(t) - \langle i(t) \rangle) / \langle i(t) \rangle$, would have an amplitude determined by the average number, $\langle N \rangle$, of molecules in the observation volume and that it would decay at a rate, $F(\tau)$, determined by the dynamics of these molecules:

$$g(\tau) = \frac{\langle (i(t) - \langle i(t) \rangle)(i(t+\tau) - \langle i(t) \rangle) \rangle}{\langle i(t) \rangle^2} = \frac{1}{\langle N \rangle} F(\tau) \tag{1}$$

where the angular brackets indicate an average over the time course of the observation. The decay function, $F(\tau)$, depends on the details of the dynamics and on the geometry of illumination and detection of the fluorescence in the volume being observed. For example, if the molecules are diffusing in two dimensions and they are illuminated with a focused laser beam in the TEM_{00} mode of characteristic width w, then the decay function is $F(\tau) = 1/ (1 + \tau / \tau_D)$, where $\tau_D = w^2/4D$ represents the characteristic diffusion time in two dimension for a diffusion coefficient of D. Correspondingly, if the molecules are flowing through the same laser beam at a velocity v, then the decay function is $F(\tau) = \exp(-(\tau / \tau_F)^2)$, where $\tau_F = w/v$ (Magde et al., 1978).

In recent years, FCS has reemerged as a major tool for studying intermolecular interactions in solutions at low concentrations (Rigler and Elson, 2001). In fact, commercial systems are now available from several microscope manufacturers and suppliers. The application of FCS to work in cellular systems is more difficult, but progress is being made (Elson, 2001; Thompson et al., 2002). Still, applications to studies of intermolecular interactions in membranes have been hampered by the fact that the fluctuations are inherently slow because the diffusion and kinetics are slow (diffusion coefficients in membranes tend to be at least two orders of magnitude smaller than that in solution, leading to characteristic fluctuation times of seconds to minutes).

In 1984, Petersen introduced the concept of scanning fluorescence correlation spectroscopy (S-FCS) as a means of imposing a flow on the sample at a known velocity and thereby capturing the amplitude as a means of estimating the number of molecules on a cell surface. Subsequently, he demonstrated that the amplitude of the correlation function could be used to estimate the density of clusters of proteins on the surface as well as the number of molecules per cluster (Petersen, 1986). While the approach did not provide information about the natural dynamics (diffusion or flow), it was argued that this could be obtained by complementary fluorescence photobleaching experiments. The S-FCS experiment was the first approach to obtain quantitative information about the state of aggregation of proteins in cell membranes and was applied successfully to studies of several protein systems in fixed cells (Petersen et al., 1986; St. Pierre and Petersen, 1990, 1992). However, the measurements were slow and cumbersome and required high-precision scanners that could move the sample over 100 μm distances with tens of nm precision (Petersen and McConnaughey, 1981).

3. IMAGE CORRELATION AND IMAGE CROSS-CORRELATION SPECTROSCOPY

Image correlation spectroscopy (ICS) was introduced (Petersen et al., 1993) as a two-dimensional implementation of the S-FCS approach to take advantage of the rapid image acquisitions provided by laser scanning confocal microscope systems developed in the late 1980s. Rather than scanning the sample, the laser beam moves linearly across the stationary sample in the x-direction and this is repeated at different y-positions to generate a two-dimensional image. At each position in the image (defined as a pixel) the intensity of fluorescence is measured as a photon count, $i(x,y)$, reflecting the number of molecules excited within the beam area when the beam is centered at that x,y position. The image represents a large number of samples of fluorescence across the surface as compared to a large number of samples of fluorescence measured in a single position as a function of time. For an ergodic system, the two types of measurements will provide the same estimate of the average properties of the system, such as the average number of molecules in the illumination area by the laser beam. Image correlation spectroscopy has been demonstrated to be a powerful tool to measure quantitatively the average cluster density of receptor molecules on a cell surface. It is currently the best demonstration of the application of the general concept of spatial FCS (Petersen, 2001).

3.1 Acquisition of Data

Figure 16–1 illustrates the critical elements of data acquisition for ICS measurements. An image is collected from a section of the membrane of an adherent cell under conditions where the image (in our work, 512 × 512 pixels) covers a small area of the surface (in our work, 15.5 × 15.5 μm; Fig. 16–1A). Under these conditions, the nominal resolution is ~30 nm per pixel whereas the actual resolution is on the order of 300 nm (the width of the laser beam at the focal plane). Thus the data are sampled at about 10-fold higher resolution—they are oversampled. This suggests that all clusters or domains smaller than about 300 nm will appear as spots with a 300 nm radius (Fig. 16–1B). Correspondingly, a trace of the intensity as a function of distance (Fig. 16–1C) will show intensity variations that are close to Gaussian in shape and with a width determined by the laser beam intensity profile. The amplitude of each of these peaks will depend on the number of molecules present in the cluster or domain being detected.

Figure 16–2A is an image of the distribution of epidermal growth factor receptors on A431 cells. There are a number of bright spots, each representing a cluster of proteins. The spots have the same dimensions, but differ in brightness, indicating that the clusters are all smaller than the beam size and that they contain different numbers of receptors. The receptor distribution is not uniform.

3.2. Image Correlation Spectroscopy

Image correlation spectroscopy involves calculating the autocorrelation function of a digital image, such as a confocal image in which the intensity is given by $i(x,y)$ at the pixel located at positions x and y in the image.

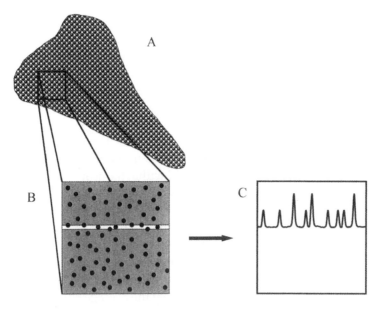

FIGURE 16–1 Schematic representation of image acquisition for image correlation spectroscopy. For further details, see text.

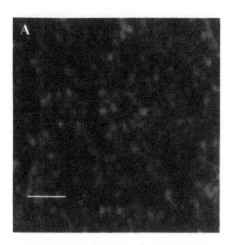
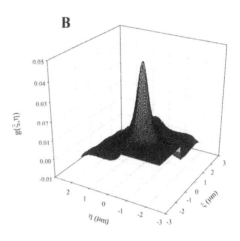

FIGURE 16–2 Example of a confocal microscope image (*A*) and the corresponding autocorrelation function (*B*). Scale bar represent 3μm

$$g(\xi,\eta)=\frac{\left\langle\left(i(x,y)-\left\langle i(x,y)\right\rangle\right)\left(i(x+\xi,y+\eta)-\left\langle i(x,y)\right\rangle\right)\right\rangle}{\left\langle i(x,y)\right\rangle^{2}} \qquad (2)$$

This two-dimensional autocorrelation function, $g(\xi,\eta)$, will decay in the spatial lag coordinates, ξ and η, as a two-dimensional Gaussian function (reflecting the beam shape, equation (3)) with a width corresponding to the width of the laser beam, w. The correlation function for the image in Figure 16–2A is depicted in Figure 16–2B. The front quadrant in Figure 16–2B shows the fit to the expected function (equation (3)) with three adjustable variables: the amplitude, $g(0,0)$, the width, w, and the baseline offset, g_0:

$$g(\xi,\eta)=g(0,0)e^{-(\xi^{2}+\eta^{2})/w^{2}}+g_0 \qquad (3)$$

The amplitude of the autocorrelation function has the same interpretation as in FCS and thus in the first instance, $g(0,0) = 1/\langle N\rangle$. However, when the system consists of a mixture of clusters with a distribution of the number of monomers in each cluster, as seems to be the case in Figure 16–2A, the interpretation is less obvious. Magde et al. (1978) demonstrated that for a mixture of molecules with different extinction coefficients, ε, quantum yield, Q, and concentration, C, the amplitude of the correlation function is given by equation (4):

$$g(0,0)=\frac{1}{A}\frac{\displaystyle\sum_{j=1}^{N}\left(\varepsilon_j Q_j\right)^2\overline{C}_j}{\left(\displaystyle\sum_{j=1}^{N}\left(\varepsilon_j Q_j\right)\overline{C}_j\right)^2} \qquad (4)$$

Later it was shown (Petersen, 1986, 1993, 2001; Wiseman and Petersen, 1999) that for a single distribution, $f(j)$, with a mean, μ, and a variance, σ^2, the amplitude of the correlation function can be rewritten as equation (5):

$$g(0,0) = \frac{1}{\overline{N_c}} \frac{\sum\limits_{j=1}^{n} j^2(j)}{\left(\sum\limits_{j=1}^{n} jf(j)\right)^2} = \frac{1}{\overline{N_m}}\left[\frac{\mu^2 + \sigma^2}{\mu}\right] \tag{5}$$

where N_c is the number of clusters, N_m is the number of monomers, and the bar indicates an average over all of the image. It is clear from equation (5) that if the mean of the distribution is significantly greater than the variance, then the amplitude of the autocorrelation function can be approximated as the inverse of the average number of clusters observed in the observation area.

$$g(0,0) \approx \frac{1}{\overline{N_m}}(\mu) \approx \frac{1}{\overline{N_c}} \tag{6}$$

Accordingly, we have defined two useful parameters: the cluster density, (CD) and the degree of aggregation (DA):

$$CD = \frac{1}{g(0,0)\pi w^2} \approx \frac{\overline{N_c}}{\mu m} \tag{7}$$

$$DA = \langle i(x,y) \rangle g(0,0) \approx c\overline{N_m}\frac{1}{\overline{N_m}} = c\mu = c\frac{\overline{N_m}}{\overline{N_c}} \tag{8}$$

where the constant c is an instrumental parameter that relates the measured intensity in a particular pixel, $i(x,y)$, to the number of monomers, N_m, in the beam area at that location.

Equations (2–8) illustrate that the autocorrelation function (Fig. 16–2B) of a confocal image (Fig. 16–2A) can serve to provide a measure of the average number of clusters per unit area and the average number of monomers per cluster in that area of the cell. The key to the success of ICS is that it is relatively simple to make measurements of many images on many cells in a population, calculate the autocorrelation function of each image, estimate the CD and DA values, and average these for the population of cells. More importantly, these measurements are quite sensitive and thus provide good quantitative estimates of the desired parameters.

Figure 16–3 illustrates the sensitivity for a pair of images in which the average number of molecules per unit area is the same (the average intensities are 15.9 and 16.0, respectively) while the degree of aggregation differs by a factor of four (DA = 1.50 for Fig. 16–3A and 0.36 for Fig. 16–3B). The latter difference is a direct reflection of the difference in the amplitude of the correlation functions (0.07 in Fig. 16–3C and 0.014 in Fig. 16–3D).

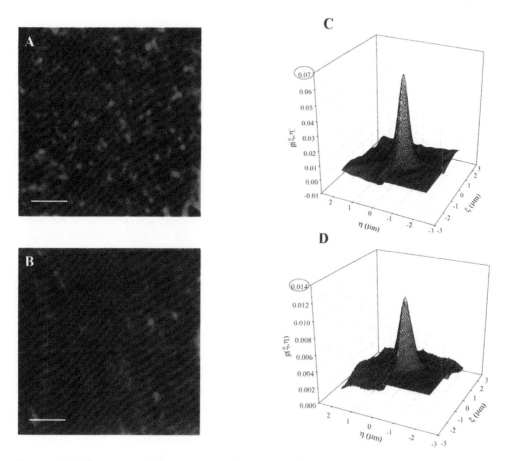

FIGURE 16–3 Illustration of the sensitivity of image correlation spectroscopy to the state of aggregation. Scale bar represents 3μm. For further explanation, see text.

3.3 Image Cross-Correlation Spectroscopy

Image correlation spectroscopy provides excellent estimates of the average number of clusters and the degree of aggregation of a particular molecule on the cell surface from a set of images of the distribution of that molecule. To obtain information about the interaction between different molecules, we must collect separate images for each of the molecules on the same areas of the cell and calculate the cross-correlation functions.

Figure 16–4 illustrates the principle. Two images are collected from the identical region of the same cell. Cross-multiplication of the intensities of corresponding pixels in the two images will yield a positive contribution to the cross-correlation function only if there are spatially coincident fluctuations in the two intensities. The spatially coincident fluctuations will arise if the two molecular species are located in the same region of the cell.

The most general cross-correlation function is calculated for a pair of images represented by $i(x,y;t)$ and $j(x,y;t)$, where the time parameter indicates that the two images could be collected at different times, as illustrated in equation (9):

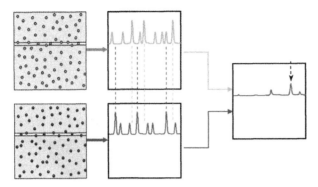

FIGURE 16–4 Schematic representation of the cross-correlation function calculation. Spatially coincident fluctuations will give positive results in the cross-multiplications inherent in the cross-correlation function calculation.

$$g_{ijt}(\chi,\xi;\tau) = \frac{\left\langle \left(i(x,y;t) - \langle i(x,y;t)\rangle \right)\left(j(x+\chi,y+\xi;t+\tau) - \langle j(x,y;t)\rangle \right)\right\rangle}{\langle i(x,y;t)\rangle\langle j(x,y;t)\rangle} \tag{9}$$

It has been shown (Srivastava and Petersen, 1996, 1998; Srivastava et al., 1998; Brown et al., 1999) that this cross-correlation function will decay in the spatial parameters as a Gaussian function and in the time domain as the $F(\tau)$ described in equation (3). The amplitude of the cross-correlation function at zero lag distances and lag time, $g(0,0;0)$, can be approximated (Schwille et al., 1997, Petersen, 2001) by

$$g_{ijt}(0,0;0) = \lim_{\chi\to 0,\xi\to 0;\tau\to 0} g_{ijt}(\chi,\xi;\tau) = \frac{\overline{N}_{ijt}}{\left(\overline{N}_{it}+\overline{N}_{ijt}\right)\left(\overline{N}_{jt}+\overline{N}_{ijt}\right)} \tag{10}$$

where N_{it} is the number of clusters with species i only, N_{jt} is the number of clusters with species j only, and N_{ijt} is the number of clusters with both species i and j present. Thus, the sums $(N_{it} + N_{ijt})$ and $(N_{jt} + N_{ijt})$ represent the total number of clusters containing species i and j, respectively. Each of these can be measured directly from the amplitudes of the corresponding autocorrelation functions as given by equation (6). Equation (10) was derived for pairs of individual molecules in solution (Schwille et al., 1997), but it has been applied with reasonable success to clusters of molecules (Brown et al., 1999) and beads in solution (Wiseman et al., 2000). We believe that equation (10) will remain a reasonable approximation as long as the distribution of sizes of clusters is reasonably narrow (as required for equation (6)).

Figure 16–5 is an example of two images collected for two proteins labeled with different antibodies. The image in Figure 16–5A yields the autocorrelation function in Figure 16–5C, from which the density of clusters of molecule A and its degree of aggregation can be inferred (equations (7) and (8)). Correspondingly, the image in Figure 16–5B yields the autocorrelation function in Figure 16–5D from which the density of clusters of molecule B and its degree of aggregation can be inferred. The cross-correlation function for the two images is shown in Figure 16–5E, from which the number of clusters with both molecule A and B can be determined.

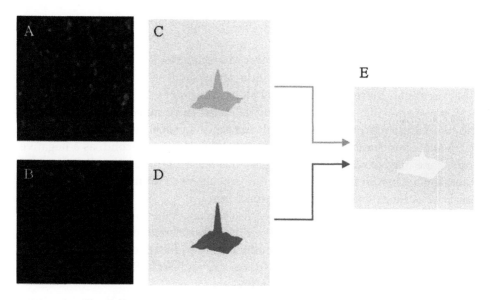

FIGURE 16–5 Illustration of the image cross-correlation spectroscopy principle. For further explanation, see text.

Combining equations (6) and (10) yields the average density of clusters with both molecules present

$$CD_{ij} = \frac{g_{ijt}(0,0;0)}{g_{it}(0,0;0)g_{jt}(0,0;0)\pi w_i^2} = \frac{\overline{N}_{ijt}}{\pi w_i^2} \tag{11}$$

Thus measurement of two images (Figs. 16–5A and 16–5B) provides the data to calculate the amplitudes of the two autocorrelation functions (Figs. 16–5C and 16–5D) and the cross-correlation function (Fig. 16–5E), which combine in equation (11) to give the density of mixed clusters.

   For convenience, we have defined (equation (12)) a pair of fractions that provide estimates of what proportion of clusters that contain one molecule also contains the other. Crudely, this can be thought of as the fraction of co-localization: the fraction of one molecule that is co-located with the other and visa versa.

$$F(i|j) = \frac{CD_{ijt}}{CD_{it}} = \frac{g_{ijt}(0,0;0)}{g_{jt}(0,0;0)\pi w_j^2}$$

and $\qquad (12)$

$$F(j|i) = \frac{CD_{ijt}}{CD_{jt}} = \frac{g_{ijt}(0,0;0)}{g_{it}(0,0;0)\pi w_i^2}$$

For example, $F(i\,|\,j)$ is the ratio of all the clusters that contain both species i and j to all the clusters that contain i (whether alone or not). If all of species i is co-located with j, then this fraction is expected to be one. Note that even if all of species i is co-located with species j, there is no requirement that all of species j be completely co-located with species i. The two fractions in equation (12) are therefore not necessarily directly related to each other.

It is possible to make reasonable control experiments that demonstrate the validity of the approximations inherent in the above analysis and to get an estimate of how accurately the image cross-correlation spectroscopy (ICCS) experiments can be. First, as a positive control, one may label the same molecule simultaneously with two different probes. In that case, one expects complete co-localization and the fractions in equation (12) should both be close to one. Second, as a negative control, one may cross-correlate images collected from different cells measured under similar conditions. In this case, one expects no co-localization and the fractions in equation (12) should both approach zero.

Table 16–1 illustrates one example of the estimate of the co-localization of pairs of receptors (Petersen et al., 1998). The positive control experiment, in which transferrin receptors (Tf-R) were labeled simultaneously with antibodies that have green fluorescent probes (Tf-R-G1) and red fluorescent probes (Tf-R-R1), shows that the two fractions are close to one (within ~10%).

Correspondingly, the negative control experiment, in which the images from green fluorescent probes (Tf-R-G1) were randomly cross-correlated with images from red fluorescent probes (Tf-R-R2), shows that these fractions are close to zero (within a few percent). This suggests that the ICCS experiments can provide estimates with an accuracy approaching 10%. The actual experiments show that there is no co-localization between Tf-R and epidermal growth factor receptors (EGF-R), whereas there is strong co-localization between Tf-R and platelet-derived growth factor receptors (PDGF-R). Approximately 80% of the Tf-R are located in clusters that also contain the PDGF-R, which suggests that about 20% of the Tf-R are not associated with PDGF-R. Correspondingly, about 70% of the PDGF-R are in clusters that also contain the Tf-R, which suggests that about 30% are not associated with Tf-R. There are therefore at least two populations of each receptor type—those in common clusters and those in separate clusters. Table 16–1 also shows (CD_{ij}) that there are about four to five clusters with both receptors per square micrometer. Whether these clusters in turn represent the coated pits in cells (Brown et al., 1999) remains to be determined.

TABLE 16–1 Illustration of Co-Localization of Receptors on a Cell Surface

Probe i	Probe j	Pairs of Images (N)	% Fit[a]	CD_{ij}	F(i | j)	F(j | i)
Tf-R-Gl	Tf-R-R1	25	100	5.4 ± 0.7	1.1 ± 0.1	0.92 ± 0.1
Tf-R-G1	Tf-R-R2	25	6	NA	<0.01	<0.01
Tf-R	EGF-R	45	<10	0.37	0.02	0.006
Tf-R	PDGF-R	80	>80	4.1 ± 0.6	0.8 ± 0.1	0.68 ± 0.1

EGF-R, epidermal growth factor receptor; NA, not available; PDGF-R, platelet-derived growth factor receptor; Tf-R, Transferrin receptor; Tf-R-G1, Tf-R with green fluorescent probes; Tf-R-R1,-R2, Tf-R with red fluorescent probes.

[a]Percentage of cross-correlated images that provide a fit at the origin of the cross-correlation function and hence have meaningful amplitudes. The other parameters are explained in the text.

Data courtesy of Mamta Srivastava.

4. Dynamic Image Correlation Spectroscopy

The ICS experiments provide quantitative estimates of receptor cluster distributions and the numbers of monomers in a cluster. The ICCS[1] experiments provide quantitative estimates of the extent of co-localization of pairs of molecules in clusters or domains. In both measurements, the dynamic information is lost. It is, however, possible to collect images in multiple spectral ranges as a function of time, which provides the opportunity to capture some of the dynamic processes within the timescale defined by the speed of image acquisition. This leads to three distinct applications of dynamic image correlation spectroscopy (DICS): ICS measurements as a function of time, ICCS measurements as a function of time, and calculation of cross-correlation functions of images separated in time.

4.1 Time Dependence of Image Correlation and Image Cross-Correlation Spectroscopy Experiments

The simplest approach to the dynamics is to calculate the time dependence of the cluster density, $CD(t)$, and the degree of aggregation, $DA(t)$, from the autocorrelation function amplitudes of each of the images in the time series. In principle, this can lead to measurements of the kinetics of assembly into clusters or disassembly of clusters into monomers on the surface. Figure 16–6 illustrates how the temporal variation of the density of clusters and the degree of aggregation can be used to follow the disassembly of $\alpha 5$–integrin receptors in CHO B2 cells on the timescale of minutes. As is evident from the images of the cells shown at four selected times (Fig. 16–6A-D), there are a number of bright adhesion complexes that disappear in region 1. The average degree of aggregation (Fig. 16–6E) decreases about fourfold over a period of ~200 s, whereas there is no change in the state of aggregation in the neighboring area (region 2). Correspondingly, the cluster density (Fig. 16–6F) increases by a corresponding factor over a similar timescale, although it appears that there is a delay between the decrease in the degree of aggregation and the increase in the number of clusters. It is evident that the number and the size of the receptor clusters vary on the minute timescale for this system and that the DICS measurements provide a convenient tool to obtain quantitative kinetic data for rather complex systems. To date, there are only a few examples of this type of analysis (Petersen et al., 1986; Nohe and Petersen, 2002, 2004), but it can provide useful insight into the mechanisms whereby protein or lipid domains assemble or disappear. For example, the dispersal of viral lipids occurs more rapidly than the dispersal of viral proteins during the fusion of a virus to a cell membrane. This finding suggests that there are intermediate states in which the lipids can transfer from the virus to the cell membrane, but the proteins cannot (Rocheleau and Petersen, 2001).

It is also possible to monitor the cross-correlation function as a function of time to obtain the kinetics of association or dissociation of different molecules in the clusters or domains. So far, this type of experiment has not been explored systematically, even though it can provide important insight into the mechanism of formation of clusters and domains in the membrane.

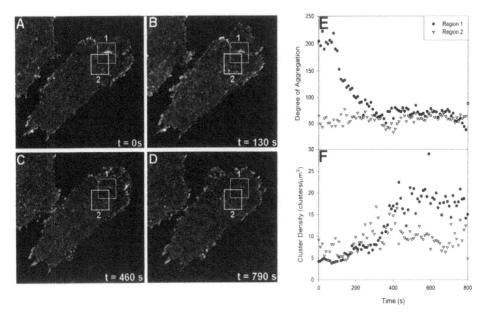

FIGURE 16–6 A CHO B2 cell expressing a5-integrin-GFP. *A–D*: Sample images from a time-lapse movie. In region 1, a row of adhesions disassembles whereas in region 2 there is no significant change in the overall protein distribution. *E, F,*: The degree of aggregation (DA) (E) decrease and the cluster density (CD) (F) increase as the adhesion complex disassembles (filled circles). Note that there is little change in the DA for region 2 and no really systematic change in the CD over the same time interval (open triangles). It should be noted that the DA and the CD can become somewhat misleading when there are large organized protein structures such as in region 1, but they can still be used to follow the overall changes in protein aggregation.

4.2 Dynamic Image Cross-Correlation Spectroscopy

The general expression for the cross-correlation function (equation (9)) contains both the space and time coordinates. If the two images, $i(x,y;t)$ and $j(x,y;t)$ are collected for the same molecules from the same area of the same cell but at different times, then the cross-correlation function will measure the extent to which the correlation of clusters with themselves is disappearing because of the movement of the clusters. This is illustrated in Figure 16–7. A series of images were collected for EGF-R over time from the same region of a single A431 cell (Fig. 16–7A). All the images were created with a red look-up table so that each of the fluorescent clusters appears red. Then the first image, corresponding to $t = 0$, was created using a green look-up table and superimposed on each of the other images. Figure 16–7B shows the result when the first image ($t = 0$) is superimposed on itself. As expected, there is complete "co-localization" of all of the clusters (all of the spots are yellow since the red and the green images are identical). Figures 16–7C to 16–7F show the result when the first image ($t = 0$) is superimposed on images collected at $t = 24$, 128, 160, and 400 s. It is evident that as the time separation between the first and subsequent images increases, the extent of co-localization decreases (the number of yellow spots decreases) and at 400 s there is no co-localization at all. The lateral movement of the clusters causes a loss of correlation of the clusters with themselves.

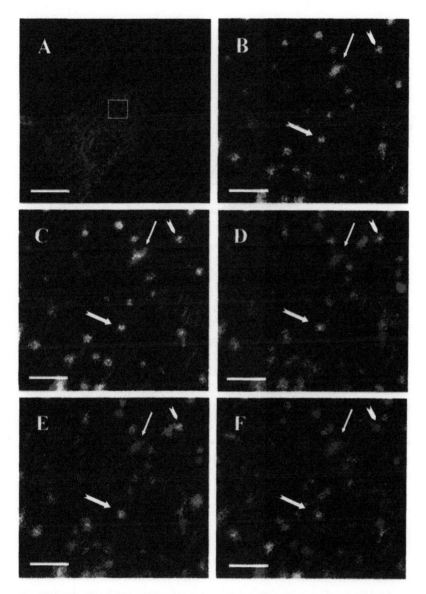

FIGURE 16–7 Illustration of the movement of clusters relative to their original positions (see text). The box in A represents the region measured in images B–F. The arrows in B–F point to representative clusters that move at slightly different speeds. Scale bar represents 3μm.

We expect that the amplitude of the cross-correlation functions for pairs of images separated in time will decrease correspondingly.

For a series of images collected from the same region as a function of time, we may express the time variation of the amplitude of the cross-correlation function as

$$g_x(0,0,\tau) = \lim_{\chi \to 0, \xi \to 0} = \frac{\langle\langle i(x,y) - \langle i(x,y)\rangle\rangle\langle i(x+\chi,y+\xi;t+\tau) - \langle i(x,y)\rangle\rangle\rangle}{\langle i(x,y)\rangle^2} = \frac{1}{N_i}F(\tau) \quad (13)$$

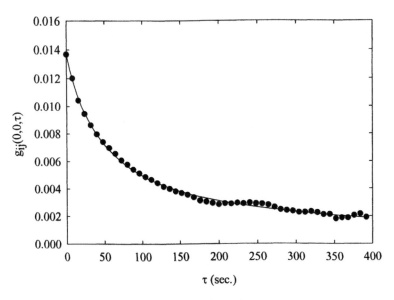

FIGURE 16–8 Decay of the amplitude of the cross-correlation function between images collected at different times.

where the decay function, $F(\tau)$, has the same functional form as that in equation (1) (Srivastava and Petersen, 1996; Wiseman et al., 2000). In effect, this experiment is an ensemble averaged FCS experiment in which the dynamics of all of the regions of the surface of the cell are measured simultaneously. Figure 16–8 shows an example of the decay of the amplitude of the cross-correlation functions as a function of the time separation between images. The solid line represents a fit to the data according to

$$g(0,0,\tau) = \frac{g(0,0;0)}{1+\dfrac{\tau}{\tau_d}} + g_0 \tag{14}$$

which suggests that the dynamic process is a two-dimensional diffusion process with a diffusion coefficient $D \sim 10^{-11}$ cm^2 s^{-1}. This is consistent with the changes in position of the clusters illustrated in Figures 16–7B–F. This analysis provided information about the dynamics of proteins within the cell membrane. If it is extended to the cross-correlation spectroscopy domain with two different proteins, the analysis will provide dynamic information that can arise only from co-movement of proteins within a common complex (Wiseman et al., 2000).

5. DISCUSSION

The systematic analysis of confocal images described above provides a set of tools to analyze quantitatively the extent of formation of clusters, the extent of association between different molecules, and the dynamics of this interaction. While these approaches are relatively new, there are several applications emerging (Fire et al., 1997; Brown and Petersen, 1998, 1999; Brown et al., 1999, Boyd et al., 2002; Nohe and Petersen, 2002, 2004; Nohe et al., 2003; Rocheleau et al., 2003) and it appears that these approaches can provide new and useful information.

The strengths of the ICS approaches are that the measurements are relatively easy with modern confocal imaging tools, the data analysis is straightforward, and it is possible to make many measurements on a large number of cells in a population. The individual measurements are quantitative and it is simple to obtain good statistics for comparison of changes that may occur as a result of perturbing the cell system.

The greatest weaknesses of ICS are that one must carry out significant numbers of controls and the data must be corrected for systematic effects (Wiseman and Petersen, 1999). Because the measurements are simple, it is easy to collect apparently meaningful data, so vigilance is also required to ensure that the results are physically meaningful.

The ICCS measurements have great potential for delivering quantitative information about the extent to which pairs of membrane molecules interact in clusters or domains on the cell surface. In principle, ICS and ICCS will capture all of the molecular interactions that persist on the timescale of the measurement, which is the time needed to capture a particular cluster in the image. For a typical laser scanning confocal microscope, this corresponds to microseconds in the x-direction and milliseconds in the y-direction. These tools should therefore measure strong as well as weak interactions, and persistent as well as transient interactions.

It is important to recognize that the correlation analysis only requires that the molecules being measured be organized in some fashion on the surface. They need not be in molecular contact as long as they are part of the same cluster. Thus, without other evidence from immunoprecipitation or FRET measurements, for example, it is impossible to determine the nature of the intermolecular interactions. For example, all of the components in a domain such as a coated pit will be seen to colocalize, regardless of whether they are in fact interacting directly with each other or only with a common component, such as an adaptor protein. However, interactions between molecules that are part of a common protein complex but not in molecular contact would be undetectable without the ICCS approach. These types of interactions could be central to understanding the mechanism of formation and disassembly of protein complexes within the cell, such as coated pits, focal adhesions, and tight junctions, which in turn may lead to significant new insight into mechanisms underlying activation of specific signal transduction pathways.

To further advance the utility of ICS, ICCS, or FRET, we propose that it may be possible to perform the correlation analysis on FRET images. If properly collected, the FRET image should represent the fluorescence intensity arising from the acceptor only, which should in turn derive only from those clusters or domains in which the donor–acceptor interactions are within the appropriate distance. The ICS analysis will then yield an estimate of the number of clusters on the surface in which there are direct intermolecular interactions. Comparison with the regular ICS could then reveal the fraction of clusters in which this direct interaction occurs. We are currently working on FRET-ICS experiments, but have so far seen very few systems in which the FRET is sufficiently strong to make the ICS analysis robust. Perhaps this is an indication that for the receptors we have studied thus far, the molecular contact between the receptors in these clusters and domains is relatively rare.

NOTE

1. The term *temporal ICCS* has also been used to describe these experiments (C.M.B.).

REFERENCES

Boyd, N.D., B.M. Chan, and N.O. Petersen. Adaptor protein-2 exhibits $\alpha_1\beta_1$ or $\alpha_6\beta_1$ integrin dependent redistribution in rhabdomyosarcoma cells grown on collagen or laminin relative to fibronectin. *Biochemistry* 41:7232–7240, 2002.

Brown, C.M. and N.O. Petersen. Distribution of clathrin associated AP-2. *J. Cell Sci.* 111:271–281, 1998.

Brown, C.M. and N.O. Petersen. Free clathrin triskelions are required for the stability of AP-2 coated pit nucleation sites. *Biochem. Cell Biol.* 77:439–448, 1999.

Brown, C.M., M.G. Roth, Y.I. Henis, and N.O. Petersen. Co-localization of AP-2 and internalization competent HA proteins in clathrin free clusters on intact cells *Biochemistry* 38:15166–15173, 1999.

Elson, E.L. Fluorescence correlation spectroscopy measures molecular transport in cells. *Traffic* 2:789–796, 2001.

Fire, E., C.M. Brown, Y.I. Henis, M.G. Roth, and N.O. Petersen. Partitioning of proteins into plasma membrane microdomains: clustering of mutant influenza virus hemagglutinins into coated pits depends on the internalization signal. *J. Biol. Chem.* 272:29438–29545, 1997.

Magde, D., E.L. Elson, and W.W. Webb. Thermodynamic fluctuations in a reacting system: measurement by fluorescence correlation spectroscopy. *Phys. Rev. Lett. (USA)* 29:705–711, 1972.

Magde, D., E. Elson, and W.W. Webb. Fluorescence correlation spectroscopy I, an experimental realization. *Biopolymers* 13:29–61, 1974.

Magde, D., W.W. Webb, and E.L. Elson. Fluorescence correlation spectroscopy III, uniform translation and laminar flow. *Biopolymers* 17:361–376, 1978.

Nohe, A., E. Keating, T.M. Underhill, P. Knaus, and N.O. Petersen. Effect of the distribution and clustering of BMP receptors on the activation of signaling pathways. *J. Cell Sci.* 116:3277–3284, 2003.

Nohe, A. and N.O. Petersen. Exploring the early events in signal transduction by image correlation spetroscopy. *Biophotonics* 9:39–52, 2002.

Nohe, A. and N.O. Petersen. Analyzing protein–protein interactions in cell membranes. *Bioessays* 26:196–203, 2004.

Petersen, N.O. Diffusion and aggregation in biological membranes *Can. J. Biochem. Cell Biol.* 62:1158–1166, 1984.

Petersen, N.O. Scanning fluorescence correlation spectroscopy I: theory and simulation of aggregation measurements. *Biophys. J.* 49:809–815, 1986.

Petersen, N.O. FCS and spatial correlations on biological surfaces. In: *Fluorescence Correlation Spectroscopy*, edited by R. Rigler and E.L. Elson, Berlin: Springer–Verlag, 2001, pp. 2–35.

Petersen, N.O., C. Brown, A. Kaminski, J. Rocheleau, M. Srivastava, and P.W. Wiseman. Analysis of membrane protein cluster densities and sizes in situ by image correlation spectroscopy. *Faraday Discuss.* 111:289–305, 1998.

Petersen, N.O., P.L. Höddelius, P.W. Wiseman, O. Seger, and K.-E. Magnusson. Quantitation of membrane receptor distributions by image correlation spectroscopy: concept and application. *Biophys. J.* 65:1135–1146, 1993.

Petersen, N.O., D.C. Johnson, and M.J. Schlesinger. Scanning fluorescence correlation spectroscopy II: applications to virus glycoprotein aggregation. *Biophys. J.* 49:817–820, 1986.

Petersen, N.O. and W.B. McConnaughey. Effects of multiple membranes on measurements of cell surface dynamics by fluorescence photobleaching. *J. Supramol. Struct. Cell. Biochem.* 17:213–221, 1981.

Rigler, R. and E.L. Elson, eds. *Fluorescence Correlation Spectroscopy*, Berlin: Springer-Verlag, 2001, pp. 56–72.

Rocheleau, J.V. and N.O. Petersen. The sendai virus membrane fusion mechanism studied using image correlation spectroscopy European. *J. Biochem.* 268:1–8, 2001.

Rocheleau, J.V., P.W. Wiseman, and N.O. Petersen. Isolation of bright aggregate fluctuations in a multi-population image correlation spectroscopy system using intensity subtraction. *Biophys. J.* 84:4011–4022, 2003.

Schwille, P., E.J. Meyer-Almes, and R. Rigler. Dual-color fluorescence cross-correlation spectroscopy for multicomponent diffusion analysis in solution. *Biophys. J.* 72:1878–1886, 1997.

Srivastava, M. and N.O. Petersen. Image cross-correlation spectroscopy: a new experimental biophysical approach to measurement of slow diffusion of fluorescent molecules. *Methods Cell Sci.* 18:1–8, 1996.

Srivastava, M. and N.O. Petersen. Diffusion of transferrin receptor clusters. *Biophys. Chem.* 75:201–211, 1998.

Srivastava, M., N.O. Petersen, G. Mount, D. Kingston, N.S. McIntyre. Analysis of 3D SIMS images using image cross-correlation spectroscopy. *Surf. Interface Anal.* 26:188–194, 1998.

St.-Pierre, P.R. and N.O. Petersen. Relative ligand binding to small or large aggregates measured by scanning correlation spectroscopy. *Biophys. J.* 58:503–511, 1990.

St.-Pierre, P.R. and N.O. Petersen. Average cluster density and size of epidermal growth factor receptors on A431 cells. *Biochemistry* 31:2459–2463, 1992.

Thompson, N.L., A.M. Lieto, and N.W. Allen. Recent advances in fluorescence correlation spectroscopy. *Curr. Opin. Struct. Biol.* 12:634–641, 2002.

Webb, D.J., C.M. Brown, and A.F. Horwitz. Illuminating adhesion complexes in migrating cells: moving towards a bright future. *Curr. Opin. Cell Biol.* 15:614–620, 2003.

Wiseman, P.W. and N.O. Petersen. Image correlation spectroscopy. II. Optimization for ultrasensitive detection of preexisting platelet-derived growth factor-α receptor oligomers on intact cells. *Biophys. J.* 76:963–977, 1999.

Wiseman, P.W., J.A. Squier, M.H. Ellisman, and K.R. Wilson. Two-photon image correlation spectroscopy and image-cross correlation spectroscopy. *J. Microsc.* 200:14–25, 2000.

Index

Printed and bound by CPI Group (UK) Ltd, Croydon, CR0 4YY

08/05/2025

01865004-0001